Physics *of* Buoyant Flows

From Instabilities to Turbulence

Physics *of* Buoyant Flows

From Instabilities to Turbulence

Mahendra K Verma

Indian Institute of Technology Kanpur, India

W⊕ World Scientific

NEW JERSEY · LONDON · SINGAPORE · BEIJING · SHANGHAI · HONG KONG · TAIPEI · CHENNAI · TOKYO

Published by

World Scientific Publishing Co. Pte. Ltd.

5 Toh Tuck Link, Singapore 596224

USA office: 27 Warren Street, Suite 401-402, Hackensack, NJ 07601

UK office: 57 Shelton Street, Covent Garden, London WC2H 9HE

British Library Cataloguing-in-Publication Data

A catalogue record for this book is available from the British Library.

PHYSICS OF BUOYANT FLOWS
From Instabilities to Turbulence

ISBN 978-981-3237-79-7

For any available supplementary material, please visit
http://www.worldscientific.com/worldscibooks/10.1142/10928#t=suppl

Desk Editor: Christopher Teo

To my mother for her
love, forbearance, and sacrifice
&
To my father for his
uprightness, hard work, and discipline

Reviews

The last three decades have seen an amazing amount of activity in buoyancy driven flows in the regime where the nonlinear terms in the dynamics play a dominant role. These flows are common in the earth's atmosphere, in the earth's mantle and a large number of realistic situations. The nonlinear terms can lead to a variety of instabilities as well as turbulence. The present text by Prof. Mahendra Verma, who is an acknowledged leader in this field, is a very comprehensive account of the large number of issues that are involved. The text starts with the basic set up and discusses in great detail the different instabilities in a stratified fluid, the very interesting issues with the convective transport and the spectral properties and anisotropies associated with the turbulence in these flows. From students who want to learn about buoyancy driven flows to researchers who want to learn a calculation technique and the specialists who want to look up an obscure issue-this text should be ideal for the entire range.

Jayanta K. Bhattacharjee,
IACS, Kolkata

In his new monograph M. K. Verma introduces a clear and rigorous mathematical framework for the description and understanding of buoyancy driven flows. Step by step and with a remarkable pedagogical effort the author leads us into a large variety of phenomena encompassing internal waves, instabilities, patterns, chaos and turbulence. Not only M. K. Verma is a prominent contributor in the field, but also an outstanding guide. I highly recommend this book to researchers and post-graduate students.

Franck Plunian,
Université Grenoble Alpes

The book "Physics of Buoyant Flows: From Instabilities to Turbulence" by M. K. Verma is a highly required contribution in the existing series of books on buoyancy driven flows. This book give a good combination of basic information on convective instabilities and convective flows (from supercritical transition to developed turbulence) with really new methods for studying buoyancy driven flows (i.e. helical decompositions and shell models, which are not widely known in this domain of fluid mechanics), and recent results on large-scale dynamics, small-scale convective turbulence, influence of global rotation and magnetic field. The book will be useful to the students and experts in fluid dynamics.

Peter Frick,
Inst. Cont. Media Mech., Perm

Foreword

Buoyant flows are common in planetary and stellar atmospheres as well as their interiors, and in numerous engineering contexts. Since buoyancy often coexists with rotation, strongly inhomogeneous material properties, magnetic fields, etc., buoyant flows host a variety of instabilities, patterns, waves, and turbulence. Research in this area is a rich tapestry of dimensional theories, numerical computations, and experiments as well as observations—and draws together engineers, physicists, geophysicists, astrophysicists, mathematicians, computational experts, and others. We know much about the subject but our ignorance is still profound.

The book "Physics of Buoyant Flows: From Instabilities to Turbulence" is an up-to-date account of buoyant flows. It contains a discussion of basic topics such as thermal instabilities, pattern formation, internal gravity waves, rotating convection, magnetoconvection, and exact relations. It is partly pedagogical and can serve as an excellent resource for a graduate course on buoyancy driven flows, thermal convection, and related topics; it is also an account of current research and thus can serve equally as a useful reference for researches in the field. In particular, the author does not shy away from covering some "controversial topics" such as the kinetic and thermal energy spectrum in thermal convection, Nusselt and Reynolds number scaling, temperature fluctuations, wind reversals, and so forth. He brings many new ideas to the fore, in hopes of addressing these complex issues of thermal convection. Considering the enormity of topics covered, some of the details are presented in only enough detail to capture the attention of the reader.

I don't intend to stand for too long between the reader and the book itself, but do wish to refer to the co-called Bolgiano-Obukhov scaling as an example of the "controversial" issues that the author considers. The Bolgiano-Obukhov scaling was originally conceived for moderately stable buoyant flows. Among its consequences is that the density spectrum rolls off with a power law exponent of -1.4. However, this power-law has been observed also for the temperature field in thermal convection, which is buoyantly unstable. It was thus deemed that the Bolgiano-Obukhov result is applicable to unstable stratification as well. The author argues on the basis of variable energy flux that this is not so and that the appearance of the -1.4 slope in thermal convection is largely accidental. The book also tackles other items of conventional wisdom—for example, the Nusselt-Rayleigh scaling in a certain range of these parameters.

It is my fond hope that the book will be successful by inspiring young entrants to this wonderful subject.

K. R. Sreenivasan
New York University

Preface

Gravity pervades the whole universe; hence buoyancy drives fluids everywhere including those in the atmospheres and interiors of planets and stars. Prime examples of such flows are mantle convection, atmospheric flows, solar convection, dynamo process, heat exchangers, airships, and hot air balloons. In this book we present fundamentals and applications of buoyancy-driven flows.

Buoyancy-driven flows can be categorised into two classes—stably stratified and unstably stratified flows. Thermal convection belongs to the latter class. These types of flows exhibit very different behaviour. In the linear limit, stably stratified flows yield waves, while unstably stratified ones exhibit instability. In the turbulent regime, the kinetic energy flux of these systems have very different properties. In the present book, we treat the two classes of Buoyancy-driven flows using common framework to contrast them.

In the first part of the book, we setup the relevant equations, and present a unified treatment of linear theory that yields waves and instabilities for stably and unstably stratified flows respectively. Then we describe how nonlinearity saturates the unstable growth, and how it helps in pattern formation. We also introduce boundary layer and exact relations.

The second part of the book is dedicated to the buoyancy-driven turbulence, both in stably stratified flows and in thermal convection. We describe the spectral theory and show that the thermally-driven turbulence is similar to hydrodynamic turbulence. We also describe flow anisotropy, flow structures, flow reversals, and the scaling of large-scale quantities—Reynolds and Nusselt numbers. In the third part of the book, we introduce the effects of rotation and magnetic field in Buoyancy-driven flows; these discussions are kept at a preliminary level to limit the size of the book.

Thus, Buoyancy-driven flows exhibit extremely rich phenomena including waves, instabilities, patterns, chaos, boundary layer, and turbulence. Last two decades have witnessed resolutions of some of the outstanding problems of the field, specially the energy spectrum and fluxes, boundary layers, scaling of large-scale quantities such as Reynolds and Nusselt numbers, and flow reversals. Some of the topics covered in the book are quite advanced, yet, we have tried to develop them in a simple and intuitive manner. For example, Craya-Herring and helical basis simplify the linear stability analysis. Also, the framework of variable energy flux helps us contrast the energetics in stably stratified flows, thermal convection, and hydrodynamic turbulence. The discussion on the scaling of Reynolds and Nusselt numbers is quite exhaustive. Of course, significant number of important problems still remain unsolved, for example, elusive ultimate regime, physics of flow reversals, quantification of fluctuations, etc. We list the open problems of the field at appropriate places. We hope that the reader would find these discussions attractive.

Despite richness of buoyancy-driven flows, both in application and in theory, there are only a handful of books written on this topic. We hope that our book fills this gap, and it is useful to the graduate students and researchers. The present book may also act as a reference material for graduate courses on buoyancy-driven flows, hydrodynamic instabilities, and turbulence. Though, I strove hard to make the book error-free, I am sure that there are deficiencies in the presentation. I would be glad to receive your feedback and suggestions at email: mkv@iitk.ac.in.

Mahendra Verma

Acknowledgments

A significant part of the book contains contributions from my collaborators and colleagues. It is a great pleasure to acknowledge these contributions.

A significant fraction of discussion on the energy spectrum and flux in Chapters 12, 13, and 16 is based on the PhD thesis work of Abhishek Kumar. Also, Abhishek painstakingly drew many figures of the book and provided enormous feedback and suggestions on the manuscript.

The contents of Chapter 11 on the scaling of large-scale quantities of RBC is based on the PhD thesis work of Ambrish Pandey. The energy spectrum and flux for large-Pr convection, discussed in Chapters 13, were derived during this collaboration.

Pankaj Mishra and I started to work on energy spectrum and flux and on flow reversals when Pankaj was a Ph.D. student in our laboratory. The numerical results on the energy spectrum of small-Pr and zero-Pr convection, as well as on the flow reversals in a cylinder, were obtained during collaborative work with Pankaj Mishra.

Many ideas of Chapter 17 originated during collaborations with Stephan Fauve, Mani Chandra, Sagar Chakraborty, Vinayak Eswaran, Gaurav Dar, Pankaj Mishra, Ambrish Pandey, Rodion Stepanov, Peter Frick, Andre Sukhanovskii, Abhishek Kumar, and Manu Mannattil. I thank all of them for those fruitful and enjoyable discussions.

On convective instability and pattern formation, I collaborated with Pinaki Pal, Krishna Kumar, Supriyo Pal, Pankaj Mishra, Pankaj Wahi, and Stephan Fauve. I am greatly indebted to them for ideas and discussions, some of which reflect in Chapter 8.

For proof-read and important suggestions on the manuscript, I thank Sumit Kumar, Ambrish Pandey, Avishek Ranjan, Jai Sukhatme, Abhishek Kumar, and Shashwat Bhattacharya. I thank Shashwat Bhattacharya for discussions on viscous dissipation in RBC, and Manohar Sharma for discussions on inertial waves. I am grateful to Manmohan Dewbanshi and Roshan Bhaskaran for assistance in formatting of the manuscript, and drawing some of the figures.

I owe a great deal to my collaborators—Katepalli Sreenivasan (Sreeni), Jayant Bhattacharjee (JKB), Stephan Fauve (Stephan), Abhishek Kumar, Ambrish

Pandey, Mani Chandra, Jai Sukhatme, Sagar Chakraborty, Ravi Samtaney, Rodion Stepanov, Franck Plunian, and Jörg Schumacher—for discussions, ideas, references, fun talks, coffee and dinner sessions, plans, successes, support, and encouragement. I am particularly indebted to senior colleagues Sreeni, JKB, and Stephan who guided me to the wonderful world of thermal convection, as well as for constant feedback and encouragement. I wish to state that I started my research in RBC after an introduction to the field by Sreeni during my visits to ICTP in 2003 and 2004 as an ICTP associate; I am thankful to Sreeni and ICTP, Trieste for the same.

I learnt many important things on rotating flows and instabilities from Anirban Guha and Avishek Ranjan. I thank K.-Q. Xia for discussions and ideas. I also benefitted greatly from discussions with colleagues—Jaywant Arakeri, Baburaj Puthenveettil, Peter Frick, Peter Davidson, Eric Lindborg, Detlef Lohse, Richard Steven, Prateek Sharma, Anurag Gupta, Anindya Chatterjee, Edger Knobloch, Raja Lakkaraju, and Joe Niemela. I thank the past and present set of lab students who contributed to this book, directly or indirectly. I am sure I missed some names...apologies to those whom I missed.

Our spectral code, TARANG, was used to solve several problems discussed in the book. Success of TARANG is due to major efforts of its developers—Anando Chatterjee, Subhadeep Sadhukhan, and Abhishek Kumar—and its many testers. I am grateful to King Abdullah University of Science and Technology (KAUST) for providing computational time on SHAHEEN I and SHAHEEN II through project K1052. We also used the computational clusters HPC2010 and HPC2013 of IIT Kanpur, and PARAM YUVA of CDAC. The computations of no-slip RBC were performed using the open-source code OpenFOAM.

For our work on buoyancy-driven flows, we benefitted from the financial and computational support through the following research projects: Swaranajayanti fellowship from Department of Science and Technology; research grants from DoRD and DoRA, IIT Kanpur; PLANEX/PHY/2015239 from ISRO; SERB/F/3279/2013-14 from Science and Engineering Research Board; and Indo-Russian project (DST-RSF) INT/RUS/RSF/P-03 and RSF-16-41-02012.

For my work on buoyancy-driven turbulence, I benefitted from the discussions, and open and intellectual environment provided during the following scientific meetings: *Summer school and discussion meeting on Buoyancy-driven Flows* and *Turbulence from Angstorms to light years* held at ICTS Bengaluru, and *Turbulence Mixing and Beyond 2017* held at ICTP Trieste.

Mahendra Verma

Notation

Gravity is along $-\hat{z}$.

Shorthand:
KE	Kinetic energy
PE	Potential energy
SST	Stably stratified flows
RBC	Rayleigh-Bénard convection
OB	Oberbeck-Boussinesq
BO	Bolgiano-Obukhov
FFT	Fast Fourier transform
RTI	Rayleigh-Taylor instability
RTT	Rayleigh-Taylor turbulence

Fields and Fourier transform:
$\mathbf{r}$	Real space coordinate
$\mathbf{k}$	Wavenumber in Fourier pace
k	Magnitude of $\mathbf{k}$
$f(\mathbf{k})$	Fourier transforms of $f(\mathbf{r})$ with no hat
$\hat{f}(\mathbf{k})$	Fourier transforms of $f(\mathbf{r})$ in free-slip basis (with hat)
$\hat{f}(\mathbf{k})$	In Chapter 17, Fourier transforms of $f(\mathbf{r})$ (exception)

Density:
$\varrho(\mathbf{r})$	Total density in real space
$\bar{\rho}(z)$	Density stratification along the verbal direction (z)
$\rho(\mathbf{r})$	Fluctuations in real space (also see b)
$\rho(\mathbf{k})$	Fluctuations in Fourier space
$b(\mathbf{r})$	Fluctuations in real space in velocity units ($g\rho/(N\rho_m)$)
$b(\mathbf{k})$	Fluctuations in Fourier space in velocity units
ρ_b, ρ_t	ρ at the bottom and top plates or surfaces
ρ_m	Mean total density

Temperature:

$T(\mathbf{r})$ Total temperature in real space

$\bar{T}(z)$ Temperature stratification

$T_m(z)$ Planar average ($\langle T(x,y,z)\rangle_{xy}$)

$\theta(\mathbf{r})$ Fluctuations in real space

$\theta(\mathbf{k})$ Fluctuations in Fourier space

$\theta_{\text{res}}(\mathbf{r})$ Fluctuations over the mean temperature $T_m(z)$

$\theta_{\text{res}}(\mathbf{k})$ Fourier transforms of $\theta_{\text{res}}(\mathbf{r})$

θ_b, θ_t θ at the bottom and top plates or surfaces

Δ Temperature difference between the bottom and top plates

Θ Large scale temperature fluctuations ($= \theta_{\text{rms}}$)

Θ_{res} Large scale temperature fluctuations without $\theta(0,0,2n)$ Fourier modes

Velocity, pressure, and magnetic fields:

$\mathbf{u}(\mathbf{r})$ Velocity field in real space

$\mathbf{u}(\mathbf{k})$ Velocity field in Fourier space

$\boldsymbol{\omega}(\mathbf{r})$ Vorticity field in real space

$\boldsymbol{\omega}(\mathbf{k})$ Vorticity field in Fourier space

$\mathbf{b}(\mathbf{r})$ Magnetic field in real space

$\mathbf{b}(\mathbf{k})$ Magnetic field in Fourier space

U Large scale velocity ($= U_{\text{rms}}$)

Energy spectrum:

$E_u(\mathbf{k})$ Modal kinetic energy ($|\mathbf{u}(\mathbf{k})|^2/2$)

$E_\rho(\mathbf{k})$ Modal potential energy ($|\rho(\mathbf{k})|^2/2$)

$\mathcal{F}(\mathbf{k})$ Kinetic energy supply to $\mathbf{u}(\mathbf{k})$ by the external force

$\mathcal{F}_B(\mathbf{k})$ Kinetic energy supply to $\mathbf{u}(\mathbf{k})$ by buoyancy

$E_b(\mathbf{k})$ Modal potential energy ($|b(\mathbf{k})|^2/2$)

$E_\theta(\mathbf{k})$ Modal entropy ($|\theta(\mathbf{k})|^2/2$), also called modal thermal energy

$E_\omega(\mathbf{k})$ Modal enstrophy ($|\boldsymbol{\omega}(\mathbf{k})|^2/2$)

$E_f(k)$ One-dimensional energy spectrum of f

Energy transfers:

$\Pi_u(k)$ Kinetic energy flux

$\Pi_b(k)$ Potential energy flux

$\Pi_\theta(k)$ Entropy flux

$\Pi_\omega(k)$ Enstrophy flux

$T_{u,B}^{u,A}$ Kinetic energy transfer from region A to region B

$T_{u,n}^{u,m}$ Kinetic energy transfer from shell m to shell n.

$T_{(u,n,\beta)}^{(u,m,\alpha)}$ Ring-to-ring kinetic energy transfer from ring (m,α) to (n,β) (spherical)

$T^{(u,m,h_1)}_{(u,n,h_2)}$ Ring-to-ring kinetic energy transfer from ring (m, h_1) to (n, h_2)
        (cylindrical)

$\mathcal{P}_\|(\mathbf{k}')$ Kinetic energy gained by $u_\|(\mathbf{k})$ from $\mathbf{u}_\perp(\mathbf{k})$ via pressure

$\mathcal{P}_\perp(\mathbf{k}')$ Kinetic energy gained by $\mathbf{u}_\perp(\mathbf{k})$ from $u_\|(\mathbf{k})$ via pressure

Global quantities:

E_u Average kinetic energy ($\int d\mathbf{r}[|\mathbf{u}^2|/2]/\text{Vol.} = \sum_\mathbf{k} |\mathbf{u}(\mathbf{k})|^2/2$)

E_ρ Average potential energy

E_b Average potential energy

E_θ Average entropy, also called average thermal energy

E_ω Average enstrophy (for 2D)

U Large scale velocity $(= U_\text{rms})$

Θ Large scale temperature fluctuations $(= \theta_\text{rms})$

Θ_res Large scale temperature fluctuations without $\theta(0, 0, 2n)$ Fourier modes

L Integral length scale

Anisotropy quantification:

$u_\|$ Velocity field parallel to the anisotropy direction

$\mathbf{u}_\perp$ Velocity field perpendicular to the anisotropy direction

$k_\|, k_z$ Component of $\mathbf{k}$ along $\hat{z}$

$k_\perp$ $\sqrt{k_x^2 + k_y^2}$

$L_\|$ Integral length scale parallel to the anisotropy direction

$L_\perp$ Integral length scale perpendicular to the anisotropy direction

$E_{u,\|}$ Kinetic energy parallel to the anisotropy direction $(|u_\||^2/2)$

$E_{u,\perp}$ Kinetic energy perpendicular to the anisotropy direction $(|\mathbf{u}_\perp|^2/2)$

$E_{u,\|}(k)$ Kinetic energy spectrum parallel to the anisotropy direction

$E_{u,\perp}(k)$ Kinetic energy spectrum perpendicular to the anisotropy direction

$E_u(k, \beta)$ Ring spectrum for the kinetic energy in spherical system
        ($\beta=$ sector index)

$E_u(k, i)$ Ring spectrum for the kinetic energy in cylindrical system
        ($i =$ height index)

$T^{(u,m,\alpha)}_{(u,n,\beta)}$ Ring-to-ring energy transfer from ring (m, α) to (n, β) (spherical)

$T^{(u,m,h_1)}_{(u,n,h_2)}$ Ring-to-ring energy transfer from ring (m, h_1) to (n, h_2)
        (cylindrical)

$\mathcal{P}_\|(\mathbf{k}')$ Kinetic energy gained by $u_\|(\mathbf{k})$ from $\mathbf{u}_\perp(\mathbf{k})$ via pressure

$\mathcal{P}_\perp(\mathbf{k}')$ Kinetic energy gained by $\mathbf{u}_\perp(\mathbf{k})$ from $u_\|(\mathbf{k})$ via pressure

System parameters:

d Distance between thermal plates

Ω Angular velocity of the rotating system

μ Dynamic viscosity

ν Kinematic viscosity

κ	Thermal diffusivity
κ	Diffusivity of passive scalar
α	Thermal expansion coefficient
g	Acceleration due to gravity
R	Gas constant
C_p, C_v	Heat capacities at constant pressure and volume respectively

System parameters 2:

Re	Reynolds number (Ud/ν)
Pe	Péclet number (Ud/κ)
Pr	Prandtl number (ν/κ)
Sc	Schmidt number (ν/κ)
Nu	Nusselt number
Ra	Rayleigh number $(\alpha g \Delta d^3/(\nu\kappa))$
Ro	Rossby number $(U/(\Omega d))$
Ta	Taylor number $(4\Omega^2 d^4/\nu^2))$
E	Ekman number $(\nu/(2\Omega d^2))$
N	Brunt Väisälä frequency

Miscellaneous parameters:

ζ	Polar angle in spherical coordinate system
ϕ	Azimuthal angle in spherical coordinate system
ϕ	Polar angle in 2D coordinate system
$\hat{e}_{1,2,3}$	Unit vectors of Craya-Herring basis
$u_{1,2}$	The components of the velocity field in Craya-Herring basis
$\hat{e}_{\pm}$	Unit vectors of helical basis
$u_{\pm}$	The components of the velocity field in helical basis
ω	Frequency of a wave
β	Sector index

Shell model:

k_n	Wavenumber of shell n
u_n	Velocity of shell n
θ_n	Temperature fluctuation of shell n
b_n	Density of shell n
f_n	Forcing on the shell n

Contents

PART 1

Basic Formulation, Patterns, and Chaos

This part covers

(1) basic formulation
(2) waves and instabilities
(3) patterns and chaos
(4) some exact relations

of buoyancy-driven flows.

Chapter 1

Introduction

Hydrodynamic flows exhibit turbulent behaviour which is not well understood till date. When a fluid is in a gravitational field, the flow becomes even more complex due to the gravity, confining walls, and two vector fields–velocity and temperature. Despite such complexities, researchers have been able to construct models and theories for buoyancy-driven flows that will be discussed in this book.

We start our discussion on buoyancy-driven flows with an example. Imagine a cold winter night, and you are sitting one meter away from a heater. How long will it take for the heat to reach you? From the heat diffusion equation:

$$\frac{\partial \theta}{\partial t} = \kappa \nabla^2 \theta \implies \frac{\theta}{T} \sim \kappa \frac{\theta}{L^2}, \tag{1.1}$$

where θ is the temperature fluctuation, $L = 1$ m is the distance concerned, T is the required time, and κ is the heat diffusivity of air, whose value at the normal pressure and temperature is approximately 10^{-5} m^2/s. Substitution of the above values in Eq. (1.1) yields

$$T \approx \frac{L^2}{\kappa} \approx 10^5 \text{ s}, \tag{1.2}$$

which is an incorrect estimate. A reasonable estimate is obtained by treating air as a fluid in which heat is convected by the velocity field. The relevant equation for this description is

$$\frac{\partial \theta}{\partial t} + \mathbf{u} \cdot \nabla \theta = \kappa \nabla^2 \theta, \tag{1.3}$$

where $\mathbf{u}$ is the fluid velocity. In this equation, for a typical large-scale velocity $U \approx 0.1$ m/s, the ratio

$$\frac{\mathbf{u} \cdot \nabla \theta}{\kappa \nabla^2 \theta} \approx \frac{UL}{\kappa} \approx \frac{0.1 \times 1}{10^{-5}} \approx 10^4 \gg 1. \tag{1.4}$$

Hence, nonlinear term $\gg$ diffusion term. Therefore, for an estimate of the timescale T, we need to equate $\partial\theta/\partial t$ with the nonlinear term $\mathbf{u} \cdot \nabla \theta$ that yields

$$\frac{\partial \theta}{\partial t} \approx -\mathbf{u} \cdot \nabla \theta \implies \frac{\theta}{T} \sim \frac{\theta U}{L}. \tag{1.5}$$

Hence

$$T \approx \frac{L}{U} \approx 1/0.1 \approx 10 \text{ s},$$ (1.6)

which is a reasonable estimate.

The above example demonstrates that many natural processes are governed by fluid flows that involves nonlinear term (e.g., $\mathbf{u} \cdot \nabla\theta$ of Eq. (1.3)). Such nonlinear systems are more complex than the diffusion equation, which is an exactly solvable linear equation. The timescale of Eq. (1.1) can be obtained from its exact solution, but such analytical solution or approach is unavailable at present for fluid equations like Eq. (1.3). Such equations have many complex and unresolved issues, some of which will be covered in this book.

Gravity pervades everywhere and it affects fluid flows from micro scales to astrophysical scales. Hence, we need to account for the gravitational force or buoyancy in such flows. In this book, we classify buoyancy-driven flows into two categories: *stably stratified flows* in which a lighter fluid is above denser fluid with gravity acting downwards (along $-\hat{z}$); and *unstably stratified flows* in which a denser fluid is above a lighter fluid. Examples of stably stratified flows are

(1) Planetary and stellar atmospheres.
(2) Oceanic flow: In oceans, denser salty water is below lighter pure water.

The examples of unstably stratified flows are

(1) Thermal convection: cold and dense fluid is above warm and light fluid. This unstable configuration yield convective motion.
(2) Rayleigh-Taylor instability and turbulence: A dense fluid is above a light fluid.
(3) Unstably stratified flows: Similar to Rayleigh-Taylor instability.
(4) Taylor-Couette turbulence: A viscous fluid confined between two coaxial and rotating cylinders. The centrifugal force that acts like a gravitational force destabilises the flow.
(5) Bubbly flow: Fast air bubbles supply energy to the flow and make it unstable.

When the nonlinearity is weak, stably stratified flows exhibit waves, while unstably stratified flows show instability with fluctuations growing in time. Typically, nonlinearity saturates the growth of the instability. For moderate nonlinearity, both stably and unstably stratified flows exhibit patterns and chaos. These systems however become turbulent for large nonlinearity. Based on the work done so far, including recent ones, the present status of the field is as follows[1]:

(1) stably stratified flows

    (a) Waves: Good understanding.
    (b) Weak turbulence: Good understanding.
    (c) Turbulence: Partial understanding. Unresolved issues in quasi2D turbulence.

---

[1]The assessment is somewhat personal, and may not be perceived in a similar light by others.

(2) Unstably stratified flows

    (a) Instability: Good understanding.

    (b) Patterns: Good understanding, but quantitative predictions are lacking.

    (c) Chaos: Reasonable understanding. Detailed routes to chaos not worked out.

    (d) Turbulence: Reasonable understanding.

In this book, we will discuss the above issues in a reasonable detail. We will focus on the instabilities and waves, the scaling of large-scale quantities like Reynolds and Nusselt numbers, and the spectra and fluxes of the velocity, density, and temperature fields. Though we cover both stably and unstably stratified buoyancy-driven flows, the balance is tilted towards thermal convection.

Throughout the book, we attempt to present the phenomena using common frameworks applicable to both stably and unstably stratified flows. For example, we solve the linear equations of such flows using a common scheme, and show how waves and thermal instability appear in the two systems. Since Fourier transform helps describe waves and instability, as well as multiscale phenomena of turbulence, we employ it throughout the book. On many occasions, buoyant flows are also affected by rotation and magnetic field, for example in the outer core of the Earth, and in the solar convection zone. Such flows are very complex, and they are beyond the scope of this book. Here we analyse only the linearised version of such systems. We also cover the linearized version of double diffusive systems.

There are a number of excellent textbooks and review articles on buoyancy-driven flows. Chandrasekhar (2013)'s classic tome covers instabilities in Rayleigh-Bénard convection, rotating convection, magnetoconvection, and rotating magnetoconvection, and Rayleigh-Taylor flow. Kundu *et al.* (2015) and Bhattacharjee (1987) describe instabilities and chaos in convective flows. Tritton (1988), Kundu *et al.* (2015), and Landau and Lifshitz (1987) provides a general overview of buoyancy-driven flows in the general framework of fluid mechanics. Turner (2009) deals with gravity waves and instabilities in great detail. Getling (1998) and Lappa (2010) describe the patterns in thermal convection; Davidson (2013) and Chassignet *et al.* (2012) focus on stably stratified flows; and Lappa (2012) covers rotating convection. For magnetoconvection, not covered in this book in detail, we refer the readers to a recent book by Weiss and Proctor (2014). Books by Manneville (2004), Cross and Greenside (2009), and Hoyle (2006) are important references on amplitude equations and their applications to patterns and chaos in thermal convection.

Among review articles, Siggia (1994) discusses scaling arguments for high Rayleigh number convection. The reviews by Ahlers *et al.* (2009), Lohse and Xia (2010), Chillà and Schumacher (2012), and Bodenschatz *et al.* (2000) cover recent developments in Rayleigh-Bénard convection. Cross and Hohenberg (1993) cover patterns in thermal convection using amplitude equations. Behringer (1985), Turner (1985), and Sherman and Imberger (1978) review specialised topics of buoyancy-driven flows.

In the present book, we cover almost all aspects stably stratified flows and thermal convection—waves, instabilities, patterns, chaos, turbulence, scaling of large-scale quantities like Rayleigh and Nusselt numbers, anisotropy, turbulent structures and their dynamics, and flow reversals. We cover results obtained till date including some works from the year 2017. Thus, the present book is an up-to-date account of buoyancy-driven flows. Of course, to make the book thematic and to save space, we have minimised our discussions on many topics including amplitude equation, rotating convection, magnetoconvection, etc. We do hope you will enjoy reading this book.

Further reading

For introduction to buoyancy-driven flows, refer to the books and review articles described above. According to me, the simplest ones among them are Chandrasekhar (2013), Tritton (1988), and Kundu *et al.* (2015).

Chapter 2

Basic Framework

Buoyancy is induced by density variations that may occur due to gravity, as in planetary and stellar atmosphere; or due to temperature variation, as in thermal convection; or due to concentration gradient, as in oceans. In this chapter, we introduce the framework and equations of buoyancy-driven flows. In later chapters we will describe complex behaviour exhibited by such flows.

2.1 Equations in terms of density variation

A fluid flow is described using velocity, density, and energy variables. In this book we focus on systems in which the density fluctuations are much smaller than the average density. We write the local density ϱ as a sum of the density stratification $\bar{\rho}(z)$ and density fluctuation ρ:

$$\varrho(x, y, z) = \bar{\rho}(z) + \rho(x, y, z). \tag{2.1}$$

It is customary to assume a linear density profile for $\bar{\rho}(z)$:

$$\bar{\rho}(z) = \rho_b + \frac{d\bar{\rho}}{dz}z = \rho_b + \frac{\rho_t - \rho_b}{d}z, \tag{2.2}$$

where ρ_b, ρ_t[1] are respectively the densities at the bottom and top layers that are separated by distance d (see Fig. 2.1). These top and bottom layers may be embedded in a larger system, e.g. in the Earth's atmosphere, or they could be plates. The latter configuration has boundary layers near the plates, but former one does not.

We assume that the gravity is downward, along $-\hat{z}$ (see Fig. 2.1). Hence the gravitational force density on the fluid is

$$\mathbf{F}_g = -g\varrho\hat{z} = -g(\bar{\rho} + \rho)\hat{z}$$
$$= -g\nabla \left(\int^z \bar{\rho}(z')dz' \right) - \rho g\hat{z}. \tag{2.3}$$

The second term of the above expression, $-\rho g\hat{z}$, occurs due to the density variation of the fluid with relative to the local surrounding, and it is called *buoyancy*. It is

---

[1]The subscripts b and t stand for *bottom* and *top*.

7

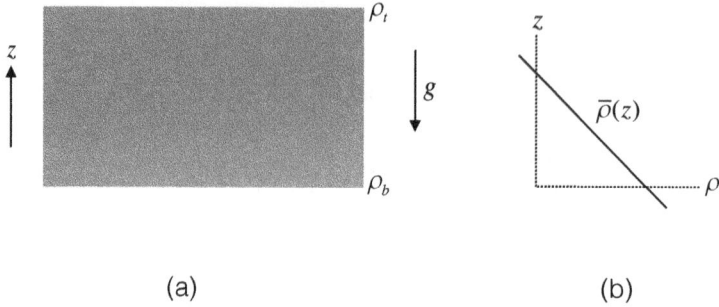

Fig. 2.1   (a) Fluid contained between two layers whose densities are ρ_b (bottom) and ρ_t (top). (b) The fluid is subjected to a density stratification $\bar{\rho}(z)$.

along $-\hat{z}$ for $\rho > 0$, but along $\hat{z}$ for $\rho < 0$. We remark that the density variation could be caused by strong gravity, for example in planetary and stellar atmospheres; or by temperature variations, as in thermal convection in which hotter (lighter) fluid ascend, and colder (heavier) fluid descends; or by infusion of soluble material, for example, salt in ocean. We will discuss these systems in detail in later parts of the book.

A fluid flow is described by Navier Stokes equation

$$\varrho \left[\frac{\partial \mathbf{u}}{\partial t} + (\mathbf{u} \cdot \nabla)\mathbf{u} \right] = -\nabla p + \mathbf{F}_g + \mu \nabla^2 \mathbf{u} + \mathbf{F}_u, \tag{2.4}$$

where $\mathbf{u}, p$ are respectively the velocity and pressure fields, $\mathbf{F}_u$ is the external force field in addition to the buoyancy, and μ is the dynamic viscosity of the fluid. Substitution of Eq. (2.3) in Eq. (2.4) yields

$$\varrho \left[\frac{\partial \mathbf{u}}{\partial t} + (\mathbf{u} \cdot \nabla)\mathbf{u} \right] = -\nabla \sigma - \rho g \hat{z} + \mu \nabla^2 \mathbf{u} + \mathbf{F}_u, \tag{2.5}$$

where

$$\sigma = p + g \left(\int^z \bar{\rho}(z') dz' \right) \tag{2.6}$$

is the modified pressure. The continuity equation for the density yields

$$\frac{\partial \varrho}{\partial t} + \nabla \cdot (\varrho \mathbf{u}) = \nabla \cdot (\kappa \nabla \varrho), \tag{2.7}$$

where κ is the diffusivity of the density. We assume that κ is constant in space and time. Equation (2.7) can be rewritten as

$$\nabla \cdot \mathbf{u} = -\frac{1}{\varrho} \frac{d\varrho}{dt} + \frac{1}{\varrho} \kappa \nabla^2 \varrho. \tag{2.8}$$

Now we employ Oberbeck-Boussinesq (OB) approximation according to which the density of the fluid is treated as a constant, i.e., $(d\varrho/dt)/\varrho = 0$. Therefore, the relative magnitude of $\nabla \cdot \mathbf{u}$ would be

$$\frac{\nabla \cdot \mathbf{u}}{U/L} \approx \frac{L}{\varrho U} \kappa \nabla^2 \varrho \approx \frac{\kappa}{UL} = \frac{1}{\mathrm{Pe}}, \tag{2.9}$$

where L, U are the large length and velocity scales respectively, and Pe is the Péclet number. Typically, buoyancy-driven flows have large Péclet number[2], hence for such flows, we can assume that $\nabla \cdot \mathbf{u} = 0$ and replace ϱ of Eq. (2.5) with the mean density, ρ_m. Hence the governing equations for the buoyancy-driven flows are

$$\frac{\partial \mathbf{u}}{\partial t} + (\mathbf{u} \cdot \nabla)\mathbf{u} = -\frac{1}{\rho_m}\nabla\sigma - \frac{\rho}{\rho_m}g\hat{z} + \nu\nabla^2\mathbf{u} + \mathbf{f}_u, \qquad (2.10)$$

$$\frac{\partial \rho}{\partial t} + (\mathbf{u} \cdot \nabla)\rho = -\frac{d\bar{\rho}}{dz}u_z + \kappa\nabla^2\rho, \qquad (2.11)$$

$$\nabla \cdot \mathbf{u} = 0, \qquad (2.12)$$

where $\nu = \mu/\rho_m$ is the kinematic viscosity, and $\mathbf{f}_u = \mathbf{F}_u/\rho_m$. An assumption that ν, κ are constants in space and time is also considered to be a part of the OB approximation. It is important to note that even though fluid density is assumed to be an approximate constant in OB approximation, density variation is retained in the momentum equation because buoyancy is comparable to the other terms of the momentum equation.

We will perform linear stability analysis of buoyancy-driven flows in Chapters 5 and 6 and show that the system is stable when $d\bar{\rho}/dz < 0$, and unstable when $d\bar{\rho}/dz > 0$. Flows with $d\bar{\rho}/dz < 0$ are called *stably stratified flows*, and they have a lighter fluid above a heavier fluid (see Fig. 2.2(a)). Such systems support waves in the linear limit, and nonlinear waves and turbulence in the nonlinear regime. In the presence of viscosity, the total energy of stably stratified flows decays without an external force, hence, $\mathbf{f}_u$ is employed to maintain a steady state. Examples of stably stratified systems are the Earth's atmosphere and oceans. Note however that the density profile of the ambient planetary atmosphere need not be linear; for the Earth, the density profile is exponential. A linear profile however is a good approximation for a small region of the atmosphere.

The second class of flows have $d\bar{\rho}/dz > 0$ (see Fig. 2.2(c)). Here a denser fluid sits above a lighter one. Such flows become unstable when $d\bar{\rho}/dz$ exceeds a critical value, hence they are called *unstably stratified flows*. For larger nonlinearity, such flows exhibit various patterns, chaos, and turbulence. Examples of such flows are Rayleigh-Bénard convection (RBC) and Rayleigh-Taylor instability. Note however that the background density of RBC varies linearly, but that of Rayleigh-Taylor instability exhibits a sharp jump at the interface between the light and dense fluids.

As described earlier, temperature induces density variation. In the next section, we will discuss the governing equations of buoyancy-driven flows in terms of velocity and temperature fields.

2.2 Equations in terms of temperature

In a heated fluid, hotter elements are lighter than colder ones. Suppose that the background temperature profile is $\bar{T}(z)$. A fluid parcel with $T(z) > \bar{T}(z)$ will be

---

[2]When Pe $\ll 1$, the mass diffusivity κ is very large that leads to strong mass diffusion. Hence, the constant density approximation is not likely to hold for such flows.

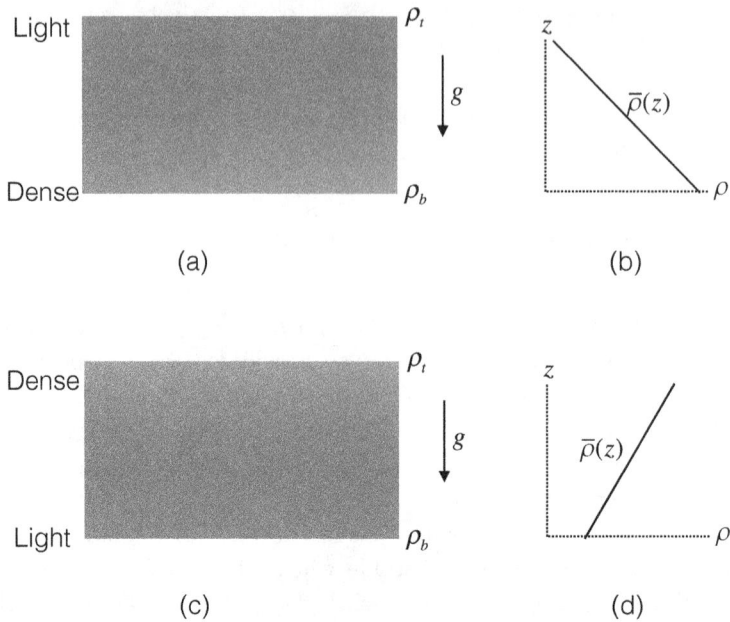

Fig. 2.2 (a) Schematic diagram of stably stratified system in which a lighter fluid is above a denser one with the density stratification $\bar{\rho}(z)$ as shown in (b). (c) Schematic diagram of unstable stratification in which a denser fluid is above a lighter one with the density stratification $\bar{\rho}(z)$ as shown in (d).

lighter that its background, and hence it will rise due to buoyancy. Opposite effect will occur to a fluid element with $T(z) < \bar{T}(z)$. In this section we will describe the equation of motion of a fluid element in terms of its temperature.

Here we consider a layer of fluid confined between two surfaces that are kept at temperature T_b and T_t as shown in Fig. 2.3. If the system is inside a larger system, the top and bottom surfaces would be imaginary, and there would be no boundary layer. However, when the bounding surfaces are plates, boundary layers appear near the plates.

We relate the temperature field T to the density ϱ using a linear relationship:

$$\varrho(x, y, z) = \rho_b \left[1 - \alpha\{T(x, y, z) - T_b\}\right], \tag{2.13}$$

where α is the thermal expansion coefficient, which is assumed to be constant in space and time[3]. In analogy with $\bar{\rho}$, we assume a linear temperature profile:

$$\bar{T}(z) = T_b + \frac{d\bar{T}}{dz}z = T_b - \frac{T_b - T_t}{d}z. \tag{2.14}$$

It is customary to separate the temperature into ambient mean value $\bar{T}(z)$ and fluctuation $\theta(x, y, z)$:

$$T(x, y, z) = \bar{T}(z) + \theta(x, y, z). \tag{2.15}$$

---

[3]The linear approximation of Eq. (2.13) holds for water, oil, and air in terrestrial environment, but it may not hold for exotic fluids or for very steep temperature gradients.

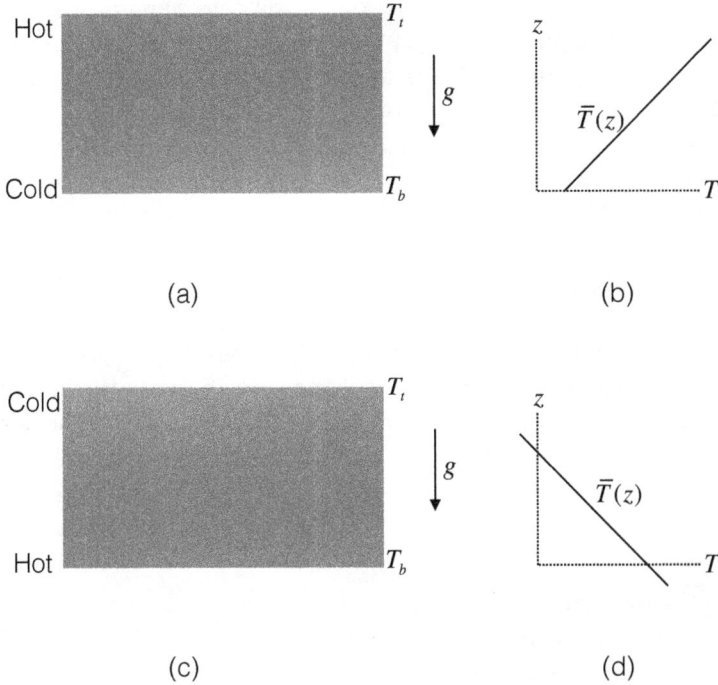

Fig. 2.3 (a) Schematic diagram of stably stratified system in which a hotter fluid is above a colder one with the temperature stratification $\bar{T}(z)$ as shown in (b). (c) Schematic diagram of unstable stratification in which a colder fluid is above a hotter one with the temperature stratification $\bar{\rho}(z)$ as shown in (d).

A comparison of Eqs. (2.1, 2.13–2.15) yields

$$\rho = -\rho_m \alpha \theta; \quad \frac{d\bar{\rho}}{dz} = -\alpha \frac{d\bar{T}}{dz}, \tag{2.16}$$

substitution of which in Eqs. (2.10, 2.11) with $\mathbf{f}_u = 0$ yields

$$\frac{\partial \mathbf{u}}{\partial t} + (\mathbf{u} \cdot \nabla)\mathbf{u} = -\frac{1}{\rho_m}\nabla\sigma + \alpha g\theta \hat{z} + \nu\nabla^2 \mathbf{u}, \tag{2.17}$$

$$\frac{\partial \theta}{\partial t} + (\mathbf{u} \cdot \nabla)\theta = -\frac{d\bar{T}}{dz}u_z + \kappa\nabla^2 \theta, \tag{2.18}$$

where κ is the *thermal diffusivity*. These equations coupled with

$$\nabla \cdot \mathbf{u} = 0 \tag{2.19}$$

are the governing equations for RBC. Note that substitution of Eq. (2.13) in Eq. (2.7) yields

$$\frac{\partial T}{\partial t} + (\mathbf{u} \cdot \nabla)T = \kappa\nabla^2 T, \tag{2.20}$$

which is another form of Eq. (2.18) with variable T.

In the above equation, we employ Oberbeck-Boussinesq approximation described in the previous section, namely, $|\rho|/\rho_m = \alpha|\theta| \ll 1$, and that κ is constant in space and time. The OB approximation is applicable when the temperature difference between the two surfaces is not too large.

Equation (2.18) can also be derived using the energy equation for fluids. The energy density per unit volume $\varepsilon = \rho C_v T$, and an equation for its evolution is [Kundu *et al.* (2015)]

$$\rho C_v \left[\frac{\partial T}{\partial t} + (\mathbf{u} \cdot \nabla)T \right] = -\nabla \cdot \mathbf{q} - p\nabla \cdot \mathbf{u} + \rho \epsilon_u, \qquad (2.21)$$

where $\mathbf{q}$ is the heat flux, C_v is the specific heat at constant volume, and $\rho \epsilon_u$ is the kinetic energy dissipation rate per unit volume that enhances the internal energy. Under OB approximation, $\rho \epsilon_u$ is ignored in the above equation. In the next section, we will illustrate this approximation for realistic situations. To model the heat flux, we employ Fourier's law

$$\mathbf{q} = -K\nabla T, \qquad (2.22)$$

where K is the thermal conductivity.

For liquids like water, incompressibility condition yields $\nabla \cdot \mathbf{u} = 0$, substitution of which in Eq. (2.21) yields Eq. (2.18, 2.20) with thermal diffusivity

$$\kappa = \frac{K}{\rho C_v}. \qquad (2.23)$$

For gases, $p\nabla \cdot \mathbf{u}$ is comparable to the other terms of the energy equation:

$$p\nabla \cdot \mathbf{u} = -\frac{p}{\rho}\frac{D\rho}{Dt} = -\frac{p}{\rho}\left(\frac{\partial \rho}{\partial T}\right)_p \frac{DT}{Dt} = p\alpha\frac{DT}{Dt}. \qquad (2.24)$$

Here $\nabla \cdot \mathbf{u} \neq 0$, contrary to OB approximations. Now we assume the gas to be ideal, hence $p = \rho R T$, where R is the Rydberg's constant. Therefore,

$$\alpha = -\frac{1}{\rho}\left(\frac{\partial \rho}{\partial T}\right)_p = \frac{1}{T}. \qquad (2.25)$$

Hence

$$p\nabla \cdot \mathbf{u} = \rho R\frac{DT}{Dt} = \rho(C_p - C_v)\frac{DT}{Dt}, \qquad (2.26)$$

where C_p, C_v are respectively the specific heats of the gas for constant pressure and constant volume respectively. Substitution of the above in Eq. (2.21) yields Eq. (2.18, 2.20) with $\kappa = K/(\rho C_p)$.

In the linear limit, Eqs. (2.17, 2.18) yield stable solution (e.g., waves) when $d\bar{T}/dz > 0$, and unstable solution when $d\bar{T}/dz < 0$. A major application with $d\bar{T}/dz < 0$ profile is Rayleigh-Bénard convection, which is the topic of the next section.

2.3 Rayleigh-Bénard convection

In general, thermal flows involve complex geometries and materials. A simplified model that captures essential features of thermal convection is called *Rayleigh-Bénard convection (RBC)* in which a fluid is confined between two thermally-conducting horizontal plates kept at constant temperatures T_b and T_t, as shown in Fig. 2.3(c). Note that $d\bar{T}/dz < 0$. We employ Eqs. (2.17–2.19) to describe RBC, which is a major subject of this book.

Depending on the temperature difference between the plates, RBC exhibits host of interesting phenomena. First, RBC exhibits thermal instability when the temperature difference between the plates, $\Delta = T_b - T_t$, exceeds a critical value. Below this value, the heat is transported by conduction with a linear temperature profile

$$\bar{T}(z) = T_b - \beta z, \tag{2.27}$$

where

$$\beta = \frac{T_b - T_t}{d}. \tag{2.28}$$

As the temperature difference Δ is increased further, nonlinearity becomes significant and it saturates the growth of the instability leading to stable convective rolls. When Δ or nonlinearity is increased further, we obtain patterns, chaos, and turbulence. We will discuss these topics in the subsequent chapters of this book.

Most of the buoyancy-driven flows in the Earth's atmosphere and in engineering applications involve water, air, and oil, whose thermal parameters are listed in Table A.1. In Tables A.1 and A.2 we also list the parameters for the Earth's mantle and outer core, the Sun's convection zone, and some other fluids. The ratio of ν and κ, called *Prandtl number* (Pr), plays an important role in the dynamics of RBC. Flows with large Pr tends to be dominated by the viscous term, and hence the velocity fluctuations are weak in such flows. On the contrary, the nonlinearity is strong in small-Pr RBCs. Hence, for same $T_b - T_t$, small-Pr RBCs are more turbulent than large-Pr.

It is mathematically convenient to study the limiting cases $\mathrm{Pr} \to \infty$ and $\mathrm{Pr} \to 0$. The former corresponds to finite κ, and $\nu \to \infty$, while the latter to finite ν, and $\kappa \to \infty$. Note that Eq. (2.9) implies that the condition $\nabla \cdot \mathbf{u} = 0$ is difficult to satisfy for $\mathrm{Pr} \to 0$ or $\kappa \to \infty$ case. Hence, OB approximation would not hold for $\mathrm{Pr} \to 0$ case. Most realistic flows, at least in terrestrial experiments and atmosphere, have $\mathrm{Pe} \gg 1$, hence OB approximation works well for such flows.

2.4 Justification of Oberbeck-Boussinesq approximation

In this section we will demonstrate how typical terrestrial atmospheric and engineering flows obey OB approximation. Here we assume that the fluid is at normal temperature and pressure.

(1) The parameters ν, κ, α are assumed to be constants. For water and air, these constants vary by a small amount when the temperature fluctuations are less than $50\,^\circ$C.

(2) If the temperature difference between the plates $\Delta \leq 30\,^\circ$C, then for typical fluids, $\rho/\rho_m \ll 1$. For example, for water that has $\alpha \approx 3 \times 10^{-4}\mathrm{K}^{-1}$ at $30\,^\circ$C, $\delta\rho/\rho_m = \alpha\Delta \approx 10^{-3}$ for $\Delta = 30\,^\circ$C. Similarly for air at room temperature, $\alpha = 1/T \approx 1/300\mathrm{K}^{-1}$ [Eq. (2.25)], hence, at $\Delta = 30\,^\circ$C, $\delta\rho/\rho_m \approx 0.1$, which is relatively small.

(3) As a result of item (2), the density of the flow is assumed to be a constant, except for the buoyancy of the momentum equation. The density variation in the fluid causes buoyancy, which is comparable to the other terms of the momentum equation (Eq. (2.17)). In Secs. 6.1, 11.3, 11.9 we will study the relative strength of buoyancy with relative to the other terms of the Navier-Stokes equation in the linear and turbulent regimes.

(4) Under the OB approximation, the viscous dissipate rate $\rho\epsilon_u$ is assumed to be small (see Eq. (2.21)). For justification of this assumption, we estimate $\epsilon_u/(C_v(DT/Dt))$. In the viscous regime,

$$\frac{\epsilon_u}{C_v(DT/Dt)} \approx \frac{\nu U^2/L^2}{C_v U\Delta/L} \approx \frac{\nu U}{C_v L\Delta}, \tag{2.29}$$

and in the turbulent regime,

$$\frac{\epsilon_u}{C_v(DT/Dt)} \approx \frac{U^3/L}{C_v U\Delta/L} \approx \frac{U^2}{C_v\Delta}. \tag{2.30}$$

We estimate $U \approx \sqrt{\alpha g\Delta d}$. For water, with $C_v = 4200$ J/kgK, $\Delta = 30\,^\circ$C, $\nu = 10^{-6}\,\mathrm{m}^2/\mathrm{s}$, $\rho = 10^3$ kg/m^3, $L = d = 0.1$ m, in the viscous regime,

$$\frac{\epsilon_u}{C_v(DT/Dt)} \approx \frac{\nu U}{C_v L\Delta} \approx \frac{10^{-6} \times 10^{-1}}{4200 \times 30 \times 10^{-1}} \approx 10^{-11}. \tag{2.31}$$

But in the turbulent regime,

$$\frac{\epsilon_u}{C_v(DT/Dt)} \approx \frac{U^2}{C_v\Delta} \approx \frac{10^{-2}}{4200 \times 30} \approx 10^{-7}. \tag{2.32}$$

Thus the viscous dissipation in the system is too small to be able to heat the fluid significantly. Hence we ignore ϵ_u term in Eq. (2.21). Physically, this is because C_v for water and air are quite large.

Thus we show that Oberbeck-Boussinesq approximation is valid for fluids at normal temperature and pressure. The above approximation however breaks down when Δ is too large, for example in planetary and stellar interiors. For such systems, we need to solve the equations of compressible fluids, which are Eqs. (2.5, 2.7), coupled with equations of state and energy.

In the next section we will rewrite the equations for stably stratified flows by transforming density to velocity variables.

2.5 Equations for the stably stratified flows simplified

The linearized version of Eqs. (2.10, 2.11) with $\nu = \kappa = 0$ and $\mathbf{f}_u = 0$ yields

$$\frac{\partial u_z}{\partial t} = -\frac{\rho}{\rho_m}g, \tag{2.33}$$

$$\frac{\partial \rho}{\partial t} = -\frac{d\bar{\rho}}{dz}u_z. \tag{2.34}$$

The above set of equations can be rewritten as

$$\frac{\partial^2 u_z}{\partial t^2} = \frac{g}{\rho_m}\frac{d\bar{\rho}}{dz}u_z, \tag{2.35}$$

whose solutions for $d\bar{\rho}/dz < 0$ are waves with *Brunt-Väisälä frequency*:

$$N = \sqrt{\frac{g}{\rho_m}\left|\frac{d\bar{\rho}}{dz}\right|}. \tag{2.36}$$

These waves are called *internal gravity waves* that will be detailed in Chapter 5.
   Researchers often write the equation for stably stratified flow using variable

$$b = \frac{g}{N}\frac{\rho}{\rho_m}, \tag{2.37}$$

which has dimension of velocity [Lindborg (2006)]. In terms of b, Eqs. (2.10, 2.11) become

$$\frac{\partial \mathbf{u}}{\partial t} + (\mathbf{u} \cdot \nabla)\mathbf{u} = -\frac{1}{\rho_m}\nabla\sigma - Nb\hat{z} + \nu\nabla^2\mathbf{u}, \tag{2.38}$$

$$\frac{\partial b}{\partial t} + (\mathbf{u} \cdot \nabla)b = Nu_z + \kappa\nabla^2 b. \tag{2.39}$$

The nondimensional parameters used for describing stably stratified flows are

$$\text{Reynolds number Re} = \frac{u_{\text{rms}}d}{\nu}, \tag{2.40}$$

$$\text{Prandtl number Pr} = \frac{\nu}{\kappa}, \tag{2.41}$$

$$\text{Péclet number Pe} = \frac{u_{\text{rms}}d}{\kappa}, \tag{2.42}$$

$$\text{Froude number Fr} = \frac{u_{\text{rms}}/d}{N} = \frac{u_{\text{rms}}}{dN}, \tag{2.43}$$

$$\text{Richardson number Ri} = \frac{N|b|_{\text{rms}}d}{u_{\text{rms}}^2}, \tag{2.44}$$

where u_{rms} is the rms velocity of flow, and b_{rms} is the rms value of b. The Reynolds number is typically defined as the ratio of the nonlinear term $\mathbf{u} \cdot \nabla\mathbf{u}$ and the viscous term[4], Péclet number as the ratio of the nonlinear term $\mathbf{u} \cdot \nabla\rho$ and the diffusion term, and the Richardson number as the ratio of the buoyancy and the nonlinear

---

[4]However, in Sec. 11.2, we show that in RBC, Re differs from the ratio of the nonlinear term and the viscous term by a factor that depends on Ra.

term. Froude number is the ratio of the frequencies associated with u_{rms} and gravity waves respectively.

Note that

$$\mathrm{Ri} = \frac{N|b|_{\mathrm{rms}}d}{u_{\mathrm{rms}}^2} = \frac{1}{\mathrm{Fr}^2}\frac{|b|_{\mathrm{rms}}}{Nd} = \frac{1}{\mathrm{Fr}^2}\frac{\rho_{\mathrm{rms}}/d}{|d\bar{\rho}/dz|} \approx \frac{1}{\mathrm{Fr}^2}. \tag{2.45}$$

This is an important relation which is used for describing various regimes of stably stratified flows. We will describe later that strong gravity induces strong anisotropy that requires more parameters, which will be described in Chapter 12.

In the next section we will describe the boundary conditions employed for buoyancy-driven flows.

2.6 Boundary conditions, Box geometry

Boundary conditions play an important role in buoyancy-driven flows. We employ the following set of boundary conditions for the velocity field:

(1) *Periodic*: Imagine that the system of Fig. 2.2 or Fig. 2.3 is embedded in a larger box. For such cases, we can employ periodic boundary condition with

$$\mathbf{u}(\mathbf{r} + lL_x\hat{x} + mL_y\hat{y} + nL_z\hat{z}) = \mathbf{u}(\mathbf{r}), \tag{2.46}$$

where l, m, n are integers, and the box has dimension of $L_x \times L_y \times L_z$.

(2) *No-slip*: A fluid is often confined in a box with top, bottom, and side walls. For such flows, the generic boundary condition employed for the velocity field is no-slip,

$$\mathbf{u} = 0 \tag{2.47}$$

at all the walls. Sometimes, for convenience of computation, no-slip boundary conditions are employed at the top and bottom walls, but periodic boundary condition at the side walls.

(3) *Free-slip*: When a fluid under consideration flows over another fluid, for example air over water. In such cases, we employ free-slip boundary condition, which is

$$\mathbf{u}_\perp = 0; \quad \frac{\partial \mathbf{u}_\parallel}{\partial n} = 0, \tag{2.48}$$

where $\mathbf{u}_\perp, \mathbf{u}_\parallel$ are respectively the velocity components perpendicular and parallel to the concerned surface, and n is the coordinate perpendicular to the wall. Note that $\mathbf{u} = \mathbf{u}_\perp + \mathbf{u}_\parallel$. For example, for a horizontal wall perpendicular to the z axis,

$$u_z = 0; \quad \frac{\partial u_x}{\partial z} = \frac{\partial u_y}{\partial z} = 0. \tag{2.49}$$

We will show later that in computer simulations, the free-slip boundary condition is easier to implement than the no-slip boundary condition.

The boundary conditions for the temperature or density are similar. Here we describe the boundary conditions for the temperature field.

(1) *Periodic*: The temperature fluctuations are periodic along all the three directions:

$$\theta(\mathbf{r} + lL_x\hat{x} + mL_y\hat{y} + nL_z\hat{z}) = \theta(\mathbf{r}), \tag{2.50}$$

where l, m, n are integers, and the box has dimension of $L_x \times L_y \times L_z$.

(2) *Conducting*: A constant temperature field at the walls:

$$\theta = 0. \tag{2.51}$$

(3) *Insulating*: The heat flux at the wall, $K\nabla\theta$, is zero. To illustrate, for a vertical surface perpendicular to the x axis, the boundary condition on θ is

$$\frac{\partial\theta}{\partial x} = 0. \tag{2.52}$$

The other important factors that affect the flow properties are the box geometry and size. An important geometrical parameter is *aspect ratio*, which is the ratio of the box width and the box hight. We will show in this book that some of the important properties of turbulent flows are weakly dependent on the box geometry. For example, turbulent thermal convection exhibits Kolmogorov's spectrum independent of the box geometry. However, large-scale flow structures are affected by the box geometry. For example, the rolls in a cylinder can freely rotate along the azimuthal direction, which is not the case in a cube.

In the next section, we will describe Nondimensionalized equations of RBC.

2.7 Nondimensionalized RBC equations

The equations for Rayleigh-Bénard convection are often presented in several nondimensionalized forms. These equations capture the relative strengths of various terms of the equations. Also, they help reduce the number of computational parameters of a system, which is quite useful for analysis, as well as for the numerical simulations and experiments.

RBC has two diffusive parameters ν and κ that yield two diffusive time scales: $\tau_\nu = d^2/\nu$ and $\tau_\kappa = d^2/\kappa$. Clearly $\tau_\nu < \tau_\kappa$ for $\mathrm{Pr} > 1$, but $\tau_\nu > \tau_\kappa$ for $\mathrm{Pr} < 1$. It is customary, specially for numerical simulations, to resolve lower of the two time scales: τ_ν for $\mathrm{Pr} > 1$, and τ_κ for $\mathrm{Pr} < 1$. These choices become specially relevant for very large or very small Pr's.

For large Pr, if we use ν/d as the velocity scale, then the nondimensionalized velocity $u' = ud/\nu$ becomes too small for very large ν. Therefore, we employ κ/d as the velocity scale. The other scales are: d for the length, d^2/κ for time, $\Delta = |T_b - T_t|$ for the temperature, and $\rho_m(\kappa/d)^2$ for the pressure. These quantities yield the following nondimensional variables:

$$\mathbf{u}' = \frac{\mathbf{u}}{(\kappa/d)}, \tag{2.53}$$

$$\mathbf{r}' = \frac{\mathbf{r}}{d}, \tag{2.54}$$

$$\nabla' = d\nabla, \tag{2.55}$$

$$t' = \frac{t}{(d^2/\kappa)}, \tag{2.56}$$

$$\sigma' = \frac{\sigma}{\rho_m(\kappa/d)^2}, \tag{2.57}$$

$$\theta' = \frac{\theta}{\Delta}. \tag{2.58}$$

In terms of the nondimensional variables, Eqs. (2.17, 2.18) get converted to

$$\frac{\partial \mathbf{u}'}{\partial t'} + (\mathbf{u}' \cdot \nabla')\mathbf{u}' = -\nabla'\sigma' + \mathrm{RaPr}\theta'\hat{z} + \mathrm{Pr}\nabla'^2\mathbf{u}' \tag{2.59}$$

$$\frac{\partial \theta'}{\partial t'} + (\mathbf{u}' \cdot \nabla')\theta' = u_z' + \nabla'^2\theta', \tag{2.60}$$

where

$$\mathrm{Ra} = \frac{\alpha g d^3 \Delta}{\nu \kappa} \tag{2.61}$$

is the Rayleigh number[5], and the Prandtl number $\mathrm{Pr} = \nu/\kappa$, as defined in Eq. (2.41). Note that the ratio of the buoyancy and the viscous term is

$$\frac{\mathrm{RaPr}\theta_{\mathrm{rms}}}{\mathrm{Pr}\nabla^2 u_{\mathrm{rms}}} \approx \frac{\mathrm{Ra}}{k_{\mathrm{min}}^2} \frac{\theta_{\mathrm{rms}}}{u_{\mathrm{rms}}}. \tag{2.63}$$

We will show in Sec. 6.1 that near the onset of Rayleigh-Bénard instability, the aforementioned ratio is approximately 3. However, in the turbulent regime, the ratio of the buoyancy and the viscous term is more complex. Hence, broadly speaking, the Rayleigh number is an approximate measure of the strength of the buoyancy with relative to the viscous term.

When $\mathrm{Re} \ll 1$, the nonlinear term $(\mathbf{u}' \cdot \nabla')\mathbf{u}'$ of Eq. (2.59) becomes much smaller than the other terms, and hence

$$-\nabla'\sigma' + \mathrm{RaPr}\theta'\hat{z} + \mathrm{Pr}\nabla'^2\mathbf{u}' = 0. \tag{2.64}$$

Now let us consider a case when $\mathrm{Pe} \gg 1$, or $\mathrm{Pr} = \mathrm{Pe}/\mathrm{Re} \gg 1$. In this case, the nonlinear term of Eq. (2.60) is significant. The incompressibility condition implies that

$$-\nabla'^2\sigma' + \mathrm{RaPr}\partial_z\theta' = 0, \tag{2.65}$$

---

[5]Note that we can also define the Rayleigh number in terms of density difference using the relation $|\rho_b - \rho_t|/\rho_m \approx \alpha\Delta$, which yields

$$\text{Rayleigh number } \mathrm{Ra} = \frac{g d^3}{\nu \kappa} \frac{|\rho_b - \rho_t|}{\rho_m}. \tag{2.62}$$

and hence $\sigma' \propto \mathrm{Pr}$. Therefore, for $\mathrm{Pr} = \infty$, it is customary to define $\sigma'' = \sigma'/\mathrm{Pr}$. Consequently, the dynamical equations for $\mathrm{Pr} = \infty$ are

$$-\nabla'\sigma'' + \mathrm{Ra}\theta'\hat{z} + \nabla'^2 \mathbf{u}' = 0, \tag{2.66}$$

$$\frac{\partial \theta'}{\partial t'} + (\mathbf{u}' \cdot \nabla')\theta' = u_z' + \nabla'^2\theta'. \tag{2.67}$$

For RBC with very small Pr, we take ν/d as the velocity scale, and d^2/ν as the time scale. In this limit, the diffusion term dominates Eq. (2.18), and hence

$$\frac{\Delta}{d}u_z \approx \kappa\nabla^2\theta \implies \frac{\nu\Delta}{d^2} \approx \frac{\kappa}{d^2}\theta. \tag{2.68}$$

Therefore

$$\theta \sim \frac{\nu}{\kappa}\Delta \to 0, \tag{2.69}$$

which is due to strong diffusion of temperature fluctuations by large κ. Consequently, for small Pr, we employ $(\nu/\kappa)\Delta$ as the scale for temperature fluctuations. In summary, for flows with small Pr, the nondimensionalized variables are

$$\mathbf{u}' = \frac{\mathbf{u}}{(\nu/d)}, \tag{2.70}$$

$$t' = \frac{t}{(d^2/\nu)}, \tag{2.71}$$

$$\sigma' = \frac{\sigma}{\rho_m(\nu/d)^2}, \tag{2.72}$$

$$\theta' = \frac{\theta}{\mathrm{Pr}\Delta}, \tag{2.73}$$

and the nondimensionalized equations in terms of these variables are

$$\frac{\partial \mathbf{u}'}{\partial t} + (\mathbf{u}' \cdot \nabla)\mathbf{u}' = -\nabla\sigma' + \mathrm{Ra}\theta'\hat{z} + \nabla^2\mathbf{u}', \tag{2.74}$$

$$\mathrm{Pr}\left(\frac{\partial \theta'}{\partial t} + (\mathbf{u}' \cdot \nabla)\theta'\right) = u_z' + \nabla^2\theta'. \tag{2.75}$$

Now let us take the case of $\mathrm{Pe} \to 0$ and $\mathrm{Re} \gg 1$, for which $\mathrm{Pr} \to 0$. In this case, Eq. (2.75) becomes

$$u_z' + \nabla^2\theta' = 0, \tag{2.76}$$

while the momentum equation remains the same. As described in Sec. 2.3, zero-Pr RBC does not satisfy OB approximation.

For large Ra, however, the aforementioned nondimensional velocity becomes very large [derivation in Chapter 11]. As a result, for a grid size of δx, the time step for numerical integration,

$$\delta t = \frac{\delta x}{U_{\max}}, \tag{2.77}$$

becomes very small. Hence, in numerical simulations, it is customary to employ $\sqrt{\alpha g \Delta d}$ as the velocity scale that yields the following set of equations for moderate Pr:

$$\frac{\partial \mathbf{u}'}{\partial t'} + (\mathbf{u}' \cdot \nabla)\mathbf{u}' = -\nabla \sigma' + \theta' \hat{z} + \sqrt{\frac{\mathrm{Pr}}{\mathrm{Ra}}} \nabla^2 \mathbf{u}', \qquad (2.78)$$

$$\frac{\partial \theta'}{\partial t'} + (\mathbf{u}' \cdot \nabla)\theta' = u_z' + \frac{1}{\sqrt{\mathrm{RaPr}}} \nabla^2 \theta'. \qquad (2.79)$$

As described earlier, for small Pr, temperature fluctuations are scaled using $\mathrm{Pr}\Delta$. Therefore, for small Pr, Eqs. (2.78, 2.79) get converted to

$$\frac{\partial \mathbf{u}'}{\partial t'} + (\mathbf{u}' \cdot \nabla)\mathbf{u}' = -\nabla \sigma' + \mathrm{Pr}\theta' \hat{z} + \sqrt{\frac{\mathrm{Pr}}{\mathrm{Ra}}} \nabla^2 \mathbf{u}', \qquad (2.80)$$

$$\mathrm{Pr} \left(\frac{\partial \theta'}{\partial t'} + (\mathbf{u}' \cdot \nabla)\theta' \right) = u_z' + \sqrt{\frac{\mathrm{Pr}}{\mathrm{Ra}}} \nabla^2 \theta'. \qquad (2.81)$$

In future discussion, for brevity we will drop the primes from the nondimensional equations.

In the next section we will discuss various quadratic quantities of buoyancy-driven flows.

2.8   Kinetic and potential energies, entropy

The kinetic energy of a small fluid element of volume $d\mathbf{r}$ is $\rho(u^2/2)d\mathbf{r}$. In an incompressible fluid, the fluid density is constant, hence we drop ρ from the above expression. Hence the local kinetic energy density is defined as

$$E_u(\mathbf{r}) = \frac{u^2}{2}, \qquad (2.82)$$

which is the local kinetic energy at location $\mathbf{r}$, averaged over the tiny volume. We start from Eq. (2.38) and derive the equation for the kinetic energy, which is

$$\frac{\partial}{\partial t} \frac{u^2}{2} + \nabla \cdot \left[\frac{u^2}{2} \mathbf{u} \right] = -\nabla \cdot (\sigma \mathbf{u}) - N b u_z + \mathbf{u} \cdot \nabla^2 \mathbf{u} \qquad (2.83)$$

where N is the Brunt-Väisälä frequency, b is the density fluctuation in the velocity units, and $\rho_m = 1$. Now, using [Spiegel (2010)][6]

$$\mathbf{u} \cdot \nabla^2 \mathbf{u} = -\omega^2 + \nabla \cdot [\mathbf{u} \times \boldsymbol{\omega}] \qquad (2.85)$$

where $\boldsymbol{\omega} = \nabla \times \mathbf{u}$ is the vorticity field, we obtain

$$\frac{\partial}{\partial t} \frac{u^2}{2} + \nabla \cdot \left[\frac{u^2}{2} \mathbf{u} \right] = -\nabla \cdot (\sigma \mathbf{u} - \nu \mathbf{u} \times \boldsymbol{\omega}) - N b u_z - \nu \omega^2. \qquad (2.86)$$

---

[6]Proof:

$$-\omega^2 + \nabla \cdot [\mathbf{u} \times \boldsymbol{\omega}] = -\epsilon_{ijk}\epsilon_{ilm}(\partial_j u_i)(\partial_l u_m) + \epsilon_{kij}\epsilon_{klm}\partial_i[u_j(\partial_l u_m)]$$
$$= u_i \partial_j \partial_j u_i - u_i \partial_i \partial_j u_j$$
$$= \mathbf{u} \cdot \nabla^2 \mathbf{u} \qquad (2.84)$$

for incompressible flows.

The corresponding equation for RBC is

$$\frac{\partial}{\partial t}\frac{u^2}{2} + \nabla \cdot \left[\frac{u^2}{2}\mathbf{u}\right] = -\nabla \cdot (\sigma\mathbf{u} - \nu\mathbf{u} \times \boldsymbol{\omega}) + \alpha g\theta u_z - \nu\omega^2. \tag{2.87}$$

Similarly, we construct quadratic quantities for the density ρ, normalised density b, and temperature fluctuation θ:

$$E_\rho(\mathbf{r}) = \frac{\rho^2}{2}, \tag{2.88}$$

$$E_b(\mathbf{r}) = \frac{b^2}{2}, \tag{2.89}$$

$$E_\theta(\mathbf{r}) = \frac{\theta^2}{2}. \tag{2.90}$$

In literature, $E_b(\mathbf{r})$ is called the *local potential energy*, while $E_\theta(\mathbf{r})$ is called *local entropy* or *local thermal energy*. Note that $\theta^2/2$ is very different from the thermodynamic entropy or internal energy. Potential energy is a useful quantity for stably stratified flows, while entropy is used in RBC. Using Eq. (2.39), we derive the equation for the potential energy as

$$\frac{\partial}{\partial t}\frac{b^2}{2} + \nabla \cdot \left[\frac{b^2}{2}\mathbf{u} - \kappa b\nabla b\right] = Nbu_z - \kappa(\nabla b)^2. \tag{2.91}$$

The corresponding equation for entropy is derived using Eq. (2.18) as

$$\frac{\partial}{\partial t}\frac{\theta^2}{2} + \nabla \cdot \left[\frac{\theta^2}{2}\mathbf{u} - \kappa\theta\nabla\theta\right] = -\frac{d\bar{T}}{dz}\theta u_z - \kappa(\nabla\theta)^2. \tag{2.92}$$

Corresponding to the local kinetic energy, potential energy, and entropy, we define the following set of global quantities, which are the average energies and entropy over the whole volume. They are defined as

$$E_u = \frac{1}{2}\langle u^2 \rangle = \frac{1}{\text{Vol}}\int d\mathbf{r}\frac{u^2}{2}, \tag{2.93}$$

$$E_\rho = \frac{1}{2}\langle \rho^2 \rangle = \frac{1}{\text{Vol}}\int d\mathbf{r}\frac{\rho^2}{2}, \tag{2.94}$$

$$E_b = \frac{1}{2}\langle b^2 \rangle = \frac{1}{\text{Vol}}\int d\mathbf{r}\frac{b^2}{2}, \tag{2.95}$$

$$E_\theta = \frac{1}{2}\langle \theta^2 \rangle = \frac{1}{\text{Vol}}\int d\mathbf{r}\frac{\theta^2}{2}, \tag{2.96}$$

where Vol is the volume of the box. In literature, the above quantities are referred to as *total kinetic energy*, *total potential energy*, and *total entropy*. Another global quantity of interest is the *total kinetic helicity*, which is defined as

$$H_K = \frac{1}{2}\langle \mathbf{u} \cdot \boldsymbol{\omega} \rangle = \frac{1}{\text{Vol}}\int d\mathbf{r}\frac{1}{2}\mathbf{u} \cdot \boldsymbol{\omega}, \tag{2.97}$$

where $\boldsymbol{\omega} = \nabla \times \mathbf{u}$ is the vorticity of the flow.

When we integrate the Eqs. (2.83, 2.91) over the whole box, we obtain

$$\frac{dE_u}{dt} = -\int d\mathbf{r} N b u_z - \nu \int d\mathbf{r} \omega^2, \tag{2.98}$$

$$\frac{dE_b}{dt} = \int d\mathbf{r} N b u_z - \kappa \int d\mathbf{r} (\nabla b)^2, \tag{2.99}$$

$$\frac{dE_\theta}{dt} = -\int d\mathbf{r} \frac{d\bar{T}}{dz} \theta u_z - \kappa \int d\mathbf{r} (\nabla \theta)^2, \tag{2.100}$$

In the above derivation, we use the fact that for a scalar field f,

$$\int d\mathbf{r} \nabla \cdot (f\mathbf{u}) = \int f\mathbf{u} \cdot d\mathbf{S} = 0 \tag{2.101}$$

for periodic boundary condition, or when $\mathbf{u}_\perp$ vanishes at the boundary. Here $d\mathbf{S}$ is a elemental surface. In the inviscid and nondiffusive limit, $\nu = \kappa = 0$, we deduce that

$$\frac{d}{dt}(E_u + E_b) = 0, \tag{2.102}$$

or, that the sum of kinetic and potential energies, $E_u + E_b$, is conserved. Note that E_u and E_b are not conserved individually. There is an exchange of energy between them via buoyancy.

It is easy to derive that the corresponding conserved quantity for RBC is

$$E_u + \frac{\alpha g}{d\bar{T}/dz} E_\theta, \quad \text{or,} \quad E_u - \left|\frac{\alpha g}{d\bar{T}/dz}\right| E_\theta. \tag{2.103}$$

Note that $d\bar{T}/dz < 0$ for RBC. The above conservation laws indicate that there is an active cross transfer between $\mathbf{u}$ and ρ fields in both stably stratified flows and RBC.

Example 2.1: Consider the following flow field in a two-dimensional box of size (π, π):

$$\mathbf{u} = 4A(\hat{x} \sin x \cos y - \hat{y} \cos x \sin y)$$
$$\theta = 4B \cos x \sin y$$

Show that the fluid is incompressible. Also, compute the average kinetic energy and the total entropy of the flow.

Solution: It is easy to verify that $\nabla \cdot \mathbf{u} = 0$, thus we prove that the fluid is incompressible. The average kinetic energy of the flow is

$$E_u = \frac{1}{\pi^2} \int_0^\pi dx dy \frac{1}{2}(u_x^2 + u_y^2) = 4A^2,$$

and the total entropy of the flow is

$$E_\theta = \frac{1}{\pi^2} \int_0^\pi dx dy \frac{1}{2}\theta^2 = 2B^2.$$

Example 2.2: Consider the following flow field in a three-dimensional box of size (π, π, π):

$$\mathbf{u} = 8A(\hat{x} \sin x \cos y \cos 2z + \hat{y} \cos x \sin y \cos 2z - \hat{z} \cos x \cos y \sin 2z),$$

$$\theta = 8B \cos x \cos y \sin 2z.$$

Show that the fluid is incompressible. Compute the average kinetic energy and entropy of the flow.

Solution: We find that $\nabla \cdot \mathbf{u} = 0$, hence the flow is incompressible. Following the same procedure as Example 2.1, we find that the average kinetic energy and entropy of the flow are

$$E_u = 12A^2$$
$$E_\theta = 4B^2$$

In the next section, we will quantify heat transport in RBC.

2.9   Heat transport and Nusselt number in RBC

In RBC, hot fluid rises, hence it transports heat from the hot plate to the cold plate. In addition, heat is also transported by conduction. The equation for the temperature, Eq. (2.20), is rewritten as

$$\frac{\partial T}{\partial t} + \nabla \cdot (\mathbf{u}T - \kappa \nabla T) = 0 \tag{2.104}$$

that helps us quantify the heat transport by conduction as[7]

$$H_{\text{cond}} = -\kappa \frac{\partial T}{\partial z}, \tag{2.105}$$

and by convection as

$$H_{\text{conv}} = u_z T. \tag{2.106}$$

Since u_z and T fluctuate appreciably, we redefine H_{cond} and H_{conv} as averages over a horizontal plane. For a quantity $f(x, y, z)$, planar average is defined as

$$\langle f \rangle_{xy}(z) = \frac{1}{A} \int dx dy f(x, y, z), \tag{2.107}$$

where A is the cross section area of the plane.

Using $T(x, y, z) = T_c(z) + \theta(x, y, z)$, and $\langle u_z \rangle_{xy} = 0$, we deduce that

$$H_{\text{cond}}(z) = -\kappa \left\langle \frac{\partial T}{\partial z} \right\rangle_{xy}, \tag{2.108}$$

$$H_{\text{conv}}(z) = \langle u_z T \rangle_{xy} = \langle u_z \theta \rangle_{xy}, \tag{2.109}$$

$$H_{\text{total}}(z) = H_{\text{cond}}(z) + H_{\text{conv}}(z). \tag{2.110}$$

---

[7]Here the heat flux has the dimension of K m/sec. To convert it to usual heat flux, we need to multiply H to ρC_v, where C_v is the specific heat at constant volume.

Note that $\langle \partial T / \partial z \rangle_{xy} \neq -\Delta/d$, rather it is a function of z (see Fig. 9.3).

When we integrate Eq. (2.104) over a fluid slab confined between horizontal planes $z = z_1$ and $z = z_2$, we obtain

$$\frac{\partial}{\partial t} \int_{z_1}^{z_2} dz dx dy T(x, y, z) + H_{\text{total}}(z_2) - H_{\text{total}}(z_1) = 0. \qquad (2.111)$$

Under a steady state,

$$\frac{\partial}{\partial t} \int_{z_1}^{z_2} dz dx dy T(x, y, z) = 0 \qquad (2.112)$$

that leads to

$$H_{\text{total}}(z_2) = H_{\text{total}}(z_1). \qquad (2.113)$$

Thus, under a steady-state, the total heat flux averaged over horizontal planes are equal. Note that in the boundary layer, $H_{\text{total}}(z) \approx H_{\text{cond}}(z)$ because $u_z \approx 0$ in this region. However, in the bulk $H_{\text{total}}(z) \approx H_{\text{conv}}(z)$ because $dT/dz \approx 0$ here (see Fig. 9.3).

The volume average of the conductive heat flux is

$$H_{\text{cond}} = -\frac{1}{Ad} \kappa \int dx dy dz \frac{\partial T}{\partial z} = \kappa \frac{(T_b - T_t)}{d} = \kappa \frac{\Delta}{d}, \qquad (2.114)$$

and that of convective heat flux is

$$H_{\text{conv}} = \langle u_z T \rangle_{xyz} = \frac{1}{d} \int dz \langle u_z \theta \rangle_{xy}. \qquad (2.115)$$

A sum of Eq. (2.114) and Eq. (2.115) yields

$$H_{\text{total}} = H_{\text{cond}} + H_{\text{conv}}. \qquad (2.116)$$

Under a steady state, planar and volume averaged heat transport are equal, or

$$H_{\text{total}} = H_{\text{total}}(z). \qquad (2.117)$$

Nusselt number, Nu, is a nondimensional measure of the convective heat transport. It is defined as the ratio of the total heat flux (convective plus conductive) and the conductive heat flux:

$$\text{Nu} = \frac{H_{\text{cond}} + H_{\text{conv}}}{H_{\text{cond}}} = 1 + \frac{\langle u_z \theta \rangle_{xyz}}{\kappa \Delta/d}. \qquad (2.118)$$

Under a steady state, using Eq. (2.117), we deduce that $\text{Nu}(z)$ computed using the convective heat flux for various plane is same as Nu of Eq. (2.118). This equality however does not hold for dynamic phenomena like flow reversal, which will be discussed in Chapter 17.

With this we close our discussion on the framework of buoyancy-driven flows. In the subsequent chapters we will describe various phenomena of buoyancy-driven flows.

Further reading

For derivation of the basic equations of buoyancy-driven flows, refer to Chandrasekhar (2013), Tritton (1988), and Kundu *et al.* (2015).

Exercises

(1) Consider thermal convection in a cylinder containing water and of 50 cm height and 50 cm of diameter. The bottom and top plates are kept at 60 °C and 30 °C respectively. Test whether Oberbeck-Boussinesq approximation is valid for the above setup. What is the Rayleigh number of the above setup?

(2) Consider the setup of Exercise 1. Analyze the flow for air as the experimental fluid with the temperature of the two plates as 500 °C and 30 °C respectively.

(3) For the Earth's lower atmosphere, estimate the Froude, Richardson, and Reynolds numbers.

(4) Write down RBC equations for $\mathrm{Pr} = 0$ and ∞.

(5) List the conserved quantities for RBC and stably stratified flows.

(6) Consider the following two-dimensional flow field in a box of size (π, π),

$$\mathbf{u} = \hat{x} 2B \cos y + \hat{y} 2C \cos x + 4A(\hat{x} \sin x \cos y - \hat{y} \cos x \sin y),$$

Show that the fluid is incompressible. Compute the total energy and vorticity field $\boldsymbol{\omega}$ of the flow.

(7) Consider the following fluid flow in a three-dimensional box of size (π, π, π):

$$\mathbf{u} = 4C(\hat{x} \sin x \cos z - \hat{z} \cos x \sin z) + 4B(\hat{y} \sin y \cos z - \hat{z} \cos y \sin z)$$
$$+ 8A(-\hat{x} \sin x \cos y \cos 2z - \hat{y} \cos x \sin y \cos 2z + \hat{z} \cos x \cos y \sin 2z).$$

Show that the fluid is incompressible. Compute the total energy and the vorticity field of the flow.

Chapter 3

Fourier Space Description

Fourier space description of a flow helps us quantify structures and their dynamics at different scales. In this chapter, we will discuss how to decompose the velocity, temperature, and density fields of buoyancy-driven flows in Fourier space.

Fourier series can be used to represent RBC or stably stratified flows under periodic and free-slip boundary conditions. In this chapter we will detail various aspects of this description. We begin this chapter with the definition of Fourier transform.

3.1 Definitions

Fourier basis is suitable for describing flows in a periodic box, say of size $L_x \times L_y \times L_z$. For RBC, the velocity and temperature fields are decomposed in Fourier basis as follows:

$$\mathbf{u}(\mathbf{r}, t) = \sum_{\mathbf{k}} \mathbf{u}(\mathbf{k}, t) \exp(i\mathbf{k} \cdot \mathbf{r}), \tag{3.1}$$

$$\theta(\mathbf{r}, t) = \sum_{\mathbf{k}} \theta(\mathbf{k}, t) \exp(i\mathbf{k} \cdot \mathbf{r}), \tag{3.2}$$

where $\mathbf{r}$ is the real space coordinate, $\mathbf{k} = (k_x, k_y, k_z)$ is the wavenumber with

$$k_x = \frac{2l\pi}{L_x}; \quad k_y = \frac{2m\pi}{L_y}; \quad k_z = \frac{2n\pi}{L_z}. \tag{3.3}$$

Here l, m, n are integers (from $-\infty$ to ∞). The variables $\mathbf{u}(\mathbf{k}, t)$ and $\theta(\mathbf{k}, t)$ are the Fourier amplitudes of velocity and temperature fields.

The associated inverse transforms are

$$\mathbf{u}(\mathbf{k}, t) = \frac{1}{L_x L_y L_z} \int d\mathbf{r} \mathbf{u}(\mathbf{r}, t) \exp(-i\mathbf{k} \cdot \mathbf{r}), \tag{3.4}$$

$$\theta(\mathbf{k}, t) = \frac{1}{L_x L_y L_z} \int d\mathbf{r} \theta(\mathbf{r}, t) \exp(-i\mathbf{k} \cdot \mathbf{r}), \tag{3.5}$$

where the aforementioned integrals are performed over the whole box. It is easy to show that the Fourier transform of the product of two real functions $f(\mathbf{r})$ and $g(\mathbf{r})$

is a convolution:

$$(fg)(\mathbf{k}) = \sum_{\mathbf{p}} f(\mathbf{k} - \mathbf{p})g(\mathbf{p}), \tag{3.6}$$

and the Fourier transform of the derivative of a real function is given by

$$(df/dx_j)(\mathbf{k}) = ik_j f(\mathbf{k}). \tag{3.7}$$

Using the above identities and Eqs. (2.17–2.19), we can derive equations for RBC in Fourier space. For convenience we set $\rho_m = 1$. The equations are

$$\frac{d}{dt}\mathbf{u}(\mathbf{k}) + \mathbf{N}_u(\mathbf{k}) = -ik\sigma(\mathbf{k}) + \alpha g\theta(\mathbf{k})\hat{z} - \nu k^2\mathbf{u}(\mathbf{k}), \tag{3.8}$$

$$\frac{d}{dt}\theta(\mathbf{k}) + N_\theta(\mathbf{k}) = -\frac{d\bar{T}}{dz}u_z(\mathbf{k}) - \kappa k^2\theta(\mathbf{k}), \tag{3.9}$$

$$\mathbf{k} \cdot \mathbf{u}(\mathbf{k}) = 0, \tag{3.10}$$

where the nonlinear terms are given by

$$\mathbf{N}_u(\mathbf{k}) = i\sum_{\mathbf{p}}[\mathbf{k} \cdot \mathbf{u}(\mathbf{k} - \mathbf{p})]\mathbf{u}(\mathbf{p}), \tag{3.11}$$

$$N_\theta(\mathbf{k}) = i\sum_{\mathbf{p}}[\mathbf{k} \cdot \mathbf{u}(\mathbf{k} - \mathbf{p})]\theta(\mathbf{p}). \tag{3.12}$$

The following equation for the pressure $\sigma(\mathbf{k})$ is derived by taking dot product of Eq. (3.8) with $i\mathbf{k}$ and by employing $\mathbf{k} \cdot \mathbf{u}(\mathbf{k}) = 0$:

$$\sigma(\mathbf{k}) = \frac{1}{k^2}[i\mathbf{k} \cdot \mathbf{N}(\mathbf{k}) - i\alpha g k_z \theta(\mathbf{k})]. \tag{3.13}$$

Equations (3.8–3.9) can also be rewritten in tensorial form as

$$\frac{d}{dt}u_i(\mathbf{k}) = -ik_i\sigma(\mathbf{k}) - ik_j\sum_{\mathbf{p}}u_j(\mathbf{k} - \mathbf{p})u_i(\mathbf{p}) + \alpha g\theta(\mathbf{k})\delta_{i,3} - \nu k^2 u_i(\mathbf{k}), \tag{3.14}$$

$$\frac{d}{dt}\theta(\mathbf{k}) = -\frac{d\bar{T}}{dz}u_{i,3}(\mathbf{k}) - ik_j\sum_{\mathbf{p}}u_j(\mathbf{k} - \mathbf{p})\theta(\mathbf{p}) - \kappa k^2\theta(\mathbf{k}), \tag{3.15}$$

$$k_i u_i(\mathbf{k}) = 0. \tag{3.16}$$

In Eqs. (3.14–3.16), i represents two things: $\sqrt{-1}$ in front of the $k_i\sigma(\mathbf{k})$ term, and x, y, z components in u_i.

One of the nondimensionalized versions of RBC equations is

$$\frac{d}{dt}\mathbf{u}(\mathbf{k}) + \mathbf{N}_u(\mathbf{k}) = -ik\sigma(\mathbf{k}) + \mathrm{RaPr}\theta(\mathbf{k})\hat{z} - \mathrm{Pr}k^2\mathbf{u}(\mathbf{k}), \tag{3.17}$$

$$\frac{d}{dt}\theta(\mathbf{k}) + N_\theta(\mathbf{k}) = u_z(\mathbf{k}) - \kappa k^2\theta(\mathbf{k}), \tag{3.18}$$

$$\mathbf{k} \cdot \mathbf{u}(\mathbf{k}) = 0. \tag{3.19}$$

The corresponding equations for the stably stratified flows are

$$\frac{d}{dt}\mathbf{u}(\mathbf{k}) + \mathbf{N}_u(\mathbf{k}) = -ik\sigma(\mathbf{k}) - Nb(\mathbf{k})\hat{z} - \nu k^2\mathbf{u}(\mathbf{k}), \tag{3.20}$$

$$\frac{d}{dt}b(\mathbf{k}) + N_b(\mathbf{k}) = Nu_z(\mathbf{k}) - \kappa k^2 b(\mathbf{k}), \tag{3.21}$$

$$\mathbf{k} \cdot \mathbf{u}(\mathbf{k}) = 0, \tag{3.22}$$

where N is the Brunt-Väisälä frequency[1], and the nonlinear terms are given by

$$\mathbf{N}_u(\mathbf{k}) = i \sum_{\mathbf{p}} [\mathbf{k} \cdot \mathbf{u}(\mathbf{k} - \mathbf{p})] \mathbf{u}(\mathbf{p}), \tag{3.23}$$

$$N_b(\mathbf{k}) = i \sum_{\mathbf{p}} [\mathbf{k} \cdot \mathbf{u}(\mathbf{k} - \mathbf{p})] b(\mathbf{p}). \tag{3.24}$$

Here the pressure $\sigma(\mathbf{k})$ is computed using

$$\sigma(\mathbf{k}) = \frac{1}{k^2}[i\mathbf{k} \cdot \mathbf{N}_u(\mathbf{k}) + iNk_z b(\mathbf{k})]. \tag{3.25}$$

In the next section, we will define energy and entropy in Fourier space.

3.2 Energy and entropy

The energy of a velocity Fourier mode, called *modal energy*, is given by

$$E_u(\mathbf{k}) = \frac{1}{2}|\mathbf{u}(\mathbf{k})|^2. \tag{3.26}$$

Using Eq. (3.2) and the orthogonality relation, we can show that the average kinetic energy of a flow field is

$$E_u = \frac{1}{2}\langle u^2 \rangle = \frac{1}{L_x L_y L_z} \int d\mathbf{r} \frac{1}{2}|\mathbf{u}(\mathbf{r})|^2 = \sum_{\mathbf{k}} \frac{1}{2}|\mathbf{u}(\mathbf{k})|^2. \tag{3.27}$$

This is the *Parseval's theorem*. Note however that E_u of the above equation is typically referred to as *total kinetic energy* because it is sum over all the Fourier modes. Similarly, the total kinetic helicity (strictly speaking, average) is given by

$$H_K = \frac{1}{2}\langle \mathbf{u} \cdot \boldsymbol{\omega} \rangle = \frac{1}{L_x L_y L_z} \int d\mathbf{r} \frac{1}{2}\mathbf{u} \cdot \boldsymbol{\omega} = \sum_{\mathbf{k}} \frac{1}{2}\Re[\mathbf{u}^*(\mathbf{k}) \cdot \boldsymbol{\omega}(\mathbf{k})] \tag{3.28}$$

Similarly, *modal entropy*, $E_\theta(\mathbf{k})$, is defined as

$$E_\theta(\mathbf{k}) = \frac{1}{2}|\theta(\mathbf{k}')|^2, \tag{3.29}$$

and the *total entropy* is

$$E_\theta = \frac{1}{2}\langle \theta^2 \rangle = \frac{1}{L_x L_y L_z} \int d\mathbf{r} \frac{1}{2}\theta(\mathbf{r})^2 = \sum_{\mathbf{k}} \frac{1}{2}|\theta(\mathbf{k})|^2. \tag{3.30}$$

Note that the above definition of entropy differs from the thermodynamic entropy.

To derive a dynamical equation for $|\mathbf{u}(\mathbf{k})|^2/2$, we multiply Eq. (3.14) with $u_i^*(\mathbf{k})$ and sum over i that yields

$$u_i^*(\mathbf{k})\frac{d}{dt}u_i(\mathbf{k}) = -ik_i u_i^*(\mathbf{k})\sigma(\mathbf{k}) - ik_j \sum_{\mathbf{p}} u_j(\mathbf{k} - \mathbf{p})u_i(\mathbf{p})u_i^*(\mathbf{k})$$

$$+ag\theta(\mathbf{k})u_z^*(\mathbf{k}) - \nu k^2 |\mathbf{u}(\mathbf{k})|^2. \tag{3.31}$$

---

[1]The symbol N is used for many quantities. N is the Brunt-Väisälä frequency, and $\mathbf{N}_u$, N_θ, and N_b (with subscripts) represent nonlinear terms.

Note that $-ik_i u_i^*(\mathbf{k})\sigma(\mathbf{k}) = 0$ due to the incompressibility condition, $\mathbf{k} \cdot \mathbf{u}(\mathbf{k}) = 0$. Hence, the pressure does not contribute to the kinetic energy transfers. Now we add the above equation with its complex conjugate that yields

$$\frac{d}{dt}E_u(\mathbf{k}) = \Im\left[\sum_{\mathbf{p}}[\mathbf{k} \cdot \mathbf{u}(\mathbf{q})]\,[\mathbf{u}(\mathbf{p}) \cdot \mathbf{u}^*(\mathbf{k})]\right] + \alpha g\Re[\theta(\mathbf{k})u_z^*(\mathbf{k})] - 2\nu k^2 E_u(\mathbf{k}),$$

(3.32)

where $\mathbf{q} = \mathbf{k} - \mathbf{p}$, and $\Re, \Im$ stand respectively for the real and imaginary parts of the argument. In the above equation, the rate of change of $E_u(\mathbf{k})$ occurs due to (a) the nonlinear energy transfers from the Fourier modes, quantified by the first term of right hand side, (b) the buoyancy $\alpha g\Re[\theta(\mathbf{k})u_z^*(\mathbf{k})]$, and (c) the dissipation term $-2\nu k^2 E_u(\mathbf{k})$. Following a similar procedure, we can derive an equation for the entropy spectrum:

$$\frac{d}{dt}E_\theta(\mathbf{k}) = \Im\left[\sum_{\mathbf{p}}[\mathbf{k} \cdot \mathbf{u}(\mathbf{q})]\,[\theta(\mathbf{p})\theta^*(\mathbf{k})]\right] - \frac{d\bar{T}}{dz}\Re[\theta(\mathbf{k})u_z^*(\mathbf{k})] - 2\kappa k^2 E_\theta(\mathbf{k}).$$

(3.33)

For stably stratified flows, we define modal energies of the density field, ρ, and its variant b, as

$$E_\rho(\mathbf{k}) = \frac{1}{2}|\rho(\mathbf{k}')|^2,$$

(3.34)

$$E_b(\mathbf{k}) = \frac{1}{2}|b(\mathbf{k}')|^2.$$

(3.35)

The dynamical equations for $E_u(\mathbf{k})$ and $E_b(\mathbf{k})$ are

$$\frac{d}{dt}E_u(\mathbf{k}) = \Im\left[\sum_{\mathbf{p}}[\mathbf{k} \cdot \mathbf{u}(\mathbf{q})]\,[\mathbf{u}(\mathbf{p}) \cdot \mathbf{u}^*(\mathbf{k})]\right] + N\Re[b(\mathbf{k})u_z^*(\mathbf{k})]$$

$$-2\nu k^2 E_u(\mathbf{k}),$$

(3.36)

$$\frac{d}{dt}E_b(\mathbf{k}) = \Im\left[\sum_{\mathbf{p}}[\mathbf{k} \cdot \mathbf{u}(\mathbf{q})]\,[b(\mathbf{p})b^*(\mathbf{k})]\right] - N\Re[b(\mathbf{k})u_z^*(\mathbf{k})]$$

$$-2\kappa k^2 E_b(\mathbf{k}).$$

(3.37)

For homogeneous and isotropic turbulence, the modal kinetic energy and modal entropy of all the modes in a thin wavenumber shell are statistically equal (on ensemble or temporal averaging). Hence, it is customary to define the following one-dimensional spectra that are sums of all the modes of a shell of unit radius:[2]

$$E_u(k) = \sum_{k-1 < k' \leq k} \frac{1}{2} |\mathbf{u}(\mathbf{k}')|^2, \tag{3.39}$$

$$E_\theta(k) = \sum_{k-1 < k' \leq k} \frac{1}{2} |\theta(\mathbf{k}')|^2, \tag{3.40}$$

$$E_\rho(k) = \sum_{k-1 < k' \leq k} \frac{1}{2} |\rho(\mathbf{k}')|^2, \tag{3.41}$$

$$E_b(k) = \sum_{k-1 < k' \leq k} \frac{1}{2} |b(\mathbf{k}')|^2. \tag{3.42}$$

The dynamical equations for the above quantities can be obtained by summing the modal energy and entropy over all the Fourier modes in a shell in their corresponding equations. For example, the equation for $E_u(k)$ is

$$\frac{d}{dt} E_u(k) = \sum_{k-1 < k' \leq k} \Im \left[\sum_{\mathbf{p}} [\mathbf{k}' \cdot \mathbf{u}(\mathbf{q})] \, [\mathbf{u}(\mathbf{p}) \cdot \mathbf{u}^*(\mathbf{k}')] \right] + \sum_{k-1 < k' \leq k} \alpha g \Re[\theta(\mathbf{k}') u_z^*(\mathbf{k}')]$$

$$- \sum_{k-1 < k' \leq k} 2\nu k'^2 E_u(\mathbf{k}') \tag{3.43}$$

with $\mathbf{k}' = \mathbf{p} + \mathbf{q}$.

In the next section we define the properties of the Fourier amplitudes of a real field, e.g., the velocity field and temperature field.

3.3   Reality condition

The velocity, density, and temperature fields are real. Hence

$$\mathbf{u}(-\mathbf{k}, t) = \frac{1}{L_x L_y L_z} \int d\mathbf{r} \, \mathbf{u}(\mathbf{r}, t) \exp(-i\mathbf{k} \cdot \mathbf{r})$$

$$= \mathbf{u}^*(\mathbf{k}, t). \tag{3.44}$$

Similarly for the temperature field,

$$\theta(-\mathbf{k}, t) = \theta^*(\mathbf{k}, t). \tag{3.45}$$

We illustrate this property in Fig. 3.1. Due to the above relation, in a computer simulation, we need to store only half the modes that reduces the computational and storage complexity.

Thermal convection typically involves walls or boundaries. Generic boundary conditions employed for the velocity field at the walls are free-slip or no-slip. In the next two sections, we will discuss these boundary conditions in the framework of Fourier transform.

---

[2]For continuum k, $E_X(k)$, where $X = u, \theta$ or b, is defined as

$$\int_k^{k+dk} E_X(k') dk' \approx E_X(k') dk'. \tag{3.38}$$

Here the integral is performed over all the modes in a shell whose inner and outer radii are k and $k + dk$ respectively. Note that $dk \to 0$ and $\int_0^\infty E_X(k') dk' = E_X$.

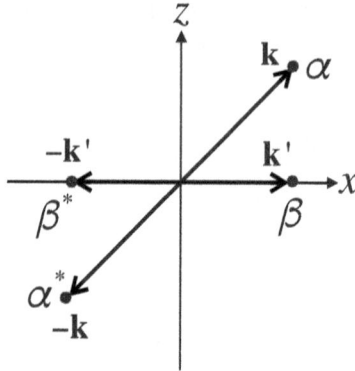

Fig. 3.1　An illustration of the reality condition. If $\theta(\mathbf{k}) = \alpha$ and $\theta(\mathbf{k}') = \beta$, then $\theta(-\mathbf{k}) = \alpha^*$ and $\theta(-\mathbf{k}') = \beta^*$.

3.4　Free-slip basis functions for RBC

The free-slip boundary condition on the velocity field is

$$\mathbf{u}_\perp = 0; \quad \frac{\partial \mathbf{u}_\parallel}{\partial n} = 0, \tag{3.46}$$

where $\mathbf{u}_\perp$ and $\mathbf{u}_\parallel$ are respectively the velocity components normal and parallel to the surface, and n is the coordinate parallel to the wall. Let us assume that the fluid is confined in a cube of size $(\pi \times \pi \times \pi)$ with the free-slip boundary condition applied to all the walls. For such a flow, the basis functions for the three components of the velocities are:

$$u_x = \hat{u}_x(k_x, k_y, k_z)(2 \sin k_x x)(\delta_{k_y,0} + 2 \cos k_y y)(\delta_{k_z,0} + 2 \cos k_z z),$$
$$\tag{3.47}$$

$$u_y = \hat{u}_y(k_x, k_y, k_z)(\delta_{k_x,0} + 2 \cos k_x x)(2 \sin k_y y)(\delta_{k_z,0} + 2 \cos k_z z),$$
$$\tag{3.48}$$

$$u_z = \hat{u}_z(k_x, k_y, k_z)(\delta_{k_x,0} + 2 \cos k_x x)(\delta_{k_y,0} + 2 \cos k_y y)(2 \sin k_z z).$$
$$\tag{3.49}$$

We term the above as *free-slip basis functions* with the hats representing the amplitudes of free-slip modes. Contrast them with the Fourier modes $\mathbf{u}(k_x, k_y, k_z)$ (without hats). Here, the components k_x, k_y, k_z are positive integers (including zero), unlike the components of a Fourier wavenumber that range from $-\infty$ to ∞. Note that the incompressibility condition for the free-slip modes are same as those for the Fourier modes, i.e.,

$$\mathbf{k} \cdot \hat{\mathbf{u}}(\mathbf{k}) = 0. \tag{3.50}$$

For the temperature field, either conducting or insulating boundary condition is employed. The conducting boundary condition is

$$T = \text{const}; \quad \text{or} \quad \theta = 0, \tag{3.51}$$

Table 3.1   Transformation rules to derive the Fourier amplitudes $\mathbf{u}(k_x, k_y, k_z)$ from the free-slip amplitudes $\hat{\mathbf{u}}(k_x, k_y, k_z)$.

mode	u_x	u_y	u_z
(k_x, k_y, k_z)	$\frac{1}{i}\hat{u}_x(k_x, k_y, k_z)$	$\frac{1}{i}\hat{u}_y(k_x, k_y, k_z)$	$\frac{1}{i}\hat{u}_z(k_x, k_y, k_z)$
$(k_x, k_y, -k_z)$	$\frac{1}{i}\hat{u}_x(k_x, k_y, k_z)$	$\frac{1}{i}\hat{u}_y(k_x, k_y, k_z)$	$-\frac{1}{i}\hat{u}_z(k_x, k_y, k_z)$
$(k_x, -k_y, k_z)$	$\frac{1}{i}\hat{u}_x(k_x, k_y, k_z)$	$-\frac{1}{i}\hat{u}_y(k_x, k_y, k_z)$	$\frac{1}{i}\hat{u}_z(k_x, k_y, k_z)$
$(k_x, -k_y, -k_z)$	$\frac{1}{i}\hat{u}_x(k_x, k_y, k_z)$	$-\frac{1}{i}\hat{u}_y(k_x, k_y, k_z)$	$-\frac{1}{i}\hat{u}_z(k_x, k_y, k_z)$
$(-k_x, k_y, k_z)$	$-\frac{1}{i}\hat{u}_x(k_x, k_y, k_z)$	$\frac{1}{i}\hat{u}_y(k_x, k_y, k_z)$	$\frac{1}{i}\hat{u}_z(k_x, k_y, k_z)$
$(-k_x, k_y, -k_z)$	$-\frac{1}{i}\hat{u}_x(k_x, k_y, k_z)$	$\frac{1}{i}\hat{u}_y(k_x, k_y, k_z)$	$-\frac{1}{i}\hat{u}_z(k_x, k_y, k_z)$
$(-k_x, -k_y, k_z)$	$-\frac{1}{i}\hat{u}_x(k_x, k_y, k_z)$	$-\frac{1}{i}\hat{u}_y(k_x, k_y, k_z)$	$\frac{1}{i}\hat{u}_z(k_x, k_y, k_z)$
$(-k_x, -k_y, -k_z)$	$-\frac{1}{i}\hat{u}_x(k_x, k_y, k_z)$	$-\frac{1}{i}\hat{u}_y(k_x, k_y, k_z)$	$-\frac{1}{i}\hat{u}_z(k_x, k_y, k_z)$

while the insulating boundary condition is

$$\frac{\partial T}{\partial n} = 0. \tag{3.52}$$

The following function

$$\theta = \hat{\theta}(k_x, k_y, k_z)(\delta_{k_x,0} + 2\cos k_x x)(\delta_{k_y,0} + 2\cos k_y y)(2\sin k_z z) \tag{3.53}$$

is suitable for conducting walls at $z = 0, \pi$, and insulating side walls at $x = 0, \pi$, and $y = 0, \pi$.

The Fourier basis function, which is suitable for a periodic box, is $\exp[i(k_x x + k_y y + k_z z)]$, where k_x, k_y, k_z take both positive or negative values. We can convert the coefficient of a free-slip basis function to the corresponding Fourier basis function using the identities: $2\cos x = [\exp(ix) + \exp(-ix)]$ and $2\sin x = [\exp(ix) - \exp(-ix)]/i$. In Table 3.1, we list the Fourier amplitudes corresponding to $\hat{\mathbf{u}}(k_x, k_y, k_z)$ of the free-slip basis. As an example,

$$
\begin{aligned}
u_x &= \hat{u}_x(1,0,1)(2\sin x)(2\cos z) \\
&= \hat{u}_x(1,0,1)\frac{1}{i}\left[\exp(ix) - \exp(-ix)\right]\left[\exp(iz) + \exp(-iz)\right] \\
&= \hat{u}_x(1,0,1)\left[\frac{1}{i}\exp\{i(x+z)\} + \frac{1}{i}\exp\{i(x-z)\}\right. \\
&\quad \left. -\frac{1}{i}\exp\{i(-x+z)\} - \frac{1}{i}\exp\{i(-x-z)\}\right]
\end{aligned}
\tag{3.54}
$$

Therefore,

$$u_x(1,0,1) = u_x(1,0,-1) = \frac{1}{i}\hat{u}_x(1,0,1), \tag{3.55}$$

$$u_x(-1,0,1) = u_x(-1,0,-1) = -\frac{1}{i}\hat{u}_x(1,0,1), \tag{3.56}$$

as shown in Table 3.1.

Example 3.1: Consider an incompressible velocity field in the xy plane whose $u_x = 4A \sin k_x x \cos k_y y$. Assume that the velocity field satisfies free-slip boundary conditions at all the confining walls. Construct the full velocity field.

Solution: When we compare the expression $u_x = 4A \sin k_x x \cos k_y y$ with Eq. (3.47), we obtain $\hat{u}_x(k_x, k_y, 0) = A$. The incompressible condition $\mathbf{k} \cdot \hat{\mathbf{u}}(\mathbf{k}) = 0$ yields

$$\hat{u}_y(k_x, k_y, 0) = -\frac{k_x}{k_y}\hat{u}_x(k_x, k_y, 0) = -\frac{k_x}{k_y}A,$$

while $\hat{u}_z = B$ can take any value since $k_z = 0$. Therefore, the velocity field is

$$\mathbf{u} = \hat{x}4A \sin k_x x \cos k_y y - \hat{y}\frac{k_x}{k_y}4A \cos k_x x \sin k_y y + \hat{z}4B \cos k_x x \cos k_y y.$$

Since the velocity field is function only of x and y, but it has three components, it is called *two-dimensional three-component (2D3C)* field. In Fig. 3.2 we illustrate the aforementioned velocity field with $B = 0$.

The above form of the velocity field is however valid when both k_x and k_y are nonzero. For modes with $k_y = 0$,

$$\mathbf{u}(x) = \hat{y}2A \cos k_x x + \hat{z}2B \cos k_x x,$$

while, for $k_x = 0$,

$$\mathbf{u}(y) = \hat{x}2A \cos k_y y + \hat{z}2B \cos k_y y.$$

When $k_x = k_y = 0$, $\mathbf{u} = \text{const.}$

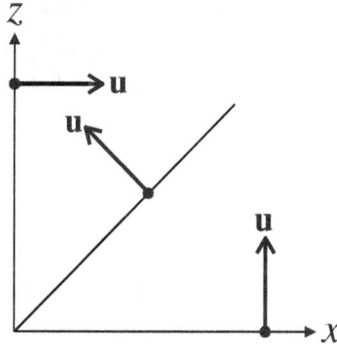

Fig. 3.2   Example 3.1: Illustration of the velocity Fourier modes. Here we assume that $B = 0$.

Example 3.2: Consider the following flow field:

$$\mathbf{u} = \hat{x}2B \cos y + \hat{y}2C \cos x + 4A(\hat{x} \sin x \cos y - \hat{y} \cos x \sin y).$$

List the active Fourier modes. Is the flow incompressible? Compute the total energy of the flow. Derive the equations of motion for A, B, and C.

Solution: In the free-slip basis, the wavenumbers of the flow are (1,0), (0,1) and (1,1), and the corresponding Fourier modes have amplitudes C, B, and A respectively. In Fourier basis, the wavenumbers are $(\pm1,0)$, $(0,\pm1)$, and $(\pm1,\pm1)$. The Fourier amplitudes of the corresponding modes are listed in Table 3.2.

Table 3.2 Example 3.2: Velocity Fourier modes. u_1 is the amplitude of the velocity field in Craya-Herring basis. Used in Example 3.4.

mode	u_x	u_y	u_1
$(1,0)$	0	C	$-C$
$(-1,0)$	0	C	C
$(0,1)$	B	0	B
$(0,-1)$	B	0	$-B$
$(1,1)$	$\frac{A}{i}$	$-\frac{A}{i}$	$\sqrt{2}A/i$
$(-1,-1)$	$-\frac{A}{i}$	$\frac{A}{i}$	$\sqrt{2}A/i$
$(1,-1)$	$\frac{A}{i}$	$\frac{A}{i}$	$-\sqrt{2}A/i$
$(-1,1)$	$-\frac{A}{i}$	$-\frac{A}{i}$	$-\sqrt{2}A/i$

In Fourier space, $\mathbf{k} \cdot \mathbf{u}(\mathbf{k}) = 0$ for each wavenumber. Therefore, the incompressibility condition is automatically satisfied in the Fourier space. The total energy of the flow is obtained by performing the following sum over all the Fourier modes:

$$E_u = \sum_{\mathbf{k}} \frac{1}{2}|\mathbf{u}(\mathbf{k})|^2 = C^2 + B^2 + 4A^2.$$

Note that above value matches with the average energy computed in real space. This is in accordance with Parseval's theorem.

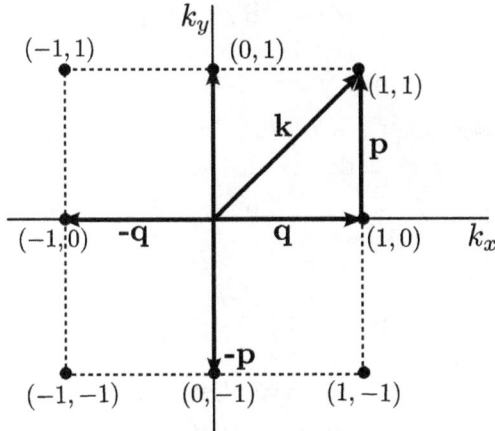

Fig. 3.3 Fourier modes of the flow field of Example 3.2.

In Fig. 3.3, we mark all the active Fourier modes and one of the four triads. Now we will derive the equations of motion for A, B, and C using the triads involving $\mathbf{q} = (1,0)$, $\mathbf{p} = (0,1)$, and $\mathbf{k} = (1,1)$. We compute the nonlinear term $\mathbf{N}_u$ using Eq. (3.11):

$$\mathbf{N}_u(\mathbf{k}) = i\mathbf{k} \cdot \mathbf{u}(\mathbf{q})\mathbf{u}(\mathbf{p}) + i\mathbf{k} \cdot \mathbf{u}(\mathbf{p})\mathbf{u}(\mathbf{q})$$
$$= iu_y(\mathbf{q})u_x(\mathbf{p})\hat{x} + iu_x(\mathbf{p})u_y(\mathbf{q})\hat{y} = iBC(\hat{x} + \hat{y}).$$
$$\mathbf{N}_u(\mathbf{q}) = i\mathbf{q} \cdot \mathbf{u}(\mathbf{k})\mathbf{u}(-\mathbf{p}) + i\mathbf{q} \cdot \mathbf{u}(-\mathbf{p})\mathbf{u}(\mathbf{k})$$
$$= iu_x(\mathbf{k})u_x^*(\mathbf{p})\hat{x} + iu_x^*(\mathbf{p})\mathbf{u}(\mathbf{k}) = AB(2\hat{x} - \hat{y}),$$
$$\mathbf{N}_u(\mathbf{p}) = AC(\hat{x} - 2\hat{y}).$$

In the above computations, we substitute the values of the velocity components from the Table 3.2. We also employ $\mathbf{q} = \mathbf{k} - \mathbf{p}$ and $\mathbf{p} = \mathbf{k} - \mathbf{q}$.

Pressure computations using Eq. (3.13) yields

$$\sigma(\mathbf{k}) = -BC$$
$$\sigma(\mathbf{p}) = -2iAC$$
$$\sigma(\mathbf{q}) = 2iAB.$$

Using the above, we derive the equations of motion for the Fourier modes as

$$\frac{d}{dt}\mathbf{u}(\mathbf{k}) = -\mathbf{N}(\mathbf{k}) - i\mathbf{k}\sigma(\mathbf{k}) - \nu k^2\mathbf{u}(\mathbf{k})$$
$$= -iBC(\hat{x} + \hat{y}) - i(\hat{x} + \hat{y})(-BC) - \nu 2\mathbf{u}(\mathbf{k})$$
$$= -2\nu\mathbf{u}(\mathbf{k}).$$

Similar analysis yields

$$\frac{d}{dt}\mathbf{u}(\mathbf{q}) = AB\hat{y} - \nu\mathbf{u}(\mathbf{q}),$$
$$\frac{d}{dt}\mathbf{u}(\mathbf{p}) = -AC\hat{x} - \nu\mathbf{u}(\mathbf{p}).$$

We consider the inviscid limit $\nu = 0$. Since $\mathbf{u}(\mathbf{k}) = (A/i)(\hat{x} - \hat{y})$, $\mathbf{u}(\mathbf{q}) = C\hat{y}$, and $\mathbf{u}(\mathbf{p}) = B\hat{x}$, the equations of motion yield

$$\dot{A} = 0,$$
$$\dot{B} = -AC,$$
$$\dot{C} = AB,$$

whose solution are

$$A = \text{constant},$$
$$B = c\cos(At),$$
$$C = c\sin(At)$$

with c as a constant. Thus, B and C oscillate in time with a frequency of A. From the above we can deduce that

$$[B(t)]^2 + [C(t)]^2 + 4[A(t)]^2 = [B(0)]^2 + [C(0)]^2 + 4[A(0)]^2 = \text{const}.$$

Hence, the total energy is conserved, as expected.

Example 3.3: Consider the velocity field

$$\mathbf{u} = 4C(\hat{x}\sin x\cos z - \hat{z}\cos x\sin z) + 4B(\hat{y}\sin y\cos z - \hat{z}\cos y\sin z)$$
$$+ 8A(-\hat{x}\sin x\cos y\cos 2z - \hat{y}\cos x\sin y\cos 2z + \hat{z}\cos x\cos y\sin 2z).$$

List the active Fourier modes. Is the flow incompressible? Compute the vorticity, the total energy, and the total kinetic helicity of the flow. Derive the equations of motion for A, B, and C.

Solution: The wavenumbers of the above velocity field are $(1,0,1)$, $(0,1,1)$, and $(1,1,2)$. The wavenumbers for the Fourier expansion are $(\pm 1, 0, \pm 1)$, $(0, \pm 1, \pm 1)$, and $(\pm 1, \pm 1, \pm 2)$. In Table 3.3 we present the Fourier amplitudes for the wavenumbers $(1,0,1)$, $(0,1,1)$, and $(1,1,2)$; the amplitudes of other Fourier modes can be written easily using the rules of Table 3.1. Table 3.3 also contains the vorticity field corresponding the the Fourier modes.

Table 3.3 Exercise 3.3: The Fourier amplitudes of the velocity and vorticity fields. u_2 is the component of $\mathbf{u}$ in Craya-Herring basis. Used in Example 3.5.

mode	u_x	u_y	u_z	ω_x	ω_y	ω_z	u_2
$(1,0,1)$	$\frac{C}{i}$	0	$-\frac{C}{i}$	0	$2C$	0	$\sqrt{2}C/i$
$(0,1,1)$	0	$\frac{B}{i}$	$-\frac{B}{i}$	$-2B$	0	0	$\sqrt{2}B/i$
$(1,1,2)$	$-\frac{A}{i}$	$-\frac{A}{i}$	$\frac{A}{i}$	$3A$	$-3A$	0	$\sqrt{3}A/i$

Note that $\mathbf{k}\cdot\mathbf{u}(\mathbf{k}) = 0$ for all the modes, hence they satisfy the incompressibility condition. The total energy of the field is

$$E_u = \sum_{\mathbf{k}}\frac{1}{2}|\mathbf{u}(\mathbf{k})|^2 = 4\times C^2 + 4\times B^2 + 8\times\frac{3}{2}A^2 = 4(C^2 + B^2 + 3A^2).$$

Here the prefactors are the number of Fourier modes in all the quadrants in the Fourier space. The vorticity mode corresponding to $\mathbf{u}(\mathbf{k})$ is

$$\boldsymbol{\omega}(\mathbf{k}) = i\mathbf{k}\times\mathbf{u}(\mathbf{k}), \tag{3.57}$$

and they are listed in Table 3.3. The total kinetic helicity of the flow is

$$H_K = \sum_{\mathbf{k}}\frac{1}{2}\Re(\mathbf{u}(\mathbf{k})\cdot\boldsymbol{\omega}^*(\mathbf{k})) = 0.$$

Hence the flow has zero kinetic helicity.

We focus on the following interacting triads

$$(1,1,2)\oplus(-1,0,-1)\oplus(0,-1,-1) = 0,$$
$$(-1,1,2)\oplus(1,0,-1)\oplus(0,-1,-1) = 0,$$
$$(1,-1,2)\oplus(-1,0,-1)\oplus(0,1,-1) = 0.$$

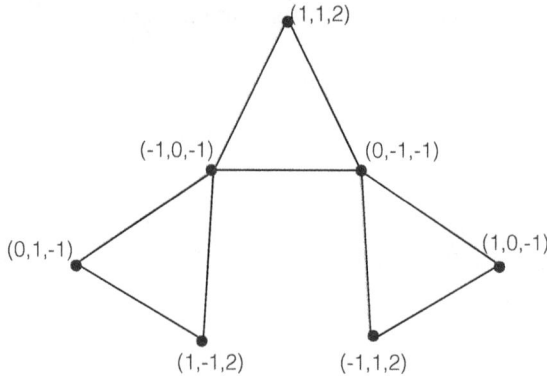

Fig. 3.4   Example 3.3. The interacting triads. The wavenumbers in a triad add up to zero.

which are depicted in Fig. 3.4. Therefore, the nonlinear interactions involving the above modes are

$$\mathbf{k} = (1, 1, 2) = (1, 0, 1) \bigoplus (0, 1, 1),$$

$$\mathbf{q} = (1, 0, 1) = (0, -1, -1) \bigoplus (1, 1, 2) + (0, 1, -1) \bigoplus (1, -1, 2),$$

$$\mathbf{p} = (0, 1, 1) = (-1, 0, -1) \bigoplus (1, 1, 2) + (1, 0, -1) \bigoplus (-1, 1, 2).$$

Using Eqs. (3.11, 3.13), we compute the nonlinear terms and the pressure as

$$\mathbf{N}(\mathbf{k}) = iBC(\hat{x} + \hat{y} - 2\hat{z}),$$

$$\mathbf{N}(\mathbf{q}) = i2AB(\hat{x} - \hat{z}),$$

$$\mathbf{N}(\mathbf{p}) = i2AC(\hat{y} - \hat{z}),$$

$$\sigma(\mathbf{q}) = \sigma(\mathbf{p}) = 0,$$

$$\sigma(\mathbf{k}) = \frac{1}{3}BC.$$

Therefore the equations of motion for Fourier modes are

$$\frac{d}{dt}\mathbf{u}(\mathbf{q}) = -i2AB(\hat{x} - \hat{z}) - 2\nu\mathbf{u}(\mathbf{q}),$$

$$\frac{d}{dt}\mathbf{u}(\mathbf{p}) = -i2AC(\hat{y} - \hat{z}) - 2\nu\mathbf{u}(\mathbf{p}),$$

$$\frac{d}{dt}\mathbf{u}(\mathbf{k}) = -i\frac{4}{3}BC(\hat{x} + \hat{y} - \hat{z}) - 6\nu\mathbf{u}(\mathbf{k}).$$

Since $\mathbf{u}(\mathbf{q}) = C/i(\hat{x} - \hat{z})$, $\mathbf{u}(\mathbf{p}) = B/i(\hat{y} - \hat{z})$, and $\mathbf{u}(\mathbf{k}) = -A/i(\hat{x} + \hat{y} - \hat{z})$, the above equations transform to the following under the $\nu = 0$ limit:

$$\dot{A} = -\frac{4}{3}BC,$$

$$\dot{B} = 2AC,$$

$$\dot{C} = 2AB.$$

The above equations have the same form as those of asymmetric top [Landau and Lifshitz (1976)]. The solutions $A(t), B(t), C(t)$, written in terms of elliptic functions, are periodic in time. Here, either B or C oscillates around a mean value, and the other two modes oscillate around zero. We will revisit this feature in Chapter 8 when we discuss the seven-mode model of thermal convection.

In the next section, we will discuss very briefly how no-slip boundary condition is handled in thermal convection.

3.5 No-slip boundary condition for RBC

For realistic fluid flows, we employ no-slip boundary condition, which is

$$\mathbf{u} = 0 \tag{3.58}$$

at the surface. Unfortunately, Fourier expansion is not suitable for this boundary condition, and more complex expansion involving Chebyshev polynomials are employed for this purpose. The discussion on Chebyshev polynomials is beyond the scope of this book; the reader is referred to Boyd (2003) and Canuto *et al.* (1988) for the same.

Typically, flows with the no-slip boundary conditions are simulated in real space using finite-difference, finite-volume, finite-element, or spectral-element methods. Near the boundary layers, the computational mesh is refined appropriately to take into account the steep variations of the velocity and temperature fields.

For turbulent flows with moderate Prandtl numbers, the viscous boundary layers are quite thin. Hence, the structures within the boundary layers are of very small size compared to the box size, and they would contribute to a large wavenumber regime of the kinetic energy spectrum $E_u(k)$. Thus, these structures do not affect the inertial range properties significantly. Therefore, the inertial-range energy and entropy spectra computed using the free-slip boundary conditions are quite close to those of realistic flows. This is a very useful observation that helps us compute the energy and entropy spectra quite easily using the Fourier basis functions.

The energy transfers among the interacting Fourier modes provide valuable diagnostics for understanding patterns, chaos, and turbulence in buoyancy-driven flows. We will show later that the energy flux provides valuable insights into the turbulence dynamics. These computations are conveniently performed using the Fourier basis functions [Verma (2004); Verma *et al.* (2017b)]. Note however that the Fourier transformation is typically performed on data sampled on uniform mesh. Since typical output of finite-difference, finite-volume, finite-element, or spectral-element computations are on nonuniform mesh, they need to be interpolated to uniform mesh for spectral analysis. Chandra and Verma (2011) and Chandra and Verma (2013) performed such computations for 2D RBC, and Kumar and Verma (2017) for 3D RBC.

In the next section we will describe Craya-Herring decomposition for incompressible flows.

3.6 Craya-Herring decomposition

Craya (1958) and Herring (1974) used incompressibility condition to reduce the three components of an incompressible velocity field to two. For the same, they proposed a set of basis vectors $\hat{e}_1$, $\hat{e}_2$, and $\hat{e}_3$ defined below:

$$\hat{e}_3(\mathbf{k}) = \hat{k}, \tag{3.59}$$

$$\hat{e}_1(\mathbf{k}) = \frac{\hat{k} \times \hat{n}}{|\hat{k} \times \hat{n}|}, \tag{3.60}$$

$$\hat{e}_2(\mathbf{k}) = \hat{e}_3 \times \hat{e}_1, \tag{3.61}$$

where $\hat{k}$ is the unit vector along the wavenumber $\mathbf{k}$, and $\hat{n}$ is along any direction (also see [Lesieur (2008); Sagaut and Cambon (2008)]). It is customary to choose $\hat{n}$ along the anisotropy direction, for example along the acceleration due to gravity, or rotation axis, or the mean magnetic field. See Figs. 3.5 and 3.6 for illustrations.

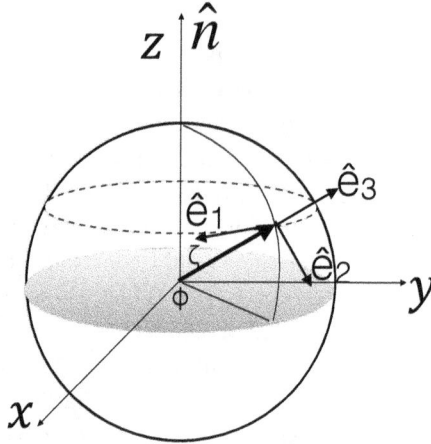

Fig. 3.5   Illustration of Craya-Herring basis vectors. The unit vector $\hat{e}_3$ is along $\mathbf{k}$, while $\mathbf{u}(\mathbf{k})$ lies in the plane formed by $\hat{e}_1$ and $\hat{e}_2$.

In the Craya-Herring basis,

$$\mathbf{u}(\mathbf{k}) = u_1(\mathbf{k})\hat{e}_1 + u_2(\mathbf{k})\hat{e}_2, \tag{3.62}$$

$$\mathbf{k} = k_3\hat{e}_3. \tag{3.63}$$

Note that $u_3(\mathbf{k})$, the velocity component along $\hat{e}_3(\mathbf{k})$ is zero because $\mathbf{k} \cdot \mathbf{u}(\mathbf{k}) = 0$. Also, $u_{1,2}(\mathbf{k})$ are in general complex number[3]. We can compute the corresponding

---

[3]Note that $u_{1,2}(\mathbf{k})$ are not the x and y components of $\mathbf{u}(\mathbf{k})$.

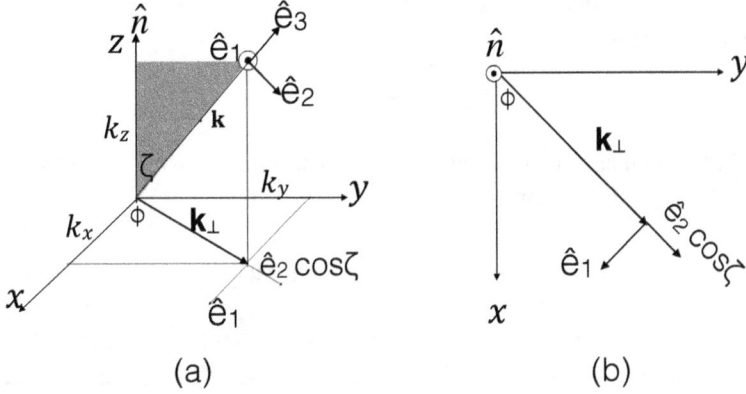

Fig. 3.6   Craya-Herring basis vectors and their relationships with the Cartesian basis vectors: (a) In 3D. (b) In k_x-k_y plane.

Cartesian components as

$$\mathbf{u}(\mathbf{k}) = \begin{pmatrix} u_x \\ u_y \\ u_z \end{pmatrix} = \begin{pmatrix} u_1(\mathbf{k}) \sin\phi + u_2(\mathbf{k}) \cos\zeta \cos\phi \\ -u_1(\mathbf{k}) \cos\phi + u_2(\mathbf{k}) \cos\zeta \sin\phi \\ -u_2(\mathbf{k}) \sin\zeta \end{pmatrix}$$

$$= \begin{pmatrix} u_1(\mathbf{k})\frac{k_y}{k_\perp} + u_2(\mathbf{k})\frac{k_x k_z}{k k_\perp} \\ -u_1(\mathbf{k})\frac{k_x}{k_\perp} + u_2(\mathbf{k})\frac{k_y k_z}{k k_\perp} \\ -u_2(\mathbf{k})\frac{k_\perp}{k} \end{pmatrix}, \tag{3.64}$$

where ζ, ϕ are respectively the polar and azimuthal angles of the wave vector $\mathbf{k}$. See Fig. 3.6 for an illustration. In Fourier space, the vorticity $\boldsymbol{\omega}(\mathbf{k})$ of the Fourier mode is

$$\boldsymbol{\omega}(\mathbf{k}) = i\mathbf{k} \times \mathbf{u}(\mathbf{k}) = ik \begin{pmatrix} -u_2(\mathbf{k}) \\ u_1(\mathbf{k}) \end{pmatrix}. \tag{3.65}$$

The modal energy $E(\mathbf{k})$ and the modal kinetic helicity $H_K(\mathbf{k})$ of the velocity mode are

$$E(\mathbf{k}) = \frac{1}{2}\mathbf{u}^*(\mathbf{k})\mathbf{u}(\mathbf{k}) = \frac{1}{2}\left[|u_1(\mathbf{k})|^2 + |u_2(\mathbf{k})|^2\right], \tag{3.66}$$

$$H_K(\mathbf{k}) = \frac{1}{2}\Re[\mathbf{u}^*(\mathbf{k}) \cdot \boldsymbol{\omega}(\mathbf{k})] = k\Im[u_1^*(\mathbf{k})u_2(\mathbf{k})]. \tag{3.67}$$

Therefore, for nonzero helicity, both u_1 and u_2 must be nonzero. In addition, u_1 and u_2 cannot be pure real or pure imaginary simultaneously.

Now let us write down the momentum equation in the Craya-Herring basis. Since $u_3(\mathbf{k}) = 0$, Eq. (3.8) transforms to

$$\frac{d}{dt}u_\alpha(\mathbf{k}) = -N_{u,\alpha}(\mathbf{k}) + F_\alpha(\mathbf{k}) \quad \text{for } \alpha = 1, 2, \tag{3.68}$$

$$\frac{d}{dt}u_3(\mathbf{k}) = 0 = -N_{u,3}(\mathbf{k}) - ik\sigma(\mathbf{k}) + F_3(\mathbf{k}), \tag{3.69}$$

where $\mathbf{F}$ is the external force (including buoyancy and viscous), and $\mathbf{N}_u$ is the nonlinear term. Note that

$$N_{u,3}(\mathbf{k}) = \mathbf{N}_u(\mathbf{k}) \cdot \hat{e}_3 \qquad (3.70)$$

need not be zero since the convolution of Eq. (3.11) involves many velocity modes. Equation (3.69) yields the pressure $\sigma(\mathbf{k})$ as

$$\sigma(\mathbf{k}) = \frac{i}{k}[N_{u,3}(\mathbf{k}) - F_3(\mathbf{k})]. \qquad (3.71)$$

Thus, the pressure is determined from the velocity and force fields. Clearly the dynamical evolution of the system is determined by Eq. (3.68) that contains a pair of equations for every $\mathbf{k}$. Also, for linear systems, $\mathbf{N} = 0$, and hence

$$\frac{d}{dt}u_\alpha(\mathbf{k}) = F_\alpha(\mathbf{k}) \text{ for } \alpha = 1, 2, \qquad (3.72)$$

$$\sigma(\mathbf{k}) = -\frac{i}{k}F_3(\mathbf{k}). \qquad (3.73)$$

In this book, we will discuss waves and instabilities of flows, hence we illustrate a wave solution along $\mathbf{k}$. If the velocity Fourier mode is aligned along $\hat{e}_2$, then the wave solution is

$$\begin{aligned}\mathbf{u}(\mathbf{r}, t) &= \hat{e}_2 \Re[u_2 \exp i(\mathbf{k} \cdot \mathbf{r} - \omega t)] \\ &= \hat{e}_2 |u_2| \cos(kz' - \omega t + \phi_k),\end{aligned} \qquad (3.74)$$

where ω is the frequency of the wave, $u_2 = |u_2| \exp(i\phi_k)$ with ϕ_k as the phase of the wave, and z' is the coordinate along $\hat{e}_3(\mathbf{k})$. Note that ϕ_k differs from ϕ, the azimuthal angle that $\mathbf{k}_\perp$ makes with the $\hat{x}$. Clearly the wave is moving along $\hat{e}_3(\mathbf{k})$ or positive z'.

Note that under the *parity transformation* $\mathbf{k} \to -\mathbf{k}$,

$$\hat{e}_1(-\mathbf{k}) = -\hat{e}_1(\mathbf{k}), \qquad (3.75)$$
$$\hat{e}_2(-\mathbf{k}) = \hat{e}_2(\mathbf{k}), \qquad (3.76)$$
$$\hat{e}_3(-\mathbf{k}) = -\hat{e}_3(\mathbf{k}). \qquad (3.77)$$

See Fig. 3.7 for an illustration. Therefore, the reality condition $\mathbf{u}(-\mathbf{k}) = \mathbf{u}^*(\mathbf{k})$ yields the following relations:

$$u_1(-\mathbf{k}) = -u_1^*(\mathbf{k}) \qquad (3.78)$$
$$u_2(-\mathbf{k}) = u_2^*(\mathbf{k}). \qquad (3.79)$$

We will use the above relations in subsequent sections.

In the next section, we will employ Craya-Herring basis to 2D flows.

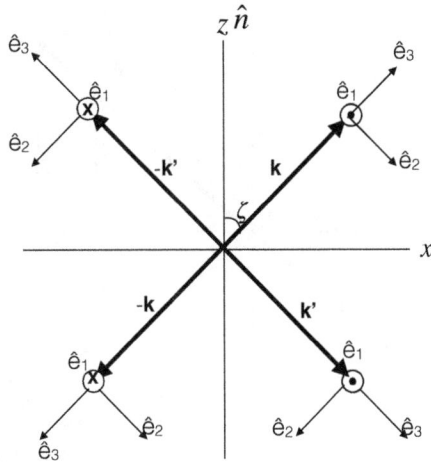

Fig. 3.7 Transformations of $\hat{e}_1$ and $\hat{e}_2$ under parity: $\mathbf{k} \to -\mathbf{k}$.

3.7 Craya-Herring basis for 2D flows

Craya-Herring decomposition simplifies the computations of two-dimensional flows, which is a special case of 3D flows. We assume $\mathbf{k}$ to lie in the k_x-k_y plane, and $\hat{n}$ along $\hat{z}$. Hence the basis vectors $\hat{e}_1$, $\hat{e}_2$, and $\hat{e}_3$ appear as shown in Fig. 3.8. A two-dimensional velocity field can be expressed as[4]

$$\mathbf{u}(\mathbf{k}) = u_1 \hat{e}_1. \tag{3.80}$$

Note that the kinetic helicity is zero for such flows since the vorticity $\boldsymbol{\omega}(\mathbf{k})$ is perpendicular to $\mathbf{u}(\mathbf{k})$. The Cartesian components of the velocity field are

$$\mathbf{u}(\mathbf{k}) = \begin{pmatrix} u_x \\ u_y \end{pmatrix} = \begin{pmatrix} u_1 \sin\phi \\ -u_1 \cos\phi \end{pmatrix} = \begin{pmatrix} (k_y/k)u_1 \\ -(k_x/k)u_1 \end{pmatrix}. \tag{3.81}$$

The reality condition is $\mathbf{u}(-\mathbf{k}) = \mathbf{u}^*(\mathbf{k})$. As shown in Fig. 3.8, $\hat{e}_1$ changes direction under $\mathbf{k} \to -\mathbf{k}$ transformation. Therefore,

$$u_1(-\mathbf{k}) = -u_1^*(\mathbf{k}). \tag{3.82}$$

If $u_1(\mathbf{k}) = A + B/i$ with A, B as real, then

$$u_1(-\mathbf{k}) = -A + B/i. \tag{3.83}$$

For a scalar field θ, the reality condition yields

$$\theta(-\mathbf{k}) = \theta^*(\mathbf{k}). \tag{3.84}$$

Hence, if $\theta(\mathbf{k}) = C + D/i$ with C and D as real, then

$$\theta(-\mathbf{k}) = C - D/i. \tag{3.85}$$

---

[4]In general, $\mathbf{u}(\mathbf{k}) = u_1 \hat{e}_1 + u_2 \hat{e}_2$ with $u_2 \hat{e}_2$ perpendicular to the k_x-k_y plane. Such two-dimensional three-component (*2D-3C*) fields are often encountered in magnetohydrodynamics. However, we will not deal with such fields in this book.

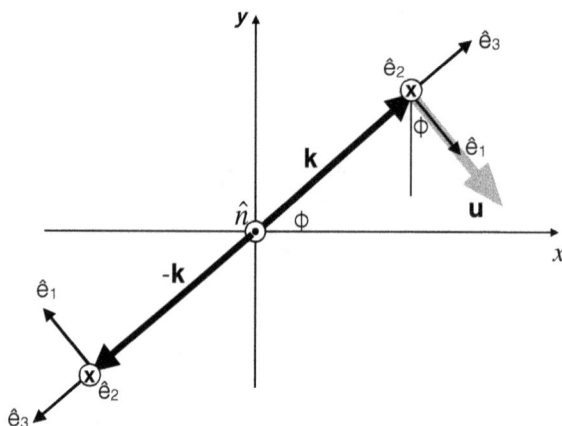

Fig. 3.8   Illustration of Craya-Herring basis vectors and $\mathbf{u}(\mathbf{k})$ in two dimensions.

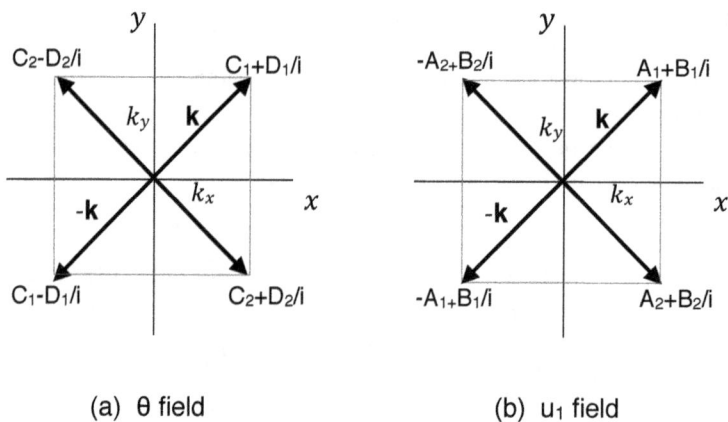

(a) θ field   (b) u_1 field

Fig. 3.9   In 2D: (a) Scalar field $\theta(\mathbf{k})$ at wavenumbers $\mathbf{k} = (\pm|k_x|, \pm|k_y|)$.   (b) Craya-Herring component u_1 at wavenumbers $\mathbf{k} = (\pm|k_x|, \pm|k_y|)$.

Craya-Herring basis can be used to represent the temperature and velocity fields for different boundary conditions including free-slip, insulating, and conducting ones. Here we describe the details for a 2D field. Let us focus on four wavenumbers $\mathbf{k} = (\pm|k_x|, \pm|k_y|)$. For the temperature field, using Eq. (3.84), we deduce that the temperature field at the above four wavenumbers are as shown in Fig. 3.9(a).

In real space, the corresponding $\theta(\mathbf{r})$ is

$$\theta(\mathbf{r}) = [(C_1 + D_1/i) \exp[i(|k_x|x) + |k_y|y)] + c.c.$$
$$(C_2 + D_2/i) \exp[i(|k_x|x) - |k_y|y)] + c.c.]$$

or

$$\theta(\mathbf{r}) = 2[(C_1 + C_2)\cos(|k_x|x)\cos(|k_y|y) + (C_2 - C_1)\sin(|k_x|x)\sin(|k_y|y)$$
$$+ (D_1 + D_2)\sin(|k_x|x)\cos(|k_y|y) + (D_1 - D_2)\cos(|k_x|x)\sin(|k_y|y)].$$

$$(3.86)$$

Therefore we have the following four possibilities:

(1) Horizontal and vertical walls are conducting: Here $D_1 = D_2 = 0$ and $C_1 = -C_2$ leading to

$$\theta(\mathbf{r}) = 4C_2 \sin(|k_x|x)\sin(|k_y|y). \qquad (3.87)$$

(2) Horizontal and vertical walls are insulating: Here $D_1 = D_2 = 0$ and $C_1 = C_2$ leading to

$$\theta(\mathbf{r}) = 4C_1 \cos(|k_x|x)\cos(|k_y|y). \qquad (3.88)$$

(3) Horizontal walls insulating and vertical walls conducting: Here $C_1 = C_2 = 0$ and $D_1 = D_2$ leading to

$$\theta(\mathbf{r}) = 4D_1 \sin(|k_x|x)\cos(|k_y|y). \qquad (3.89)$$

(4) Vertical walls insulating and horizontal walls conducting: Here $C_1 = C_2 = 0$ and $D_2 = -D_1$ leading to

$$\theta(\mathbf{r}) = 4D_1 \cos(|k_x|x)\sin(|k_y|y). \qquad (3.90)$$

Using Eq. (3.82) and arguments similar to above, the velocity component $u_1(\mathbf{k})$ for $\mathbf{k} = (\pm|k_x|, \pm|k_y|)$ are[5]:

$$u_1(|k_x|, |k_y|) = A_1 + B_1/i$$
$$u_1(-|k_x|, -|k_y|) = -A_1 + B_1/i$$
$$u_1(|k_x|, -|k_y|) = A_2 + B_2/i$$
$$u_1(-|k_x|, |k_y|) = -A_2 + B_2/i. \qquad (3.91)$$

See Fig. 3.9(b) for an illustration. We derive $u_{x,y}(\mathbf{r})$ using Eq. (3.81) that yields

$$u_x(\mathbf{r}) = 2(|k_y|/k)[(A_1 - A_2)\cos(|k_x|x)\cos(|k_y|y) - (A_1 + A_2)\sin(|k_x|x)\sin(|k_y|y)$$
$$+(B_1 - B_2)\sin(|k_x|x)\cos(|k_y|y) + (B_1 + B_2)\cos(|k_x|x)\sin(|k_y|y)] \qquad (3.92)$$
$$u_y(\mathbf{r}) = -2(|k_x|/k)[(A_1 + A_2)\cos(|k_x|x)\cos(|k_y|y) - (A_1 - A_2)\sin(|k_x|x)\sin(|k_y|y)$$
$$+(B_1 + B_2)\sin(|k_x|x)\cos(|k_y|y) + (B_1 - B_2)\cos(|k_x|x)\sin(|k_y|y)]. \qquad (3.93)$$

As expected, the above components satisfy the incompressibility condition. The aforementioned field configuration allows various kinds of boundary conditions. In the present book, we focus on the free-slip boundary condition according to which a 2D field is

$$\mathbf{u}(x, y) = \begin{pmatrix} 4(|k_y|/k)B\sin(|k_x|x)\cos(|k_y|y) \\ -4(|k_x|/k)B\cos(k_x x)\sin(|k_y|y) \end{pmatrix}. \qquad (3.94)$$

---

[5] For 2D3C fields, u_2 has the same transformation rule as θ. Hence, for such fields, $u_2(\mathbf{k})$ and $u_2(\mathbf{r})$ can be derived in manner similar to that for θ.

Equations (3.92, 3.93, 3.94) reveal that the free-slip velocity field can be obtained from the aforementioned general Craya-Herring configurations by choosing

$$A_1 = A_2 = 0; \quad B_1 = -B_2 = B. \tag{3.95}$$

Example 3.4: Consider the field configuration of Example 3.2. What are the components of the field in Craya-Herring basis? Derive the nonlinear terms and equations of motion in this basis.

Solution: Following the convention described in this section, we deduce that the velocity field has no component along $\hat{e}_2$. We denote $\mathbf{q} = (1,0)$, $\mathbf{p} = (0,1)$, and $\mathbf{k} = (1,1)$. Hence, $\hat{e}_1(\mathbf{q}) = -\hat{y}$, $\hat{e}_1(\mathbf{p}) = \hat{x}$, and $\hat{e}_1(\mathbf{k}) = (\hat{x} - \hat{y})/\sqrt{2}$. The components along $\hat{e}_1$ are computed using the rules described above, and they are listed in Table 3.2.

We compute the nonlinear term $\mathbf{N}_u$ using Eq. (3.11):

$$
\begin{aligned}
\mathbf{N}_u(\mathbf{k}) &= i\mathbf{k} \cdot \mathbf{u}(\mathbf{q})\mathbf{u}(\mathbf{p}) + i\mathbf{k} \cdot \mathbf{u}(\mathbf{p})\mathbf{u}(\mathbf{q}) \\
&= i(1,1) \cdot (0,-1)(-C)B\hat{x} + i(1,1) \cdot (1,0)B(-C)(-\hat{y}) \\
&= iBC(\hat{x} + \hat{y}).
\end{aligned}
$$

$$
\begin{aligned}
\mathbf{N}_u(\mathbf{q}) &= i\mathbf{q} \cdot \mathbf{u}(\mathbf{k})\mathbf{u}(-\mathbf{p}) + i\mathbf{q} \cdot \mathbf{u}(-\mathbf{p})\mathbf{u}(\mathbf{k}) \\
&= i(1,0) \cdot [(1,-1)/\sqrt{2}](\sqrt{2}A/i)B\hat{x} \\
&\quad + i(1,0) \cdot (-1,0)(-B)(\sqrt{2}A/i)[(\hat{x} - \hat{y})/\sqrt{2}] \\
&= AB(2\hat{x} - \hat{y}),
\end{aligned}
$$

$$\mathbf{N}_u(\mathbf{p}) = AC(\hat{x} - 2\hat{y}),$$

which are same as those obtained in Example 3.2. After this we derive equations of motion of the Fourier modes. We start with

$$\frac{d}{dt}\mathbf{u}(\mathbf{p}) = \dot{B}\hat{e}_2(\mathbf{p}) = -\mathbf{N}_u(\mathbf{p}) - i\mathbf{p}\sigma(\mathbf{p}) - \nu p^2 \mathbf{u}(\mathbf{p})$$

We take component of the above equation along $\hat{e}_2(\mathbf{p})$. The term $-i\mathbf{p}\sigma(\mathbf{p})$ disappears in the final equation since it is perpendicular to $\hat{e}_2(\mathbf{p})$. Therefore,

$$
\begin{aligned}
\dot{B} &= -\mathbf{N}_u(\mathbf{p}) \cdot \hat{e}_2(\mathbf{p}) - \nu B \\
&= -AC - \nu B.
\end{aligned}
$$

Note that the component of $-\mathbf{N}_u(\mathbf{p})$ along $\hat{e}_3$ cancels the pressure gradient. Similar computations yield

$$
\begin{aligned}
\dot{A} &= -2\nu A \\
\dot{C} &= AB - \nu C.
\end{aligned}
$$

These equations are identical to those derived in Example 3.2.

In the next section we revisit Craya-Herring for 3D vector fields.

3.8 Craya-Herring for 3D vector field revisited

For 3D velocity field, the components u_1, u_2 have the properties given by Eqs. (3.64, 3.78, 3.79). Using these conditions, the most general form for u_2 is:

$$u_2(|k_x|, |k_y|, |k_z|) = A_1 + B_1/i$$
$$u_2(|k_x|, -|k_y|, |k_z|) = A_2 + B_2/i$$
$$u_2(-|k_x|, -|k_y|, |k_z|) = A_3 + B_3/i$$
$$u_2(-|k_x|, |k_y|, |k_z|) = A_4 + B_4/i$$
$$u_2(-k_x, -k_y, -|k_z|) = u_2^*(k_x, k_y, |k_z|). \tag{3.96}$$

Similarly, the most general form of u_1 is:

$$u_1(|k_x|, |k_y|, |k_z|) = \bar{A}_1 + \bar{B}_1/i$$
$$u_1(|k_x|, -|k_y|, |k_z|) = \bar{A}_2 + \bar{B}_2/i$$
$$u_1(-|k_x|, -|k_y|, |k_z|) = \bar{A}_3 + \bar{B}_3/i$$
$$u_1(-|k_x|, |k_y|, |k_z|) = \bar{A}_4 + \bar{B}_4/i$$
$$u_1(-k_x, -k_y, -|k_z|) = -u_1^*(k_x, k_y, |k_z|).$$

For u_1 and u_2, the free-slip condition yields

$$\bar{A}_i = 0; \quad \bar{B}_1 = -\bar{B}_2 = \bar{B}_3 = -\bar{B}_4, \tag{3.97}$$
$$A_i = 0; \quad B_1 = B_2 = B_3 = B_4. \tag{3.98}$$

Example 3.5: Consider the field configuration of Example 3.3. What are the components of the field in Craya-Herring basis? Derive the nonlinear terms and equations for the coefficients.

Solution: We denote the wavenumbers as $\mathbf{q} = (1, 0, 1)$, $\mathbf{p} = (0, 1, 1)$, $\mathbf{k} = (1, 1, 2)$, and choose $\hat{n} = \hat{z}$. In Craya-Herring basis, the basis vectors for these wavenumbers are

$$\hat{e}_1(\mathbf{q}) = -\hat{y}; \qquad \hat{e}_2(\mathbf{q}) = \frac{1}{\sqrt{2}}(\hat{x} - \hat{z});$$

$$\hat{e}_1(\mathbf{p}) = \hat{x}; \qquad \hat{e}_2(\mathbf{p}) = \frac{1}{\sqrt{2}}(\hat{y} - \hat{z});$$

$$\hat{e}_1(\mathbf{k}) = \frac{1}{\sqrt{2}}(\hat{x} - \hat{y}); \qquad \hat{e}_2(\mathbf{k}) = \frac{1}{\sqrt{3}}(\hat{x} + \hat{y} - \hat{z}).$$

From the entries of Table 3.3, we can deduce that $u_1 = \mathbf{u} \cdot \hat{e}_1 = 0$ for all three wavenumbers. However, u_2's are nonzero for these wavenumbers, and they are listed in Table 3.3.

Now we compute the nonlinear term and the equation of motion for the wavenumber $\mathbf{k}$:

$$\mathbf{N}_u(\mathbf{k}) = i\mathbf{k} \cdot \mathbf{u}(\mathbf{q})\mathbf{u}(\mathbf{p}) + i\mathbf{k} \cdot \mathbf{u}(\mathbf{p})\mathbf{u}(\mathbf{q})$$
$$= i(1, 1, 2) \cdot [(1, 0, -1)/\sqrt{2}](C\sqrt{2}/i)(B\sqrt{2}/i)\hat{e}_2(\mathbf{p})$$
$$+i(1, 1, 2) \cdot [(0, 1, -1)/\sqrt{2}](B\sqrt{2}/i)(C\sqrt{2}/i)\hat{e}_2(\mathbf{q})$$
$$= i\sqrt{2}BC[\hat{e}_2(\mathbf{p}) + \hat{e}_2(\mathbf{q})].$$

Therefore, the equation of motion for the wavenumber $\mathbf{k}$ with $\nu = 0$ is

$$\frac{\dot{A}}{i}\sqrt{3}\hat{e}_2(\mathbf{k}) = -i\sqrt{2}BC(\hat{e}_2(\mathbf{p}) + \hat{e}_2(\mathbf{q})) - k\sigma(\mathbf{k})$$

$$\implies \dot{A} = \frac{4}{3}BC.$$

The equations for the other wavenumbers can be derived in a similar manner. These equations are identical to those derived in Example 3.3. Note that in the Craya-Herring formalism, the pressure gradient term automatically disappears from the equation. This is a major advantage of this method.

In the next section we will describe how Craya-Herring basis can be combined to represent helical modes.

3.9   Helical decomposition

Helical decomposition provides another set of basis vectors to expand an incompressible velocity field. Here the basis vectors are [Sagaut and Cambon (2008)]

$$\hat{e}_\pm(\mathbf{k}) = \hat{e}_2(\mathbf{k}) \mp i\hat{e}_1(\mathbf{k}), \tag{3.99}$$

and

$$\mathbf{u}(\mathbf{k}) = u_+(\mathbf{k})\hat{e}_+ + u_-(\mathbf{k})\hat{e}_-. \tag{3.100}$$

The basis functions $\hat{e}_\pm(\mathbf{k})$ have several interesting and queer properties:

$$\hat{e}_\pm^*(\mathbf{k}) = \hat{e}_\mp(\mathbf{k}), \tag{3.101}$$

$$\hat{e}_\pm \cdot \hat{e}_\pm = 0, \tag{3.102}$$

$$\hat{e}_+ \cdot \hat{e}_- = \hat{e}_- \cdot \hat{e}_+ = 2, \tag{3.103}$$

$$i\mathbf{k} \times \hat{e}_\pm(\mathbf{k}) = ik\hat{e}_3(\mathbf{k}) \times [\hat{e}_2(\mathbf{k}) \mp i\hat{e}_1(\mathbf{k})] = \pm k\hat{e}_\pm(\mathbf{k}). \tag{3.104}$$

Using the above properties of the basis vectors, we derive relationships between $u_\pm(\mathbf{k})$ and $u_{1,2}(\mathbf{k})$ as

$$u_\pm(\mathbf{k}) = \frac{1}{2}[u_2(\mathbf{k}) \pm iu_1(\mathbf{k})], \tag{3.105}$$

$$u_1(\mathbf{k}) = i[u_-(\mathbf{k}) - u_+(\mathbf{k})], \tag{3.106}$$

$$u_2(\mathbf{k}) = [u_+(\mathbf{k}) + u_-(\mathbf{k})]. \tag{3.107}$$

The vorticity field is

$$\boldsymbol{\omega}(\mathbf{k}) = i\mathbf{k} \times \mathbf{u}(\mathbf{k}) = k[u_+(\mathbf{k})\hat{e}_+(\mathbf{k}) - u_-(\mathbf{k})\hat{e}_-(\mathbf{k})]. \tag{3.108}$$

Note that $\boldsymbol{\omega}(\mathbf{k})$ lies in the plane of $\hat{e}_1$ and $\hat{e}_2$, and it has no component along $\hat{e}_3$. The modal kinetic energy and kinetic helicity are

$$E_u(\mathbf{k}) = \frac{1}{2}\mathbf{u}^*(\mathbf{k}) \cdot \mathbf{u}(\mathbf{k}) = [|u_+(\mathbf{k})|^2 + |u_-(\mathbf{k})|^2], \tag{3.109}$$

$$H_K(\mathbf{k}) = \frac{1}{2}\Re[\mathbf{u}^*(\mathbf{k}) \cdot \boldsymbol{\omega}(\mathbf{k})] = k[|u_+(\mathbf{k})|^2 - |u_-(\mathbf{k})|^2]. \tag{3.110}$$

Thus, $\mathbf{u}_+(\mathbf{k})$ is the maximal helical mode with $H_K(\mathbf{k})/(kE(\mathbf{k})) = 1$, while $\mathbf{u}_-$ is the minimal helical mode with $H_K(\mathbf{k})/(kE(\mathbf{k})) = -1$.

The helical modes have interesting properties in real space. We discuss them below.

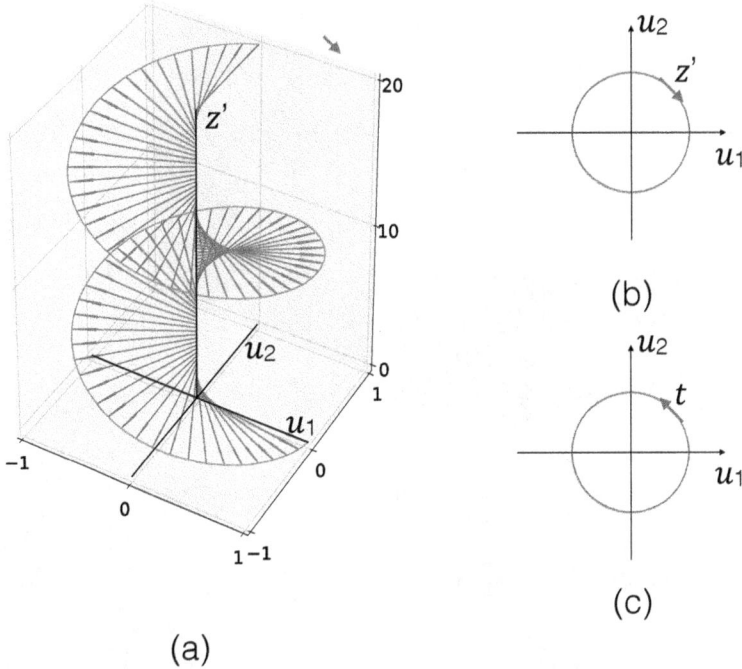

Fig. 3.10 Helical wave $\mathbf{u}_+$: (a) The static mode is left-handed. (b) The tip of $\mathbf{u}_+(\mathbf{r})$ vector rotates clockwise as we traverse along z'. (c) At a given z', with time, the tip of $\mathbf{u}_+(t)$ vector rotates in the anti-clockwise direction in time. Note that (a,b) correspond to the static case.

3.9.1   *The helical mode* $\mathbf{u}_+$

In real space, the helical mode $\mathbf{u}_+$ appears as

$$\mathbf{u}_+(\mathbf{r}) = \begin{pmatrix} u_1(\mathbf{r}) \\ u_2(\mathbf{r}) \end{pmatrix} = \Re[\hat{e}_+ u_+ \exp i(\mathbf{k} \cdot \mathbf{r})]$$

$$= |u_+|\Re[(\hat{e}_2 - i\hat{e}_1)\exp i(kz' + \phi_{k+})]$$

$$= |u_+| \begin{pmatrix} \sin(kz' + \phi_{k+}) \\ \cos(kz' + \phi_{k+}) \end{pmatrix}, \tag{3.111}$$

where $\Re$ stands for the real part, $u_+ = |u_+|\exp(i\phi_{k+})$ with ϕ_{k+} as the phase of the mode, and z' is the coordinate along $\hat{e}_3$. See Fig. 3.10 for an illustration of $\mathbf{u}_+(\mathbf{r})$. When we hold the z' axis with *left-hand*, the tip of the vector $\mathbf{u}_+$ turns along the fingers. Therefore $\mathbf{u}_+(\mathbf{r})$ is said to be *left-handed*.

Since $\cos A = \sin(A + \pi/2)$, the phase of u_2 is ahead of u_1 by $\pi/2$. This can also be deduced from the definition of the basis function. Pure $\mathbf{u}_+$ mode in $\hat{e}_1$-$\hat{e}_2$ basis is

$$\begin{pmatrix} -i \\ 1 \end{pmatrix}. \tag{3.112}$$

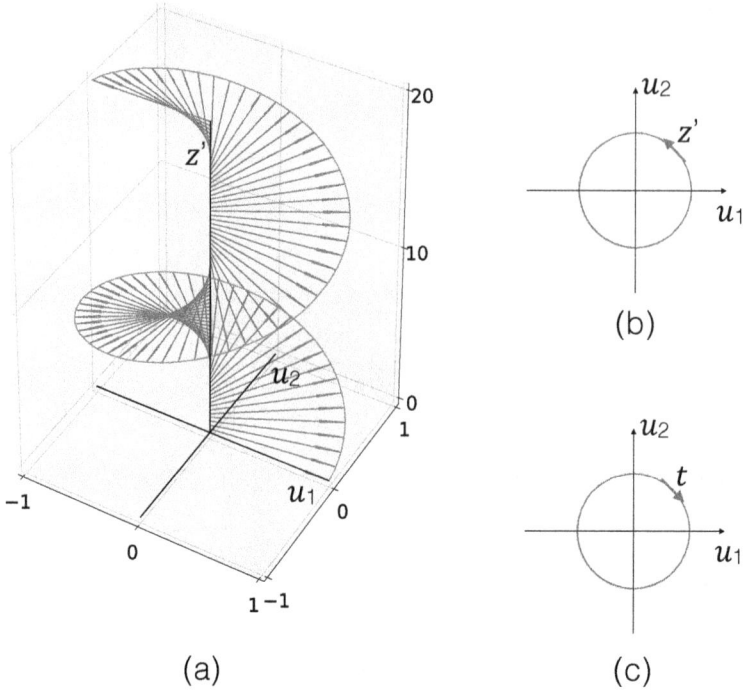

Fig. 3.11   Helical wave $\mathbf{u}_-$: (a) The static mode is right-handed. (b) The tip of $\mathbf{u}_-(\mathbf{r})$ rotates in anti-clockwise as we traverse along z'. (c) At a given z', with time, the tip of $\mathbf{u}_-(t)$ vector rotates in the clockwise direction in time. Note that (a,b) correspond to the static case.

Therefore, u_2 is ahead of u_1 by $\pi/2$. Also,

$$|u_1|^2 + |u_2|^2 = |u_+|^2. \tag{3.113}$$

Therefore, the mode $\mathbf{u}_+$ is said to be *circularly polarized*.

3.9.2   *The helical mode* $\mathbf{u}_-$

In real space, the helical mode $\mathbf{u}_-$ appears as

$$\mathbf{u}_-(\mathbf{r},t) = \begin{pmatrix} u_1(\mathbf{r}) \\ u_2(\mathbf{r}) \end{pmatrix} = \Re[\hat{e}_- u_- \exp i(\mathbf{k} \cdot \mathbf{r})]$$

$$= |u_-|\Re[(\hat{e}_2 + i\hat{e}_1) \exp i(kz' + \phi_{k-})]$$

$$= |u_-| \begin{pmatrix} -\sin(kz' + \phi_{k-}) \\ \cos(kz' + \phi_{k-}) \end{pmatrix} \tag{3.114}$$

where $u_- = |u_-| \exp(i\phi_{k-})$. We illustrate $\mathbf{u}_-(\mathbf{r},t)$ in Fig. 3.11(a,b). When we hold the z' axis with *right-hand*, the tip of $\mathbf{u}_-$ turns along the fingers. Therefore $\mathbf{u}_-(\mathbf{r})$ is said to be *right-handed*. Since $-\sin A = \cos(A + \pi/2)$, the phase of u_1 is ahead of that of u_2 by $\pi/2$. It also follows from the definition of the pure u_- mode. In addition, $|u_1|^2 + |u_2|^2 = |u_-|^2$, hence we conclude that $\mathbf{u}_-$ is *circularly polarized*.

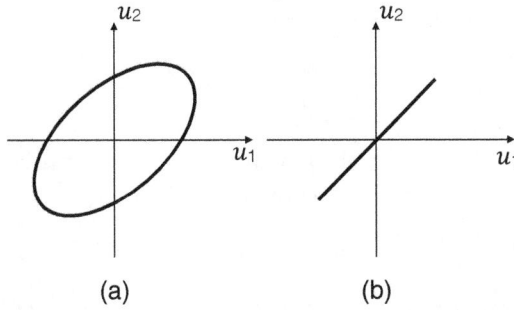

Fig. 3.12 (a) An illustration of elliptically polarized wave. (b) An illustration of linearly polarized wave.

3.9.3 *Mixture of* u₊ *and* u₋

A combination of $\mathbf{u}_+$ and $\mathbf{u}_-$ modes yields

$$
\begin{aligned}
\mathbf{u}(\mathbf{r},t) &= \Re[\{\hat{e}_+ u_+ + \hat{e}_- u_-\} \exp i(\mathbf{k} \cdot \mathbf{r})] \\
&= \Re[\{u_1 \hat{e}_1 + u_2 \hat{e}_2\} \exp i(kz')] \\
&= \Re[\{|u_1| \exp(i\phi_1)\hat{e}_1 + |u_2| \exp(i\phi_2)\hat{e}_2\} \exp i(kz')] \\
&= \begin{pmatrix} |u_1| \cos(kz' + \phi_1) \\ |u_2| \cos(kz' + \phi_2) \end{pmatrix}.
\end{aligned}
\tag{3.115}
$$

Hence

$$
u_1(\mathbf{r},t) = |u_1| \cos(kz' + \phi_1) = |u_1| \cos \Phi, \tag{3.116}
$$
$$
u_2(\mathbf{r},t) = |u_2| \cos(kz' + \phi_2) = |u_2| \cos(\Phi + \Delta\Phi), \tag{3.117}
$$

where $\Phi = kz' + \phi_1$ and $\Delta\Phi = \phi_2 - \phi_1$. Elimination of $\cos \Phi$ from the above equation yields

$$
\frac{u_1^2}{|u_1|^2} + \frac{u_2^2}{|u_2|^2} - \frac{2u_1 u_2}{|u_1||u_2|} \cos(\phi_2 - \phi_1) = \sin^2(\phi_2 - \phi_1). \tag{3.118}
$$

Thus, u_1 and u_2 form an ellipse as shown in Fig. 3.12(a). Hence the above wave is said to be elliptic polarised. When $\phi_1 = \phi_2$, the above equation yields

$$
u_2 = \frac{|u_2|}{|u_1|} u_1 \tag{3.119}
$$

indicating that u_2 is proportional u_1; this configuration corresponds to a plane polarised wave (see Fig. 3.12(b)). When $\phi_2 - \phi_1 = \pm\pi/2$,

$$
u_1^2 + u_2^2 = |u_1|^2 + |u_2|^2 = \text{const} \tag{3.120}
$$

that corresponds to circularly polarised modes—left-handed $\mathbf{u}_+$ when $\phi_2 - \phi_1 = \pi/2$, and right-handed $\mathbf{u}_-$ when $\phi_2 - \phi_1 = -\pi/2$. The aforementioned linear and circular polarisation are special cases of elliptic polarisation.

Example 3.6: Consider the flow field

$$\mathbf{u} = -\hat{y}2\sin x - \hat{z}2\cos x \qquad (3.121)$$

Analyse the flow field in the light of helical basis functions.

Solution: The flow field consists of $\mathbf{u}(\mathbf{k})$ and $\mathbf{u}(-\mathbf{k})$ where $\mathbf{k} = (1,0,0)$. The Craya-Herring basis vectors for the wavenumber $\mathbf{k}$ are

$$\hat{e}_1(\mathbf{k}) = -\hat{y}; \quad \hat{e}_2(\mathbf{k}) = -\hat{z}, \qquad (3.122)$$

hence $u_1 = 1/i$ and $u_2 = 1$. Using the conversion formulas described in Sec. 3.6, we deduce the velocity Fourier modes and its components in the helical basis as

$$\mathbf{u}(\mathbf{k}) = \frac{1}{i}\hat{e}_1(\mathbf{k}) + \hat{e}_2(\mathbf{k}),$$

$$u_+(\mathbf{k}) = \frac{1}{2}[u_2 + iu_1] = 1,$$

$$u_-(\mathbf{k}) = \frac{1}{2}[u_2 - iu_1] = 0.$$

Thus, the above flow field has left-handed polarisation (see Fig. 3.10). Note that the phase of u_2 is ahead of u_1 by $\pi/2$.

The modal energy and kinetic helicity are

$$E_u(\mathbf{k}) = \frac{1}{2}|\mathbf{u}(\mathbf{k})|^2 = 1,$$

$$H_K(\mathbf{k}) = k\Im(u_1^* u_2) = k = 1.$$

Since the flow contains two wavenumbers—$(1,0,0)$ and $(-1,0,0)$, the total energy $E_u = 2$, and the total kinetic helicity $H_K = 2$. The flow field has maximal helicity.

Example 3.7: Analyse the helicity of the following flow:

$$\mathbf{u} = \hat{y}2\sin x - \hat{z}2\cos x. \qquad (3.123)$$

Solution: Following the similar steps as the earlier example, we can show that

$$\mathbf{u}(\mathbf{k}) = -\frac{1}{i}\hat{e}_1(\mathbf{k}) + \hat{e}_2(\mathbf{k}) \qquad (3.124)$$

Hence, the modal energy and kinetic helicity of the mode $(1,0,0)$ are 1 and -1 respectively. The flow field is right-handed and it has minimal helicity.

3.10 Helical waves

The helical modes are also used to describe helical waves. In this section we provide brief description of helical waves. Here we consider waves $\mathbf{u}_+(\mathbf{r}, t)$ and $\mathbf{u}_-(\mathbf{r}, t)$ travelling along $\mathbf{k}$ direction.

Following the same procedure as that used in the previous section, the real-space configuration of helical wave $\mathbf{u}_+$ travelling along $\mathbf{k}$ is

$$
\begin{aligned}
\mathbf{u}_+(\mathbf{r}, t) &= \Re[\hat{e}_+ u_+ \exp i(\mathbf{k} \cdot \mathbf{r} - \omega t)] \\
&= |u_+| \Re[(\hat{e}_2 - i\hat{e}_1) \exp i(kz' - \omega t + \phi_{k+})] \\
&= |u_+| \begin{pmatrix} \sin(kz' - \omega t + \phi_{k+}) \\ \cos(kz' - \omega t + \phi_{k+}) \end{pmatrix},
\end{aligned}
\tag{3.125}
$$

where ω is the frequency of the wave. When we consider a frozen-in field, $\mathbf{u}_+(\mathbf{r}, t = t_0)$, the field configuration appears as in Fig. 3.10(a). This is a left-handed wavy structure.

The time evolution of $\mathbf{u}_+(\mathbf{r}, t)$ at a given position, say at $z' = 0$, is

$$
\mathbf{u}_+(z' = 0, t) = \begin{pmatrix} -\sin(\omega t - \phi_{k+}) \\ \cos(\omega t - \phi_{k+}) \end{pmatrix}.
\tag{3.126}
$$

Thus, the tip of $\mathbf{u}_+(z' = 0, t)$ rotates in anti-clockwise direction, as shown in Fig. 3.10(c). The above property however holds for any z'. Therefore, $\mathbf{u}_+$ is also referred to as *anti-clockwise or right circularly polarized (RCP)* wave.

Similar computation shows that the real-space configuration of helical wave $\mathbf{u}_-$ travelling along $\mathbf{k}$ is

$$
\begin{aligned}
\mathbf{u}_-(\mathbf{r}, t) &= \Re[\hat{e}_- u_- \exp i(\mathbf{k} \cdot \mathbf{r} - \omega t)] \\
&= |u_-| \begin{pmatrix} -\sin(kz' - \omega t + \phi_{k-}) \\ \cos(kz' - \omega t + \phi_{k-}) \end{pmatrix}.
\end{aligned}
\tag{3.127}
$$

The frozen-in configuration of $\mathbf{u}_-(\mathbf{r}, t = t_0)$ is right-handed and it appears exactly as Fig. 3.11(a). The time evolution of the velocity vector at a given position, say at $z' = 0$, is given by

$$
\mathbf{u}_-(z' = 0, t) = \begin{pmatrix} \sin(\omega t - \phi_{k-}) \\ \cos(\omega t - \phi_{k-}) \end{pmatrix}.
\tag{3.128}
$$

Hence, the tip of the velocity vector rotates in clock-wise direction, as shown in Fig. 3.11(c). Therefore, the helical wave $\mathbf{u}_-$ is referred to as *clockwise or left circularly polarized (LCP)* wave.

It is important to note that the nature of polarisation would change if the wave is travelling along $-\mathbf{k}$ direction. However, $\mathbf{u}_+$ is always left handed, and $\mathbf{u}_-$ always right handed. This is left as an exercise. We also remark that the aforementioned definitions of polarisation are equivalent to those for polarised electromagnetic waves [Zangwill (2013)].

Example 3.8: Consider a velocity vector field $\mathbf{u}(x, y) = -\hat{x} \sin \omega t + \hat{y} \cos \omega t$. Analyse the vorticity and helicity of this field.

Solution: Since $\mathbf{u}$ is not a function of x or y, its vorticity is zero. Therefore, the velocity field has no kinetic helicity. The above flow resembles solid body rotation that has zero kinetic helicity. Note that not all rotating flows have vorticity or helicity.

3.11 Helicity under parity transformation

Let us briefly discuss the properties of helicity under the parity transformation, $\mathbf{k} \to -\mathbf{k}$. The energy of $\mathbf{u}(\mathbf{k})$ and $\mathbf{u}(-\mathbf{k})$ are the same under parity, but the vorticity field for the mode $-\mathbf{k}$ is

$$\begin{aligned}
\boldsymbol{\omega}(-\mathbf{k}) &= i(-\mathbf{k}) \times \mathbf{u}(-\mathbf{k}) \\
&= -i\mathbf{k} \times \mathbf{u}^*(\mathbf{k}) \\
&= [i\mathbf{k} \times \mathbf{u}(\mathbf{k})]^* = \boldsymbol{\omega}^*(\mathbf{k}).
\end{aligned} \tag{3.129}$$

Hence

$$\begin{aligned}
H_K(-\mathbf{k}) &= \frac{1}{2}\Re[\mathbf{u}(-\mathbf{k}) \cdot \boldsymbol{\omega}^*(-\mathbf{k})] \\
&= \frac{1}{2}\Re[\mathbf{u}^*(\mathbf{k}) \cdot \boldsymbol{\omega}(\mathbf{k})] = H_K(\mathbf{k}),
\end{aligned} \tag{3.130}$$

consistent with the reality condition for $\boldsymbol{\omega}(\mathbf{r})$. The above equation shows that the kinetic helicity of the mode $-\mathbf{k}$ and $\mathbf{k}$ are the same.

We can also show the above result in Craya-Herring basis. Some of the identities that can be proved easily are

$$\hat{e}_1(-\mathbf{k}) = -\hat{e}_1(\mathbf{k}) \tag{3.131}$$
$$\hat{e}_2(-\mathbf{k}) = \hat{e}_2(\mathbf{k}) \tag{3.132}$$
$$\hat{e}_\pm(-\mathbf{k}) = \hat{e}_\mp(\mathbf{k}) \tag{3.133}$$
$$u_1(-\mathbf{k}) = -u_1^*(\mathbf{k}) \tag{3.134}$$
$$u_2(-\mathbf{k}) = u_2^*(\mathbf{k}). \tag{3.135}$$

Also, using Eq. (3.67), we can show that $H_K(-\mathbf{k}) = H_K(\mathbf{k})$. It is left as an exercise. Here we conclude our discussion on Fourier representation of fluid flows.

Further reading

For Fourier transform and its application to partial differential equations, refer to texts on mathematical physics, such as Kreyszig *et al.* (2011). Leslie (1973), McComb (1990), Davidson (2004), and Pope (2000) describe fluid equations in Fourier space.

Exercises

(1) Consider the following flow fields in a two-dimensional box of size $[\pi, \pi]$:

    (a)

$$\begin{aligned}
\mathbf{u} = {}& 4C(\hat{x}\sin 3x \cos y - \hat{y}3\cos 3x \sin y) + 4B(\hat{y}\sin 2x \cos 2y - \hat{y}\cos 2x \sin 2y) \\
& + 4A(\hat{x}\sin x \cos y - \hat{y}\cos x \sin y).
\end{aligned}$$

(b)

$$\mathbf{u} = \hat{x}2B\cos y + \hat{y}2C\cos x + 2A[\hat{x}\cos(x+y) - \hat{y}\cos(x+y)]$$

For these fields, identify the active Fourier modes and their amplitudes. Compute the total energy and the total enstrophy of the flows. Verify that these fields are incompressible.

(2) For the fields of Exercise 1, find the amplitudes of the Fourier modes in Craya-Herring basis.

(3) For the fields of Exercise 1, derive equations of motion for the amplitudes A, B, and C. Assume the viscosity of the fluid to be ν. It is more convenient to work in Craya-Herring basis.

(4) Consider Fourier wavenumbers $\mathbf{k}_1 = (1, 0, 1)$ and $\mathbf{k}_2 = (1, 1, 1)$. In Craya-Herring basis, the velocity field $\mathbf{u}(\mathbf{k}_1) = \hat{e}_2(\mathbf{k}_1) + i\hat{e}_1(\mathbf{k}_1)$ and $\mathbf{u}(\mathbf{k}_2) = \hat{e}_2(\mathbf{k}_2) + 2\hat{e}_1(\mathbf{k}_2)$. The reality condition on the velocity field demands that $\mathbf{u}(-\mathbf{k}_1) = \mathbf{u}^*(\mathbf{k}_1)$ and $\mathbf{u}(-\mathbf{k}_2) = \mathbf{u}^*(\mathbf{k}_2)$. Write down the velocity field in Cartesian basis, in both Fourier space and real space.

(5) Consider a flow field containing a triad with wavenumbers $\mathbf{k} = (1/2, \sqrt{3}/2)$, $\mathbf{q} = (1, 0)$, and $\mathbf{p} = (-1/2, \sqrt{3}/2)$. Construct the flow field and compute its energy. What kind of patterns does the flow generate?

(6) For a temperature field given by

$$\theta(x, y, z) = 8\cos(x)\cos(y)\sin(2z).$$

Construct Fourier amplitudes for the wavenumbers $(\pm 1, \pm 1, \pm 2)$.

(7) Show that under the parity operation, a left-handed circularly polarised wave becomes right-handed circularly polarised wave.

(8) Construct a source at the origin to generate helical waves along $\pm\mathbf{k}$.

Chapter 4

Energy Transfers in Buoyancy-Driven Flows

In the last chapter we introduced the equations for the buoyancy-driven flows in Fourier space. We also wrote the equations for the modal energy and entropy. In this chapter, we will revisit the energy and entropy equations, and derive formulas for the energy and entropy transfers among the Fourier modes. Using these formulas we will derive diagnostic tools like the energy flux, shell-to-shell energy transfers, etc., which are very useful for understanding buoyancy-driven flows.

We start with the derivation of mode-to-mode energy transfer in buoyancy-driven flows.

4.1 Mode-to-Mode energy transfers in RBC

First we consider nonlinear interactions among the velocity modes in wavenumber triads $(\mathbf{k}, \mathbf{p}, \mathbf{q})$ and $(-\mathbf{k}, -\mathbf{p}, -\mathbf{q})$ with a condition that $\mathbf{k} = \mathbf{p} + \mathbf{q}$ (see Fig. 4.1). The corresponding velocity Fourier modes are $\mathbf{u}(\mathbf{k})$, $\mathbf{u}(\mathbf{p})$, and $\mathbf{u}(\mathbf{q})$. Note that $\mathbf{u}(-\mathbf{k}) = \mathbf{u}^*(\mathbf{k})$ due to the reality condition. We also simplify the system by making it dissipationless ($\nu = 0$) and free of buoyancy ($g = 0$) that leads to conservation of total kinetic energy.

For convenience, we use variable $\mathbf{k}' = -\mathbf{k}$ that yields $\mathbf{k}' + \mathbf{p} + \mathbf{q} = 0$. The equations are symmetric in $\mathbf{k}', \mathbf{p}, \mathbf{q}$ variables thus providing a major simplification. Using Eq. (3.32) we derive an equation for the modal energy $E_u(\mathbf{k}') = |\mathbf{u}(\mathbf{k}')|^2/2 = E_u(\mathbf{k})$ for the above triad:

$$\frac{d}{dt} E_u(\mathbf{k}', t) = -\Im \left[(\mathbf{k}' \cdot \mathbf{u}(\mathbf{q})) (\mathbf{u}(\mathbf{p}) \cdot \mathbf{u}(\mathbf{k}')) + (\mathbf{k}' \cdot \mathbf{u}(\mathbf{p})) (\mathbf{u}(\mathbf{q}) \cdot \mathbf{u}(\mathbf{k}')) \right]$$
$$= S^{uu}(\mathbf{k}'|\mathbf{p}, \mathbf{q}), \tag{4.1}$$

where $\Im$ denotes the imaginary part of the argument. Kraichnan (1959), who derived the above equation for the first time, argued that $S^{uu}(\mathbf{k}'|\mathbf{p}, \mathbf{q})$ is the *combined energy transfer* to $\mathbf{u}(\mathbf{k})$ from $\mathbf{u}(\mathbf{p})$ and $\mathbf{u}(\mathbf{q})$. The evolution equations for $E_u(\mathbf{p})$

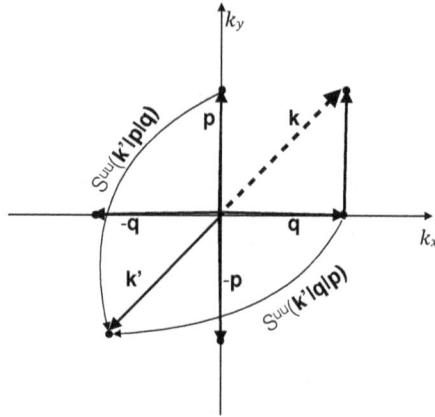

Fig. 4.1   Energy transfers in an interacting wavenumber triad: $S^{uu}(\mathbf{k'}|\mathbf{p}|\mathbf{q})$ represents the energy transfer rate from mode $\mathbf{u}(\mathbf{p})$ to mode $\mathbf{u}(\mathbf{k'})$ with $\mathbf{u}(\mathbf{q})$ acting as a helper, and $S^{uu}(\mathbf{k'}|\mathbf{q}|\mathbf{p})$ is the energy transfer rate from mode $\mathbf{u}(\mathbf{q})$ to mode $\mathbf{u}(\mathbf{k'})$ with $\mathbf{u}(\mathbf{p})$ acting as a helper.

and $E_u(\mathbf{q})$ can be derived in a similar fashion:

$$\frac{d}{dt}E_u(\mathbf{p},t) = -\Im\left[(\mathbf{p}\cdot\mathbf{u}(\mathbf{q}))\,(\mathbf{u}(\mathbf{k'})\cdot\mathbf{u}(\mathbf{p})) + (\mathbf{p}\cdot\mathbf{u}(\mathbf{k'}))\,(\mathbf{u}(\mathbf{q})\cdot\mathbf{u}(\mathbf{p}))\right],$$

(4.2)

$$\frac{d}{dt}E_u(\mathbf{q},t) = -\Im\left[(\mathbf{q}\cdot\mathbf{u}(\mathbf{k'}))\,(\mathbf{u}(\mathbf{p})\cdot\mathbf{u}(\mathbf{q})) + (\mathbf{q}\cdot\mathbf{u}(\mathbf{p}))\,(\mathbf{u}(\mathbf{k'})\cdot\mathbf{u}(\mathbf{q}))\right].$$

(4.3)

Using $\mathbf{k'}\cdot\mathbf{u}(\mathbf{k'}) = \mathbf{p}\cdot\mathbf{u}(\mathbf{p}) = \mathbf{p}\cdot\mathbf{u}(\mathbf{p}) = 0$, we derive that

$$\frac{d}{dt}\left[E_u(\mathbf{k'},t) + E_u(\mathbf{p},t) + E_u(\mathbf{q},t)\right] = 0,$$

(4.4)

or

$$E_u(\mathbf{k'},t) + E_u(\mathbf{p},t) + E_u(\mathbf{q},t) = \text{constant},$$

(4.5)

which is the *law of detailed energy conservation* [Kraichnan (1959); Lesieur (2008)]. Note that the incompressibility condition plays a critical role in the above derivation and subsequent discussion of this chapter. In compressible flows, there is an energy exchange between the kinetic energy and the internal energy, which is absent in the present formalism due to incompressibility.

Kraichnan's formula $S^{uu}(\mathbf{k'}|\mathbf{p},\mathbf{q})$ does not provide information about the individual energy transfers from mode $\mathbf{p}$ to $\mathbf{k}$, and from $\mathbf{q}$ to $\mathbf{k}$. Dar *et al.* (2001) and Verma (2004) derived a formula for the mode-to-mode energy transfer that provides the above measures. They showed that

$$S^{uu}(\mathbf{k'}|\mathbf{p}|\mathbf{q}) = -\Im\left[\{\mathbf{k'}\cdot\mathbf{u}(\mathbf{q})\}\{\mathbf{u}(\mathbf{p})\cdot\mathbf{u}(\mathbf{k'})\}\right]$$

(4.6)

is the *rate of mode-to-mode energy transferred* from mode $\mathbf{p}$ to $\mathbf{k'}$ with $\mathbf{q}$ acting as a mediator. In short, we drop the term "rate of" in the above phrase and refer to it as *mode-to-Mode energy transfer.* In the above formula, the Fourier amplitudes of the receiver and giver modes are dotted together, while the receiver wavenumber is dotted with the amplitude of the mediator mode. Since $\mathbf{k}' = -\mathbf{k}$, the above formula is also written as

$$S^{uu}(\mathbf{k}|\mathbf{p}|\mathbf{q}) = \Im\left[\{\mathbf{k} \cdot \mathbf{u}(\mathbf{q})\}\{\mathbf{u}(\mathbf{p}) \cdot \mathbf{u}^*(\mathbf{k})\}\right]. \tag{4.7}$$

The formula for the other energy transfers can be written using the above recipe. In the following we sketch the proof for the above formula:

The functions S^{uu}'s satisfy the following properties:

(1) The sum of $S^{uu}(\mathbf{k}'|\mathbf{p}|\mathbf{q})$ and $S^{uu}(\mathbf{k}'|\mathbf{q}|\mathbf{p})$ is the combined energy transferred to mode $\mathbf{k}'$ from modes $\mathbf{p}$ and $\mathbf{q}$, which is $S^{uu}(\mathbf{k}'|\mathbf{p}, \mathbf{q})$ of Eq. (4.1). Similar rules apply to the formulas for the energy transfers to modes $\mathbf{p}$ and $\mathbf{q}$. Therefore,

$$S^{uu}(\mathbf{k}'|\mathbf{p}|\mathbf{q}) + S^{uu}(\mathbf{k}'|\mathbf{q}|\mathbf{p}) = S^{uu}(\mathbf{k}'|\mathbf{p}, \mathbf{q}), \tag{4.8}$$

$$S^{uu}(\mathbf{p}|\mathbf{k}'|\mathbf{q}) + S^{uu}(\mathbf{p}|\mathbf{q}|\mathbf{k}') = S^{uu}(\mathbf{p}|\mathbf{k}', \mathbf{q}), \tag{4.9}$$

$$S^{uu}(\mathbf{q}|\mathbf{k}'|\mathbf{p}) + S^{uu}(\mathbf{q}|\mathbf{p}|\mathbf{k}') = S^{uu}(\mathbf{q}|\mathbf{k}', \mathbf{p}). \tag{4.10}$$

(2) By definition, the energy transfer from mode $\mathbf{p}$ to mode $\mathbf{k}'$, $S^{uu}(\mathbf{k}'|\mathbf{p}|\mathbf{q})$, is equal and opposite to the energy transfer from mode $\mathbf{k}'$ to mode $\mathbf{p}$, $S^{uu}(\mathbf{p}|\mathbf{k}'|\mathbf{q})$. Thus,

$$S^{uu}(\mathbf{k}'|\mathbf{p}|\mathbf{q}) + S^{uu}(\mathbf{p}|\mathbf{k}'|\mathbf{q}) = 0, \tag{4.11}$$

$$S^{uu}(\mathbf{k}'|\mathbf{q}|\mathbf{p}) + S^{uu}(\mathbf{q}|\mathbf{k}'|\mathbf{p}) = 0, \tag{4.12}$$

$$S^{uu}(\mathbf{p}|\mathbf{q}|\mathbf{k}') + S^{uu}(\mathbf{q}|\mathbf{p}|\mathbf{k}') = 0. \tag{4.13}$$

It is easy to verify that the mode-to-mode energy transfer formulas of the form of Eq. (4.6) satisfy the above conditions. Note that Eqs. (4.8–4.10) are satisfied by definition (Eq. (4.6)), while Eqs. (4.11–4.13) are satisfied due to the incompressibility condition

$$\mathbf{q} \cdot \mathbf{u}(\mathbf{q}) = \mathbf{p} \cdot \mathbf{u}(\mathbf{p}) = \mathbf{k}' \cdot \mathbf{u}(\mathbf{k}') = 0. \tag{4.14}$$

For example,

$$\begin{aligned}
&S^{uu}(\mathbf{k}'|\mathbf{p}|\mathbf{q}) + S^{uu}(\mathbf{p}|\mathbf{k}'|\mathbf{q}) \\
&= -\Im\left([\mathbf{k}' \cdot \mathbf{u}(\mathbf{q})]\,[\mathbf{u}(\mathbf{k}') \cdot \mathbf{u}(\mathbf{p})]\right) - \Im\left([\mathbf{p} \cdot \mathbf{u}(\mathbf{q})]\,[\mathbf{u}(\mathbf{k}') \cdot \mathbf{u}(\mathbf{p})]\right) \\
&= -\Im\left([(\mathbf{k}' + \mathbf{p}) \cdot \mathbf{u}(\mathbf{q})]\,[\mathbf{u}(\mathbf{k}') \cdot \mathbf{u}(\mathbf{p})]\right) \\
&= \Im\left([\mathbf{q} \cdot \mathbf{u}(\mathbf{q})]\,[\mathbf{u}(\mathbf{k}') \cdot \mathbf{u}(\mathbf{p})]\right) = 0.
\end{aligned}$$

Hence, the function $S^{uu}(\mathbf{k}'|\mathbf{p}|\mathbf{q})$ of Eq. (4.6) is a formula for the energy transfer from mode $\mathbf{u}(\mathbf{p})$ to mode $\mathbf{u}(\mathbf{k}')$ with mode $\mathbf{u}(\mathbf{q})$ acting as a mediator. The other energy transfers have a similar form. However, the determinant of the matrix formed by the linear equations (4.8–4.13) is zero. Hence, the solution consisting of $S^{uu}(\mathbf{k}'|\mathbf{p}|\mathbf{q})$ of Eq. (4.6) and similar functions for other arguments, e.g. $S^{uu}(\mathbf{p}|\mathbf{k}'|\mathbf{q})$, is not unique[1].

---

[1] Earlier Dar *et al.* (2001) and Verma (2004) showed that, in addition to the aforementioned solution, a trivial *circulating energy* flows along $\mathbf{p} \rightarrow \mathbf{k}' \rightarrow \mathbf{q} \rightarrow \mathbf{p}$ that does not alter physical observables, for example, energy flux. In this chapter, we provide an alternate derivation that does not invoke circulating transfer.

In the following discussion, we show using tensor analysis that Eq. (4.6) *uniquely* describes the energy transfer from mode $\mathbf{u}(\mathbf{p})$ to mode $\mathbf{u}(\mathbf{k}')$ with mode $\mathbf{u}(\mathbf{q})$ acting as a mediator[2].

From the energy equation, Eq. (4.1), $S^{uu}(\mathbf{k}'|\mathbf{p}|\mathbf{q})$ must satisfy the following properties in addition to Eqs. (4.8–4.13):

(1) $S^{uu}(\mathbf{k}'|\mathbf{p}|\mathbf{q})$ is real.
(2) $S^{uu}(\mathbf{k}'|\mathbf{p}|\mathbf{q})$ is a linear function of the Fourier modes $\mathbf{u}(\mathbf{k}')$, $\mathbf{u}(\mathbf{p})$, and $\mathbf{u}(\mathbf{q})$. Each of the above modes must appear in $S^{uu}(\mathbf{k}'|\mathbf{p}|\mathbf{q})$ only once.
(3) $S^{uu}(\mathbf{k}'|\mathbf{p}|\mathbf{q})$ is a linear function of the wavevector $\mathbf{k}'$ due to the derivative of the nonlinear term $\partial_j(u_j u_i)$.
(4) From the reality condition, the energy transfer from $\mathbf{u}(\mathbf{p})$ to $\mathbf{u}(\mathbf{k}')$ should be same as $\mathbf{u}(-\mathbf{p})$ to $\mathbf{u}(-\mathbf{k}')$, that is,

$$S^{uu}(-\mathbf{k}'|-\mathbf{p}|-\mathbf{q}) = S^{uu}(\mathbf{k}'|\mathbf{p}|\mathbf{q}). \qquad (4.15)$$

From the above properties, the tensor analysis yields the following form for $S^{uu}(\mathbf{k}'|\mathbf{p}|\mathbf{q})$:

$$\begin{aligned}
S^{uu}(\mathbf{k}'|\mathbf{p}|\mathbf{q}) &= c_1 \Re\left([\mathbf{k}' \cdot \mathbf{u}(\mathbf{q})]\,[\mathbf{u}(\mathbf{k}') \cdot \mathbf{u}(\mathbf{p})]\right) \\
&+ c_2 \Re\left([\mathbf{k}' \cdot \mathbf{u}(\mathbf{p})]\,[\mathbf{u}(\mathbf{k}') \cdot \mathbf{u}(\mathbf{q})]\right) \\
&+ c_3 \Re\left([\mathbf{k}' \cdot \mathbf{u}(\mathbf{k}')]\,[\mathbf{u}(\mathbf{p}) \cdot \mathbf{u}(\mathbf{q})]\right) \\
&+ c_4 \Im\left([\mathbf{k}' \cdot \mathbf{u}(\mathbf{q})]\,[\mathbf{u}(\mathbf{k}') \cdot \mathbf{u}(\mathbf{p})]\right) \\
&+ c_5 \Im\left([\mathbf{k}' \cdot \mathbf{u}(\mathbf{p})]\,[\mathbf{u}(\mathbf{k}') \cdot \mathbf{u}(\mathbf{q})]\right) \\
&+ c_6 \Im\left([\mathbf{k}' \cdot \mathbf{u}(\mathbf{k}')]\,[\mathbf{u}(\mathbf{p}) \cdot \mathbf{u}(\mathbf{q})]\right),
\end{aligned} \qquad (4.16)$$

where c_i's are constants. The fourth condition makes $c_1 = c_2 = c_3 = 0$ since the corresponding formulas change sign under the $(\mathbf{k}', \mathbf{p}, \mathbf{q}) \to (-\mathbf{k}', -\mathbf{p}, -\mathbf{q})$ transformation. In addition, the sixth term is zero since $\mathbf{k}' \cdot \mathbf{u}(\mathbf{k}') = 0$. Now we are left with

$$\begin{aligned}
S^{uu}(\mathbf{k}'|\mathbf{p}|\mathbf{q}) &= c_4 \Im\left([\mathbf{k}' \cdot \mathbf{u}(\mathbf{q})]\,[\mathbf{u}(\mathbf{k}') \cdot \mathbf{u}(\mathbf{p})]\right) \\
&+ c_5 \Im\left([\mathbf{k}' \cdot \mathbf{u}(\mathbf{p})]\,[\mathbf{u}(\mathbf{k}') \cdot \mathbf{u}(\mathbf{q})]\right).
\end{aligned} \qquad (4.17)$$

The nonlinearity in the energy equation is of the form $\partial_j(u_j u_i u_i)$. In the equation for the temperature fluctuations, θ, the nonlinear term is $\partial_j(u_j \theta \theta)$. In both these systems, u_j (the field on whom ∂_j acts) advects u_i and θ, and it does not participate directly in the energy transfer. Hence, the Fourier mode $u_j(\mathbf{q})$ that appears in the scalar product with k_j' must be the mediator in the energy transfer. Therefore, $S^{uu}(\mathbf{k}'|\mathbf{p}|\mathbf{q})$ is given by Eq. (4.6). Q.E.D.

Now, in the energy equation we introduce buoyancy and viscous force that are linear in $\mathbf{u}(\mathbf{k})$. With these terms, the energy equation becomes

$$\frac{d}{dt} E_u(\mathbf{k}) = S^{uu}(\mathbf{k}|\mathbf{p}|\mathbf{q}) + S^{uu}(\mathbf{k}|\mathbf{q}|\mathbf{p}) + \alpha g \Re[\theta(\mathbf{k}) u_z^*(\mathbf{k})] - 2\nu k^2 E_u(\mathbf{k}).$$

$$(4.18)$$

---

[2]There are some subtleties in the choice of the formula for the energy transfers. However, these discussions are beyond the scope of this book.

In RBC, hot fluid elements have positive u_z, and cold ones have negative u_z. Hence $\int d\mathbf{r}\theta(\mathbf{r})u_z(\mathbf{r}) > 0$ or $\sum \Re[\theta(\mathbf{k})u_z^*(\mathbf{k})] > 0$. From these we cannot deduce that $\Re[\theta(\mathbf{k})u_z^*(\mathbf{k})] > 0$ for each $\mathbf{k}$, yet the relation $\Re[\theta(\mathbf{k})u_z^*(\mathbf{k})] > 0$ holds for most $\mathbf{k}$'s. Hence buoyancy provides positive contribution to $dE_u(\mathbf{k})/dt$. On the other hand, viscous dissipation, $-2\nu k^2 E_u(\mathbf{k})$, yields negative input to the energy. Note that the buoyancy and viscous terms have the same $\mathbf{k}$ as $E_u(\mathbf{k})$ due to their linear nature.

When we consider all the triads connected to $\mathbf{k}$ with $\mathbf{k} = \mathbf{p} + \mathbf{q}$, we obtain

$$\frac{d}{dt}E_u(\mathbf{k}) = \sum_{\mathbf{p}} S^{uu}(\mathbf{k}|\mathbf{p}|\mathbf{q}) + \alpha g \Re[\theta(\mathbf{k})u_z^*(\mathbf{k})] - 2\nu k^2 E_u(\mathbf{k}). \qquad (4.19)$$

Using

$$\sum_{\mathbf{k}}\sum_{\mathbf{p}} S^{uu}(\mathbf{k}|\mathbf{p}|\mathbf{q}) = 0, \qquad (4.20)$$

we deduce an equation for the total kinetic energy E_u as

$$\frac{d}{dt}E_u = \alpha g \sum_{\mathbf{k}} \Re[\theta(\mathbf{k})u_z^*(\mathbf{k})] - 2\nu \sum_{\mathbf{k}} k^2 E_u(\mathbf{k}), \qquad (4.21)$$

which is equivalent to its real space counterpart:

$$\frac{d}{dt}E_u = \alpha g \langle \theta(\mathbf{r})u_z(\mathbf{r}) \rangle + 2\nu \langle \omega^2 \rangle, \qquad (4.22)$$

where $\omega = \nabla \times \mathbf{u}$, and $\langle . \rangle$ represents volume average. In the above derivation we invoke Parsevel's theorem, which is

$$\langle f(\mathbf{r})g(\mathbf{r}) \rangle = \sum_{\mathbf{k}} \Re[f(\mathbf{k})g^*(\mathbf{k})]. \qquad (4.23)$$

Note that E_u is not conserved in the inviscid limit due to the cross transfer between the kinetic energy and the entropy.

Following similar arguments as above, we can show that

$$S^{\theta\theta}(\mathbf{k}|\mathbf{p}|\mathbf{q}) = \Im\left[\{\mathbf{k} \cdot \mathbf{u}(\mathbf{q})\}\{\theta(\mathbf{p})\theta^*(\mathbf{k})\}\right] \qquad (4.24)$$

is the mode-to-mode entropy transfer from mode $\theta(\mathbf{p})$ to mode $\theta(\mathbf{k})$ with $\mathbf{u}(\mathbf{q})$ acting as a mediator. Hence,

$$\frac{d}{dt}E_\theta(\mathbf{k}) = S^{\theta\theta}(\mathbf{k}|\mathbf{p}|\mathbf{q}) + S^{\theta\theta}(\mathbf{k}|\mathbf{q}|\mathbf{p}) - \frac{d\bar{T}}{dz}\Re[\theta(\mathbf{k})u_z^*(\mathbf{k})] - 2\kappa k^2 E_\theta(\mathbf{k}). \qquad (4.25)$$

For RBC, $d\bar{T}/dz < 0$ and $\theta(\mathbf{r})u_z(\mathbf{r}) > 0$, hence, to $dE_\theta(\mathbf{k})/dt$, buoyancy and thermal diffusion term provide positive and negative contributions respectively. For all the interacting triads involving $\mathbf{k}$ with $\mathbf{k} = \mathbf{p} + \mathbf{q}$,

$$\frac{d}{dt}E_\theta(\mathbf{k}) = \sum_{\mathbf{p}} S^{\theta\theta}(\mathbf{k}|\mathbf{p}|\mathbf{q}) - \frac{d\bar{T}}{dz}\Re[\theta(\mathbf{k})u_z^*(\mathbf{k})] - 2\kappa k^2 E_\theta(\mathbf{k}). \qquad (4.26)$$

Using

$$\sum_{\mathbf{k}}\sum_{\mathbf{p}} S^{\theta\theta}(\mathbf{k}|\mathbf{p}|\mathbf{q}) = 0, \qquad (4.27)$$

we derive the following equation for the total entropy:

$$\frac{d}{dt}E_\theta = -\frac{d\bar{T}}{dz}\sum_{\mathbf{k}}\Re[\theta(\mathbf{k})u_z^*(\mathbf{k})] - 2\kappa\sum_{\mathbf{k}}k^2 E_\theta(\mathbf{k}). \qquad (4.28)$$

The corresponding equation in real space is

$$\frac{d}{dt}E_\theta = -\frac{d\bar{T}}{dz}\langle\theta(\mathbf{r})u_z(\mathbf{r})\rangle + 2\kappa\langle(\nabla\theta)^2\rangle. \qquad (4.29)$$

Using Eqs. (4.21, 4.28), and by setting $\nu = \kappa = 0$ (inviscid and nondiffusive limit), we obtain

$$\frac{d}{dt}(E_u + \frac{\alpha g}{d\bar{T}/dz}E_\theta) = 0. \qquad (4.30)$$

Thus

$$E_u + \frac{\alpha g}{d\bar{T}/dz}E_\theta = \text{const.} \qquad (4.31)$$

Thus, $E_u + E_\theta(\alpha g)/(d\bar{T}/dz)$ is a conserved quantity when $\nu = \kappa = 0$. Note that E_u and E_θ are not conserved individually due to the cross transfer between E_u and E_θ. This derivation is analogous to those of Sec. 2.8 in which we derived the conservation laws in real space. Note that $d\bar{T}/dz < 0$ for RBC, and $d\bar{T}/dz > 0$ for stably stratified flow.

Example 4.1: Consider the following flow field of Example 3.2:

$$\mathbf{u} = \hat{x}2B\cos y + \hat{y}2C\cos x + 4A(\hat{x}\sin x\cos y - \hat{y}\cos x\sin y).$$

Compute the energy transfer among the Fourier modes.

Solution: The active Fourier modes of the system are $(\pm 1, 0)$, $(0, \pm 1)$, and $(\pm 1, \pm 1)$. Among these $(a) : \{(1,0), (0,1), (-1,-1)\}$, $(b) : \{(-1,0), (0,-1), (1,1)\}$, $(c) : \{(-1,0), (0,1), (1,-1)\}$, $(d) : \{(1,0), (0,-1), (-1,1)\}$ form triads $(\mathbf{k}' + \mathbf{p} + \mathbf{q} = 0)$. The amplitudes of the modes are listed in Table 3.2. The energy transfers for the first triad (a) are

$$S^{uu}\left((-1,-1)|(0,1)|(1,0)\right) = -\Im([(-\hat{x}-\hat{y})\cdot(C\hat{y})][(B\hat{x})\cdot\frac{A}{i}(-\hat{x}+\hat{y})])$$
$$= ABC = -S^{uu}\left((0,1)|(-1,-1)|(1,0)\right)$$

$$S^{uu}\left((-1,-1)|(1,0)|(0,1)\right) = -\Im([(-\hat{x}-\hat{y})\cdot(B\hat{x})][(C\hat{y})\cdot\frac{A}{i}(-\hat{x}+\hat{y})])$$
$$= -ABC = -S^{uu}\left((1,0)|(-1,-1)|(0,1)\right)$$

$$S^{uu}\left((1,0)|(0,1)|(-1,-1))\right) = -\Im([\hat{x}\cdot\frac{A}{i}(-\hat{x}+\hat{y})][C\hat{y}\cdot B\hat{x})$$
$$= 0 = S^{uu}\left((0,1)|(1,0)|(-1,-1))\right)$$

Hence, the energy ABC flows from the mode $(0,1)$ to $(1,1)$, which is then transferred to $(1,0)$. See Fig. 4.2(a) for an illustration. In Example 3.2 we derived that $A=$ constant, $B = c\cos(At)$, and $C = c\sin(At)$. Thus ABC changes sign with a

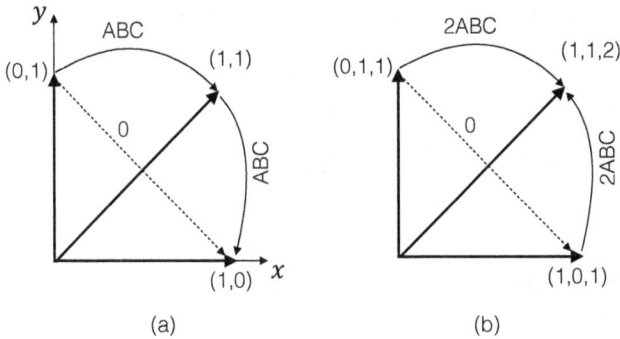

Fig. 4.2   Energy transfers in (a) Example 4.1, (b) Example 4.2.

frequency of $2A$. Hence the energy flows from $(0,1) \rightarrow (1,1) \rightarrow (1,0) \rightarrow (1,1) \rightarrow (0,1)$ periodically.

Example 4.2: Compute the energy transfer for the flow field of Example 3.3

$$\mathbf{u} = 4C(\hat{x} \sin x \cos z - \hat{z} \cos x \sin z) + 4B(\hat{y} \sin y \cos z - \hat{z} \cos y \sin z)$$
$$+ 8A(-\hat{x} \sin x \cos y \cos 2z - \hat{y} \cos x \sin y \cos 2z + \hat{z} \cos x \cos y \sin 2z).$$

Solution: The active Fourier modes are $(\pm 1, 0, \pm 1)$, $(0, \pm 1, \pm 1)$, and $(\pm 1, \pm 1, \pm 2)$, and they form a number of triads. Here we focus on the energy transfers in one of the triads whose wavenumbers are $\mathbf{q} = (1,0,1)$, $\mathbf{p} = (0,1,1)$, and $\mathbf{k} = (1,1,2)$:

$$S^{uu}(\mathbf{k}|\mathbf{p}|\mathbf{q}) = \Im[\mathbf{k} \cdot \mathbf{u}(\mathbf{q})\mathbf{u}(\mathbf{p}) \cdot \mathbf{u}^*(\mathbf{k})]$$
$$= \Im[(1,1,2) \cdot (0,1,-1)(C/i)(-2BA)] = 2ABC$$
$$S^{uu}(\mathbf{k}|\mathbf{q}|\mathbf{p}) = \Im[(1,1,2) \cdot (1,0,-1)(B/i)(-2CA)] = 2ABC$$
$$S^{uu}(\mathbf{q}|\mathbf{p}|\mathbf{k}) = 0$$

Hence, the energy flows from the modes $\mathbf{u}(\mathbf{p})$ and $\mathbf{u}(\mathbf{q})$ to the mode $\mathbf{u}(\mathbf{k})$ by an amount $2ABC$ each. See Fig. 4.2(b) for an illustration. As discussed in Example 3.3, either B or C oscillates around a mean value, while the other two oscillates around zero. Therefore, we expect that the energy transfer $2ABC$ oscillates around zero. This needs to be verified using a numerical simulation.

We will show later that the aforementioned energy transfers provide useful diagnostics for understanding pattern formation and turbulence.

4.2   Energy transfers in stably stratified flows

Starting with Eqs. (2.38, 2.39), we derive the equations for the kinetic and potential energies of the stably stratified flows as

$$\frac{d}{dt}E_u(\mathbf{k}) = \sum_{\mathbf{p}} S^{uu}(\mathbf{k}|\mathbf{p}|\mathbf{q}) - N\Re\left[u_z(\mathbf{k})b^*(\mathbf{k})\right] + \Re\left[\mathbf{f}(\mathbf{k})\cdot\mathbf{u}^*(\mathbf{k})\right] - 2\nu k^2 E_u(\mathbf{k}),$$

$$\text{(4.32)}$$

$$\frac{d}{dt}E_b(\mathbf{k}) = \sum_{\mathbf{p}} S^{bb}(\mathbf{k}|\mathbf{p}|\mathbf{q}) + N\Re\left[u_z(\mathbf{k})b^*(\mathbf{k})\right] - 2\kappa k^2 E_b(\mathbf{k}), \qquad \text{(4.33)}$$

where

$$E_b(\mathbf{k}) = \frac{1}{2}|b(\mathbf{k}')|^2 \qquad \text{(4.34)}$$

is the modal potential energy. Following the arguments similar to that of Sec. 4.1, we can derive the mode-to-mode potential energy transfer from mode $b(\mathbf{p})$ to mode $b(\mathbf{k})$ with mode $\mathbf{u}(\mathbf{q})$ acting as a mediator

$$S^{bb}(\mathbf{k}|\mathbf{p}|\mathbf{q}) = \Im\left[\{\mathbf{k}\cdot\mathbf{u}(\mathbf{q})\}\{b(\mathbf{p})b^*(\mathbf{k})\}\right]. \qquad \text{(4.35)}$$

In Eq. (4.32), the term $\Re\left[\mathbf{f}(\mathbf{k})\cdot\mathbf{u}^*(\mathbf{k})\right]$ represents the energy supply rate by the external force $\mathbf{f}$. Note that the field b, defined using Eq. (2.37), has dimension of velocity.

An interpretation of various terms for the stably stratified flows is as follows. The equation for the total energy is

$$\frac{d}{dt}(E_u + E_b) = -2\nu\sum_{\mathbf{k}} k^2 E_u(\mathbf{k}) - 2\kappa\sum_{\mathbf{k}} k^2 E_b(\mathbf{k}) + \sum_{\mathbf{k}}\Re\left[\mathbf{f}(\mathbf{k})\cdot\mathbf{u}^*(\mathbf{k})\right]. \quad \text{(4.36)}$$

Thus, in the absence of external force $\mathbf{f}$, the stably stratified flow would decay due to viscous dissipation and diffusion. Hence, we need an external force to maintain steady state in a stably stratified flow. This is unlike RBC in which the thermal plumes drive the flow and maintain a steady state. The above equation also demonstrates that $E_u + E_b$ is conserved when $\nu = \kappa = 0$ and $\mathbf{f} = 0$.

In the next two sections, using the mode-to-mode energy transfers, we derive diagnostic tools—energy flux and shell-to-shell energy transfers.

4.3   Energy flux due to nonlinear interactions

As described in the previous sections, nonlinear interactions induce energy and entropy transfers from a Fourier mode and to another. Buoyancy-driven flows involve transfers of kinetic energy, $u^2/2$, potential energy, $b^2/2$ or $\rho^2/2$, and entropy $\theta^2/2$. Fluxes of these quantities provide very useful diagnostic tools for the buoyancy-driven flow. First we focus on the kinetic energy flux.

The kinetic energy flux $\Pi_u(k_0)$ is defined as the rate of kinetic energy transfer from all the modes residing inside the wavenumber sphere of radius k_0 to the modes outside the sphere. The above quantity can be computed using $S^{uu}\left(\mathbf{k}\,|\mathbf{p}|\,\mathbf{q}\right)$ as follows:

$$\Pi_u(k_0) = \sum_{k'>k_0} \sum_{p\leq k_0} S^{uu}\left(\mathbf{k}'\,|\mathbf{p}|\,\mathbf{q}\right). \tag{4.37}$$

We illustrate the above in Fig. 4.3. Positive $\Pi_u(k)$ implies that the kinetic energy flows from small k (large length scales) to large k (small length scales), and it is termed as *forward cascade*. While negative $\Pi_u(k)$ or *inverse cascade* implies that kinetic energy flows from large k (small length scales) to small k (large length scale).

Note that

$$\sum_{k'\in A} \sum_{p\in A} S^{uu}\left(\mathbf{k}'\,|\mathbf{p}|\,\mathbf{q}\right) = 0, \tag{4.38}$$

where A represents a wavenumber region. The above sum is zero because

$$S^{uu}\left(\mathbf{k}'\,|\mathbf{p}|\,\mathbf{q}\right) + S^{uu}\left(\mathbf{p}\,|\mathbf{k}'|\,\mathbf{q}\right) = 0. \tag{4.39}$$

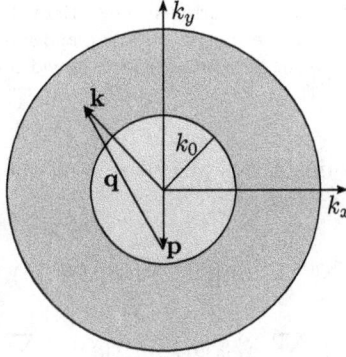

Fig. 4.3 An illustration of the energy flux $\Pi(k_0)$, which is the energy transferred from all the modes residing inside the sphere of radius k_0 to the modes outside the sphere. In the figure **p, k, q** are the giver, receiver, and mediator modes respectively.

Using the above identity, the kinetic energy flux $\Pi_u(k)$ can also be written in another useful form:

$$\Pi_u(k) = -\sum_{k'\leq k} \sum_{\mathbf{p}} S^{uu}\left(\mathbf{k}'\,|\mathbf{p}|\,\mathbf{q}\right)$$

$$= -\sum_{k'\leq k} \sum_{p\leq k} S^{uu}\left(\mathbf{k}'\,|\mathbf{p}|\,\mathbf{q}\right) - \sum_{k'\leq k} \sum_{p>k} S^{uu}\left(\mathbf{k}'\,|\mathbf{p}|\,\mathbf{q}\right)$$

$$= 0 + \sum_{p>k} \sum_{k'\leq k} S^{uu}\left(\mathbf{p}\,|\mathbf{k}'|\,\mathbf{q}\right)$$

$$= \sum_{k'>k} \sum_{p\leq k} S^{uu}\left(\mathbf{k}'\,|\mathbf{p}|\,\mathbf{q}\right), \tag{4.40}$$

which is same as Eq. (4.37). Here, $\Pi_u(k)$ can be interpreted as the rate of kinetic energy lost by the sphere of radius k due to nonlinear interactions.

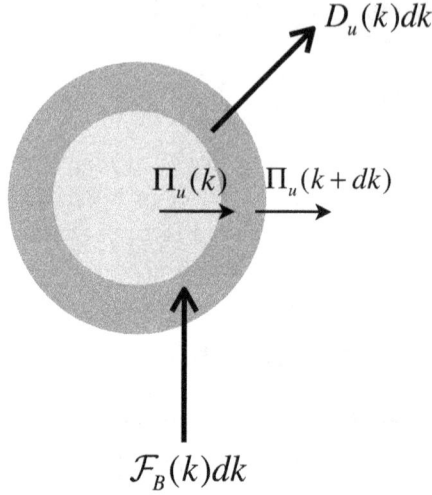

Fig. 4.4   An illustration of energetics in a turbulent flow. Here $\Pi(k)$ and $\Pi(k+dk)$ represent the kinetic energy fluxes emanating from the spheres of radii k and $k+dk$ respectively, $\mathcal{F}_B(k)dk$ is the energy supply rate by buoyancy to the wavenumber shell of width dk, and $D_u(k)dk$ is the viscous dissipation in the shell.

Equation (4.40), and sum of Eq. (4.19) over all the modes inside a sphere of radius k yield

$$\frac{d}{dt} \sum_{k' \leq k} E_u(\mathbf{k}') = \sum_{k' \leq k} \sum_{\mathbf{p}} S^{uu}(\mathbf{k}'|\mathbf{p}|\mathbf{q}) + \sum_{k' \leq k} \alpha g \Re[\theta(\mathbf{k}')u_z^*(\mathbf{k}')] - \sum_{k' \leq k} 2\nu k^2 E_u(\mathbf{k}')$$

$$= -\Pi_u(k) + \sum_{k' \leq k} \alpha g \Re[\theta(\mathbf{k}')u_z^*(\mathbf{k}')] - \sum_{k' \leq k} 2\nu k^2 E_u(\mathbf{k}')$$

$$= (a) + (b) + (c). \tag{4.41}$$

That is, the rate of gain of kinetic energy in the wavenumber sphere of radius k is a sum of (a) the rate of energy gained by the sphere due to nonlinear interactions, (b) the rate of energy supply by buoyancy, and (c) negative of the dissipation rate. We write the corresponding equation for $\sum_{k' < k+dk} E(\mathbf{k}')$ and then take a difference between the two equations, which yields

$$\frac{d}{dt} \sum_{k < k' \leq k+dk} E_u(\mathbf{k}') = [-\Pi_u(k+dk) + \Pi_u(k)] + \sum_{k < k' \leq k+dk} \alpha g \Re[\theta(\mathbf{k}')u_z^*(\mathbf{k}')]$$

$$- 2\nu \sum_{k < k' \leq k+dk} k'^2 E_u(\mathbf{k}'). \tag{4.42}$$

Now taking the limit $dk \to 0$ yields

$$\frac{d}{dt} E_u(k) = -\frac{d}{dk} \Pi_u(k) + \mathcal{F}_B(k) - D_u(k) \tag{4.43}$$

where

$$\mathcal{F}_B(k) = \sum_{|\mathbf{k}'|=k} \alpha g \Re[\theta(\mathbf{k}')u_z^*(\mathbf{k}')], \tag{4.44}$$

$$D_u(k) = 2\nu \sum_{|\mathbf{k}'|=k} k'^2 E_u(\mathbf{k}'), \tag{4.45}$$

with the sum being performed over the shell. We illustrate the above energetics in Fig. 4.4. The rate of change of kinetic energy in a shell is determined using the difference between the energy flux at the two radii, the energy supply rate by buoyancy, and the dissipation rate. In general situation, $\mathcal{F}$ represents the energy supply rate by the external force:

$$\mathcal{F}(k) = \sum_{|\mathbf{k}'|=k} \Re[\mathbf{f}(\mathbf{k}') \cdot \mathbf{u}^*(\mathbf{k}')]. \tag{4.46}$$

Using similar heuristics as described above, we define the potential energy flux as

$$\Pi_b(k_0) = \sum_{k'>k_0} \sum_{p\leq k_0} S^{bb}\left(\mathbf{k}' \,|\mathbf{p}|\, \mathbf{q}\right), \tag{4.47}$$

and the entropy flux as

$$\Pi_\theta(k_0) = \sum_{k'>k_0} \sum_{p\leq k_0} S^{\theta\theta}\left(\mathbf{k}' \,|\mathbf{p}|\, \mathbf{q}\right). \tag{4.48}$$

The energetics of the entropy is described by the following equation:

$$\frac{d}{dt}E_\theta(k) = -\frac{d}{dk}\Pi_\theta(k) + \frac{|d\bar{T}/dz|}{\alpha g}\mathcal{F}_B(k) - D_\theta(k). \tag{4.49}$$

Note that $d\bar{T}/dz < 0$ for RBC. From Equations (4.43, 4.49) we deduce that

$$\frac{d}{dt}[E_u(k) - \Xi E_\theta(k)] = -\frac{d}{dk}[\Pi_u(k) - \Xi\Pi_\theta(k)] - [D_u(k) - \Xi D_\theta(k)], \tag{4.50}$$

where

$$\Xi = \frac{\alpha g}{|d\bar{T}/dz|}. \tag{4.51}$$

Hence, under steady state, in the inertial range where the dissipative effects are negligible, we obtain

$$\Pi_u(k) - \Xi\Pi_\theta(k) = \text{const.} \tag{4.52}$$

For stably stratified turbulence, the corresponding conservation law for the fluxes is

$$\Pi_u(k) + \Pi_b(k) = \text{const.} \tag{4.53}$$

In Chapters 12 and 13 we will show that in three-dimensions, $\Pi_u(k) > 0$, $\Pi_\theta(k) > 0$, and $\Pi_b(k) > 0$. The situation however is more complex in the boundary layers of RBC, and in two-dimensional and quasi two-dimensional buoyant flows.

4.4 Shell-to-shell energy transfers

The energy flux provides information on the cumulative energy transfer from the modes inside a sphere to modes outside the sphere. For a more detailed information on the energy transfers, we compute kinetic energy transfer from a region of a wavenumber space, A, to region, B:

$$T_{u,B}^{u,A} = \sum_{k \in B} \sum_{p \in A} S^{uu} (\mathbf{k} | \mathbf{p} | \mathbf{q}). \tag{4.54}$$

For homogeneous and isotropic turbulence, it is customary to use wavenumber shells as regions A and B. The shell-to-shell energy transfer rate from shell m to shell n is defined as [Dar *et al.* (2001); Verma (2004)]

$$T_{u,n}^{u,m} = \sum_{k \in n} \sum_{p \in m} S^{uu} (\mathbf{k} | \mathbf{p} | \mathbf{q}), \tag{4.55}$$

and it is illustrated in Fig. 4.5.

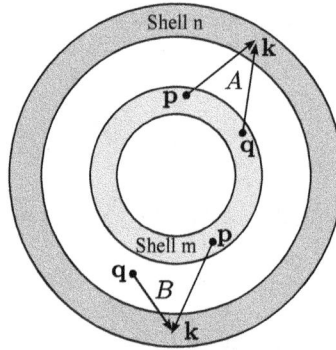

Fig. 4.5 A schematic diagram of shell-to-shell kinetic energy transfer, $T_{u,n}^{u,m}$, from shell m to shell n. The giver mode $\mathbf{p} \in m$ and the receiver mode $\mathbf{k} \in n$.

Following similar approach as above, we derive the entropy and potential energy transfer rates from region A to region B, as well as from wavenumber shell m to wavenumber shell n as

$$T_{\theta,B}^{\theta,A} = \sum_{k \in B} \sum_{p \in A} S^{\theta\theta} (\mathbf{k} | \mathbf{p} | \mathbf{q}), \tag{4.56}$$

$$T_{\theta,n}^{\theta,m} = \sum_{k \in n} \sum_{p \in m} S^{\theta\theta} (\mathbf{k} | \mathbf{p} | \mathbf{q}), \tag{4.57}$$

$$T_{b,B}^{b,A} = \sum_{k \in B} \sum_{p \in A} S^{bb} (\mathbf{k} | \mathbf{p} | \mathbf{q}), \tag{4.58}$$

$$T_{b,n}^{b,m} = \sum_{k \in n} \sum_{p \in m} S^{bb} (bk | \mathbf{p} | \mathbf{q}). \tag{4.59}$$

In Kolmogorov's theory of hydrodynamic turbulence, to be discussed in Chapter 10, the maximum shell-to-shell kinetic energy transfer occurs from shell m to shell $(m + 1)$, hence the kinetic energy transfer in hydrodynamic turbulence is *local* and *forward*. We will explore using numerical simulations whether the energy transfer in RBC is local and forward.

There are only a handful of analytical results in turbulence literature. Hence, we need to resort to numerical simulations to understand turbulent flows. In the following discussion we will briefly describe pseudo-spectral method that yields various spectral diagnostics.

4.5 Pseudo-spectral method

In pseudo-spectral method we solve Eqs. (3.8–3.9) under the constraint $\mathbf{k} \cdot \mathbf{u}(\mathbf{k}) = 0$. Equations (3.8–3.9) are time advanced using a time stepping method, e.g., Runge-Kutta scheme. In spectral space, the nonlinear terms

$$N_{u,i}(\mathbf{k}) = ik_j[u_j(\mathbf{r})u_i(\mathbf{r})](\mathbf{k}), \qquad (4.60)$$

$$N_\theta(\mathbf{k}) = ik_j[u_j(\mathbf{r})\theta(\mathbf{r})](\mathbf{k}) \qquad (4.61)$$

become convolutions (see Eqs. (3.11, 3.12)) whose computation requires $O(N^6)$ floating point operations for a N^3 grid. $O(N^6)$ computation is enormous for a large N. For example, for $N = 10^3$, the required floating point operations is 10^{18} that would take approximately 10^3 seconds on a PetaFLOP supercomputer (a supercomputer that can perform 10^{15} per second).

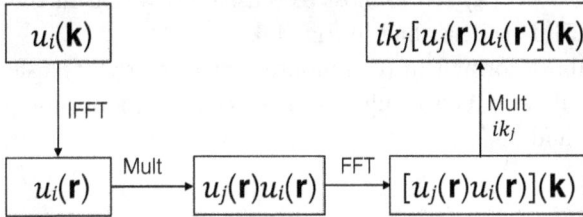

Fig. 4.6 A schematic diagram of the computation of a nonlinear term $ik_j[u_j(\mathbf{r})u_i(\mathbf{r})](\mathbf{k})$ in a pseudo-spectral solver.

Orszag circumvented the above difficulty using Fast Fourier Transform (FFT), in which a convolution is computed using $O(N^3 \log(N^3))$ operations, which is manageable. For the computation of $[u_j(\mathbf{r})u_i(\mathbf{r})](\mathbf{k})$, $\mathbf{u}(\mathbf{k})$ is transformed to real field $\mathbf{u}(\mathbf{r})$, the products $[u_j(\mathbf{r})u_i(\mathbf{r})]$ is computed in the real space, and then the products are transformed back to the Fourier space yielding $[u_j(\mathbf{r})u_i(\mathbf{r})](\mathbf{k})$. The derivative operation in Fourier space is performed by $ik_j[u_j(\mathbf{r})u_i(\mathbf{r})](\mathbf{k})$. We illustrate these operations in Fig. 4.6. Since the computation of the nonlinear terms involves multiplication in real space, this method is called *pseudo-spectral method*. The procedure however suffers from *aliasing* errors, which is solved by filling up only 2/3 part of the array in each direction. See Boyd (2003), Canuto *et al.* (1988), Verma *et al.* (2013), and Chatterjee *et al.* (2018) for details.

In numerical simulations, to compute the kinetic energy transfers from region A to region B, we adopt the following strategy.

$$T^{u,A}_{u,B} = \sum_{k \in B} \sum_{p \in A} S^{uu}(\mathbf{k}|\mathbf{p}|\mathbf{q})$$

$$= \Re \sum_{k \in B} \left\{ \sum_{p \in A} -i[\mathbf{k} \cdot \mathbf{u}(\mathbf{q})]\mathbf{u}^A(\mathbf{p}) \right\} \cdot \left\{ \mathbf{u}^B(\mathbf{k}) \right\}^*$$

$$= \Re \sum_{k \in B} \left\{ -\mathbf{N}^A(\mathbf{k}) \right\} \cdot \left\{ \mathbf{u}^B(\mathbf{k}) \right\}^*, \tag{4.62}$$

where

$$\mathbf{u}^A(\mathbf{p}) = \begin{cases} \mathbf{u}(\mathbf{p}) & \text{for } \mathbf{p} \in A \\ 0 & \text{otherwise} \end{cases}, \tag{4.63}$$

$$\mathbf{u}^B(\mathbf{k}) = \begin{cases} \mathbf{u}(\mathbf{k}) & \text{for } \mathbf{k} \in B \\ 0 & \text{otherwise} \end{cases}, \tag{4.64}$$

and $\mathbf{N}^A(\mathbf{k})$ is the nonlinear field induced by the velocity field $\mathbf{u}^A$. The sum of $\Re\left\{\mathbf{N}^A(\mathbf{k}) \cdot \mathbf{u}^{*B}(\mathbf{k})\right\}$ over the region B yields the energy transfer from the Fourier modes of region A to the Fourier modes of region B. We can also define entropy transfer rate in a similar manner [Kumar *et al.* (2014a)]. The computation of the nonlinear term $\mathbf{N}^A(\mathbf{k})$ is the most expensive operation in the code, and it is computed using FFT as described in Fig. 4.6.

We use the above formalism to compute various energy transfer diagnostics of buoyancy-driven flows. The results of these computations would be covered in Chapters 12, 13, and 15.

Further reading

Dar *et al.* (2001) and Verma (2004) first derived the formula for the mode-to-mode energy transfer in hydrodynamic and magnetohydrodynamic flows. I am not aware of simpler introduction to energy transfers than what is presented here. For spectral method, reader can refer to Canuto *et al.* (1988), Boyd (2003), Verma *et al.* (2013), and Chatterjee *et al.* (2018).

Exercises

(1) Consider the following fluid flow in a two-dimensional box of size $[\pi, \pi]$:

$$\mathbf{u} = 4D(\hat{x}\sin x \cos 3y - \frac{1}{3}\hat{y}\cos x \sin 3y) + 4C(\hat{x}\sin 3x \cos y - \hat{y}3\cos 3x \sin y)$$

$$+4B(\hat{y}\sin 2x \cos 2y - \hat{y}\cos 2x \sin 2y) + 4A(\hat{x}\sin x \cos y - \hat{y}\cos x \sin y)$$

Compute the energy transfer among the modes (1,1), (2,2), (3,1), and (1,3).

(2) Consider flow fields that contains one of the triads with wavenumbers $\mathbf{k} = (1/2, \sqrt{3}/2)$, $\mathbf{q} = (1,0)$, and $\mathbf{p} = (-1/2, \sqrt{3}/2)$. Compute the energy transfer among the above modes.

(3) Consider the flow field of Example 4.1. Compute the energy flux for wavenumber spheres of radius of 0.8 and 1.2.

(4) Consider the flow field of Example 4.2. Compute the energy flux for wavenumber spheres of radius of 2 and 8.

(5) Following arguments similar to that of Sec. 4.1, show that Eq. (4.24) is the formula for the mode-to-mode entropy transfer.

Chapter 5

Waves in Stably Stratified Flows

Stably stratified flows are linearly stable hence they support waves under linear perturbation. In this chapter we derive wave solution inside a stably stratified flow. These waves are called *internal gravity waves*.

5.1   Internal gravity waves

We start with Eqs. (3.20–3.22). A linearized version of these equations with $\nu = \kappa = 0$ is

$$\frac{d}{dt}\mathbf{u}(\mathbf{k}) = -i\mathbf{k}\frac{\sigma(\mathbf{k})}{\rho_m} - Nb(\mathbf{k})\hat{z}, \tag{5.1}$$

$$\frac{d}{dt}b(\mathbf{k}) = N\hat{u}_z(\mathbf{k}), \tag{5.2}$$

$$\mathbf{k}\cdot\mathbf{u}(\mathbf{k}) = 0, \tag{5.3}$$

where N is the *Brunt-Väisälä frequency*:

$$N = \sqrt{\frac{g}{\rho_m}\left|\frac{d\bar{\rho}}{dz}\right|}, \tag{5.4}$$

and b is the normalised density in the units of velocity:

$$b = \frac{g}{N}\frac{\rho}{\rho_m}. \tag{5.5}$$

We compute the pressure $\sigma(\mathbf{k})$ by taking a dot product between $\mathbf{k}$ and Eq. (5.1), and by employing $\mathbf{k}\cdot\mathbf{u}(\mathbf{k}) = 0$:

$$\sigma(\mathbf{k}) = iN\rho_m\frac{k_z}{k^2}b(\mathbf{k}). \tag{5.6}$$

Substitution of the above in Eq. (5.1) yields

$$\frac{d}{dt}u_z(\mathbf{k}) = -\frac{k_\perp^2}{k^2}Nb(\mathbf{k}), \tag{5.7}$$

where $\mathbf{k}_\perp = k_x\hat{x} + k_y\hat{y}$.

In a matrix form, Eqs. (5.2, 5.7) appear as

$$\frac{d}{dt}\begin{pmatrix} u_z(\mathbf{k}) \\ b(\mathbf{k}) \end{pmatrix} = \begin{pmatrix} 0 & -\frac{k_\perp^2}{k^2}N \\ N & 0 \end{pmatrix}\begin{pmatrix} u_z(\mathbf{k}) \\ b(\mathbf{k}) \end{pmatrix} = A\begin{pmatrix} u_z(\mathbf{k}) \\ b(\mathbf{k}) \end{pmatrix} \tag{5.8}$$

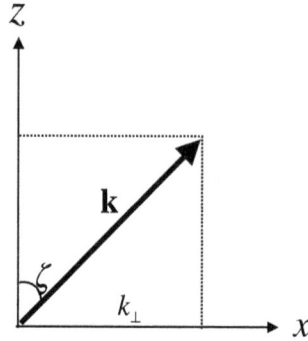

Fig. 5.1   A wavenumber **k** in Fourier space. It makes an angle ζ with $\hat{z}$.

The eigenvalues of the matrix A are $\lambda_\pm = \pm i\omega$ with

$$\omega = N\frac{k_\perp}{k} = N\sin\zeta, \tag{5.9}$$

where ζ is the angle between **k** and $\hat{z}$ (see Fig. 5.1 for an illustration). The corresponding eigenvectors are

$$\begin{pmatrix} 1 \\ -i\frac{k}{k_\perp} \end{pmatrix}; \quad \begin{pmatrix} 1 \\ i\frac{k}{k_\perp} \end{pmatrix}. \tag{5.10}$$

Hence, the solution for λ_+ is

$$u_z(\mathbf{k}, t) = A_+ \exp(i\omega t), \tag{5.11}$$

$$b(\mathbf{k}, t) = -i\frac{kA_+}{k_\perp} \exp(i\omega t), \tag{5.12}$$

where

$$A_+ = |A_+| \exp(i\phi_{k+}) \tag{5.13}$$

is the amplitude of $u_z(\mathbf{k})$. In real space, the wave appears as

$$
\begin{aligned}
u_z(\mathbf{r}, t) &= \Re[A_+ \exp(i\mathbf{k}\cdot\mathbf{r} + i\omega t)] \\
&= |A_+|\cos(\mathbf{k}\cdot\mathbf{r} + \omega t + \phi_{k+}),
\end{aligned}
\tag{5.14}
$$

$$
\begin{aligned}
b(\mathbf{r}, t) &= -\Re\left[\frac{ikA_+}{k_\perp}\exp(i\mathbf{k}\cdot\mathbf{r} + i\omega t)\right] \\
&= |A_+|\frac{k}{k_\perp}\sin(\mathbf{k}\cdot\mathbf{r} + \omega t + \phi_{k+}).
\end{aligned}
\tag{5.15}
$$

The above solution corresponds to internal gravity waves moving in the direction of $-\mathbf{k}$.

By following a similar procedure we obtain the solution corresponding to λ_- as

$$
\begin{aligned}
u_z(\mathbf{r}, t) &= \Re[A_- \exp(i\mathbf{k}\cdot\mathbf{r} - i\omega t)] \\
&= |A_-|\cos(\mathbf{k}\cdot\mathbf{r} - \omega t + \phi_{k-}),
\end{aligned}
\tag{5.16}
$$

$$
\begin{aligned}
b(\mathbf{r}, t) &= \Re\left[\frac{ikA_-}{k_\perp}\exp(i\mathbf{k}\cdot\mathbf{r} - i\omega t)\right] \\
&= -|A_-|\frac{k}{k_\perp}\sin(\mathbf{k}\cdot\mathbf{r} - \omega t + \phi_{k-}).
\end{aligned}
\tag{5.17}
$$

Here $A_- = |A_-| \exp(i\phi_-)$. The above solution represents an internal gravity wave moving along $\mathbf{k}$. A general solution of Eq. (5.8) is a superposition of waves travelling parallel and antiparallel to $\mathbf{k}$:

$$u_z(\mathbf{r}, t) = |A_+| \cos(\mathbf{k} \cdot \mathbf{r} + \omega t + \phi_+) + |A_-| \cos(\mathbf{k} \cdot \mathbf{r} - \omega t + \phi_-) \qquad (5.18)$$

$$b(\mathbf{r}, t) = \frac{k}{k_\perp} [|A_+| \sin(\mathbf{k} \cdot \mathbf{r} + \omega t + \phi_+) \qquad (5.19)$$

$$-|A_-| \sin(\mathbf{k} \cdot \mathbf{r} - \omega t + \phi_-)] . \qquad (5.20)$$

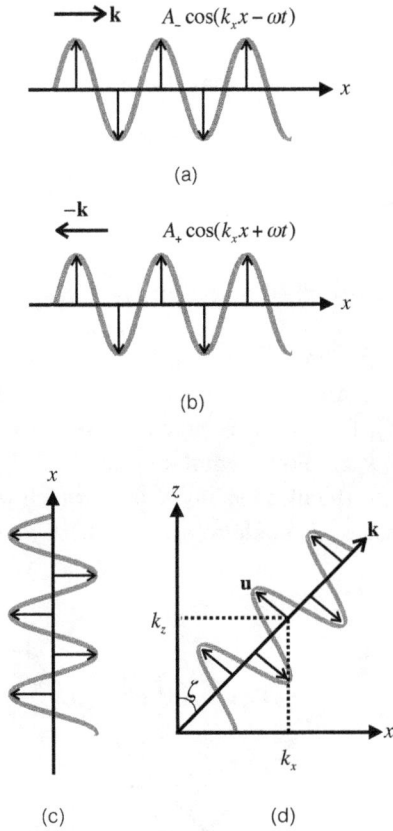

Fig. 5.2 Internal gravity waves moving along (a) $\mathbf{k} = k_x \hat{x}$; (b) $-\mathbf{k} = -k_x \hat{x}$; (d) $\mathbf{k} = k_x \hat{x} + k_z \hat{z}$. (c) For $\mathbf{k} = k_z \hat{z}$, there is no oscillation due to gravity.

The wave could be two- or three dimensional. First we discuss a 2D wave whose wavenumber is $\mathbf{k} = k_x \hat{x} + k_z \hat{z}$. The 3D waves will be discussed in the next section. Note that the incompressibility condition yields

$$u_x = -\frac{k_z}{k_x} u_z \qquad (5.21)$$

We illustrate the internal gravity waves using three limiting cases:

(1) For $\mathbf{k} = k_x \hat{x}$, Fig. 3.1(a) illustrates a wave moving along $\hat{x}$:

$$\mathbf{u}(\mathbf{r}, t) = A_- \cos(k_x x - \omega t)\hat{z}. \tag{5.22}$$

For this wave, $k_\perp = k_x = k$ or $\zeta = \pi/2$. Therefore, $\omega = N$ and $u_x = 0$ (see Eq. (5.21)). We illustrate the above wave in Fig. 5.2(a). Note that the fluid parcels are oscillating vertically due to gravity. Figure 5.2(b) depicts a wave moving along $-\mathbf{k}$ which is

$$\mathbf{u}(\mathbf{r}, t) = A_+ \cos(k_x x + \omega t)\hat{z}. \tag{5.23}$$

(2) For $\mathbf{k} = k_z \hat{z}$,

$$\mathbf{u}(\mathbf{r}, t) = A \cos(k_z z)\hat{x}. \tag{5.24}$$

Here $\omega = 0$, $u_z = 0$, and there is no gravity wave. See Fig. 5.2(c) for an illustration. The velocity field provides shear to the flow, and it is not connected to gravitational oscillations.

(3) For $\mathbf{k} = k_x \hat{x} + k_z \hat{z}$,

$$\mathbf{u}(\mathbf{r}, t) = A_- \cos(k_x x + k_z z - \omega t)\hat{n}. \tag{5.25}$$

where $\omega = N k_x / k = N \sin \zeta$ with $k = \sqrt{k_x^2 + k_y^2}$, and $\mathbf{u}$ is perpendicular to $\mathbf{k}$ as shown in Fig. 5.2(d). Note that both u_x and u_z have the same frequency ω. As described in item (1), for $\mathbf{k} = k_x \hat{x}$, u_z oscillates with the frequency of $\omega = N$, but $\omega = 0$ for $\mathbf{k} = k_z \hat{z}$. For a combination, $\mathbf{k} = k_x \hat{x} + k_z \hat{z}$, the oscillation frequency takes an intermediate value $N \sin \zeta$, which is between 0 and N. This is because the sluggish u_x is made to oscillate with u_z with the same frequency.

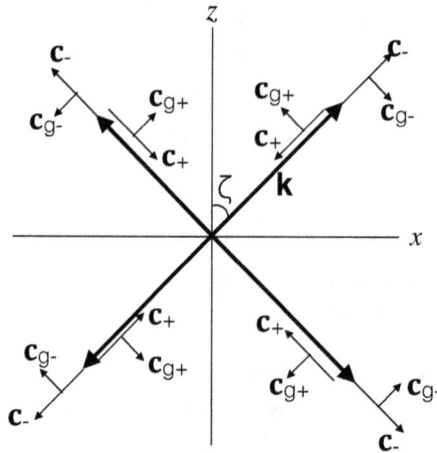

Fig. 5.3   A wave A_- travelling along $\mathbf{k}$ has phase velocity $\mathbf{c}_-$ and group velocity $\mathbf{c}_{g-}$, while a wave A_+ travelling along $-\mathbf{k}$ has phase velocity $\mathbf{c}_+$ and group velocity $\mathbf{c}_{g+}$. The figure also indicates such waves in different quadrants.

Equation (5.9) is the dispersion relation for internal gravity waves. Using this relation we derive the phase velocities of the waves associated with $A_\pm$ as

$$\mathbf{c}_+ = -\frac{N}{k}\sin\zeta\hat{k}; \quad \mathbf{c}_- = \frac{N}{k}\sin\zeta\hat{k}, \tag{5.26}$$

and the group velocities of the wave as

$$\mathbf{c}_{g\pm} = \mp\nabla_\mathbf{k}\omega = \mp\hat{\zeta}\frac{1}{k}\frac{\partial\omega}{\partial\zeta} = \mp\hat{\zeta}\frac{N}{k}\cos\zeta, \tag{5.27}$$

where $\hat{\zeta}$ is the unit vector associated with the polar angle ζ. The phase and group velocities of the waves are illustrated in Fig. 5.3. Also see Appendix B.

5.2 Derivation of waves in Craya-Herring basis

It is convenient to derive the internal gravity waves in Craya-Herring basis discussed in Sec. 3.6. In this basis, buoyancy term is

$$\mathbf{F}(\mathbf{k}) = -Nb(\mathbf{k})\hat{z}$$
$$= \hat{e}_2 Nb(\mathbf{k})\sin\zeta - \hat{e}_3 Nb(\mathbf{k})\cos\zeta, \tag{5.28}$$

where ζ is the angle between $\mathbf{k}$ and $\hat{z}$. Therefore, the dynamical equations (5.1–5.3) in Craya-Herring basis are:

$$\dot{u}_1(\mathbf{k}) = 0, \tag{5.29}$$
$$\dot{u}_2(\mathbf{k}) = \omega b(\mathbf{k}), \tag{5.30}$$
$$\dot{b}(\mathbf{k}) = -\omega u_2(\mathbf{k}), \tag{5.31}$$

where $u_{1,2}(\mathbf{k})$ are the components of the velocity mode $\mathbf{u}(\mathbf{k})$ in the Craya-Herring basis, and $\omega = N\sin\zeta$ (see Eq. (5.9)). Note that the pressure gradient balances the component of buoyancy along $\hat{e}_3$:

$$-ik\sigma(\mathbf{k}) - Nb\cos\zeta = 0. \tag{5.32}$$

Equation (5.29) admits trivial solution:

$$u_1(\mathbf{k}) = \mathrm{const}, \tag{5.33}$$

while Eqs. (5.30, 5.31) can be rewritten in a matrix form as

$$\frac{d}{dt}\begin{pmatrix} u_2(\mathbf{k}) \\ b(\mathbf{k}) \end{pmatrix} = \begin{pmatrix} 0 & \omega \\ -\omega & 0 \end{pmatrix}\begin{pmatrix} u_2(\mathbf{k}) \\ b(\mathbf{k}) \end{pmatrix} = B\begin{pmatrix} u_2(\mathbf{k}) \\ b(\mathbf{k}) \end{pmatrix}. \tag{5.34}$$

The eigenvalues of the matrix B are $\lambda_\pm = \pm i\omega$, and the corresponding eigenvectors are

$$\begin{pmatrix} 1 \\ i \end{pmatrix}; \quad \begin{pmatrix} 1 \\ -i \end{pmatrix} \tag{5.35}$$

respectively. Hence, the respective solution are

$$\begin{pmatrix} u_2(\mathbf{k}) \\ b(\mathbf{k}) \end{pmatrix} = \begin{pmatrix} 1 \\ i \end{pmatrix}B_+(\mathbf{k})\exp(i\omega t); \quad \begin{pmatrix} 1 \\ -i \end{pmatrix}B_-(\mathbf{k})\exp(-i\omega t) \tag{5.36}$$

where $B_\pm$ are the amplitudes of the waves. In real space, the wave solution are

$$\begin{pmatrix} u_2(\mathbf{r}) \\ b(\mathbf{r}) \end{pmatrix} = |B_+| \begin{pmatrix} \cos(\mathbf{k} \cdot \mathbf{r} + \omega t + \phi_{k+}) \\ -\sin(\mathbf{k} \cdot \mathbf{r} + \omega t + \phi_{k+}) \end{pmatrix}; \quad |B_-| \begin{pmatrix} \cos(\mathbf{k} \cdot \mathbf{r} - \omega t + \phi_{k-}) \\ \sin(\mathbf{k} \cdot \mathbf{r} - \omega t + \phi_{k-}) \end{pmatrix},$$

(5.37)

where $B_\pm(\mathbf{k}) = |B_\pm| \exp(i\phi_{k\pm})$. It is easy to verify that the aforementioned solutions are same as those derived in the earlier section.

Equation (5.33) yields

$$u_1(\mathbf{r}) = |u_1| \cos(\mathbf{k} \cdot \mathbf{r} + \phi_k) \tag{5.38}$$

where $u_1(\mathbf{k}) = |u_1| \exp(i\phi_k)$. Note that $u_1(\mathbf{r})$ is independent of time, and it is referred to as *vortical mode*. With u_1 and u_2, the velocity field is three dimensional, whose Cartesian components can be computed using Eq. (3.64). The aforementioned solutions are same as those of Godeferd and Cambon (1994) and Remmel *et al.* (2013).

Another important buoyancy-driven wave is surface gravity wave, which will be discussed in the next chapter.

Further reading

Vallis (2006) and Davidson (2013) discuss internal gravity waves.

Exercises

(1) Consider a stably stratified flow with gravity along $-\hat{z}$. An internal gravity wave with wavenumber $\mathbf{k} = 2\hat{x}$ is generated in the flow. Describe the fluid velocity and density fields. Give a physical mechanism to generate such waves.
(2) A point source is oscillating vertically in a stably stratified fluid. Describe the property of internal gravity waves generated by this process. What will happen when the point source oscillates horizontally?

Chapter 6

Instability in Thermal Convection

When a fluid confined between two plates is subjected to an adverse temperature gradient, as shown in Fig. 2.3(c), the flow exhibits thermal instability beyond a critical temperature gradient. In this chapter we will derive instability criteria and associated flow properties. We will study thermal instability in the framework of RBC. In addition, we will also describe Rayleigh-Taylor instability.

In the following discussion, we solve the linearised RBC equations. For the same, we drop the nonlinear terms of the RBC equations (Eqs. (2.59–2.60)) that yields the following linearised equations for u_z and θ, in addition to the incompressibility condition:

$$\frac{\partial}{\partial t}\mathbf{u} = -\nabla\sigma + \mathrm{RaPr}\theta\hat{z} + \mathrm{Pr}\nabla^2\mathbf{u}, \tag{6.1}$$

$$\frac{\partial\theta}{\partial t} = u_z + \nabla^2\theta, \tag{6.2}$$

$$\nabla\cdot\mathbf{u} = 0. \tag{6.3}$$

Here we set $\rho_m = 1$ for convenience. It is convenient to work out the instability criterion using the equation for u_z, which is

$$\frac{\partial u_z}{\partial t} = -\partial_z\sigma + \mathrm{RaPr}\theta + \mathrm{Pr}\nabla^2 u_z. \tag{6.4}$$

The u_x component is determined using the incompressibility condition, Eq. (6.3).

First we solve the above equations for the free-slip boundary condition.

6.1 Thermal instability for Free-slip boundary condition

We consider two infinite horizontal plates at $z = 0$ and $z = 1$. On these plates, we employ free-slip boundary condition for the velocity field, and conducting boundary condition for the temperature. We apply periodic boundary condition along the horizontal direction. The following basis functions satisfy the above boundary conditions:

$$u_z(x, z) = u_z(\mathbf{k})2\sin(n\pi z)\exp(i\mathbf{k}_\perp\cdot\mathbf{r}_\perp) + c.c., \tag{6.5}$$

$$\theta(x, z) = \theta(\mathbf{k})2\sin(n\pi z)\exp(i\mathbf{k}_\perp\cdot\mathbf{r}_\perp) + c.c., \tag{6.6}$$

where n is a positive integer ($n > 0$), $\mathbf{k}_\perp = k_x\hat{x} + k_y\hat{y}$, $\mathbf{r}_\perp = x\hat{x} + y\hat{y}$, and *c.c.* stands for complex conjugate. Substitution of the above in Eqs. (6.4, 6.2) yields

$$\frac{d}{dt}u_z(\mathbf{k}) = -ik_z\sigma(\mathbf{k}) + \text{RaPr}\theta(\mathbf{k}) - \text{Pr}k^2 u_z(\mathbf{k}), \tag{6.7}$$

$$\frac{d}{dt}\theta(\mathbf{k}) = u_z(\mathbf{k}) - k^2\theta(\mathbf{k}). \tag{6.8}$$

We compute the pressure $\sigma(\mathbf{k})$ by taking dot product of $\mathbf{k}$ and Eq. (6.1) and employing $\mathbf{k} \cdot \mathbf{u}(\mathbf{k}) = 0$:

$$\sigma(\mathbf{k}) = -i\frac{k_z}{k^2}\text{RaPr}\theta(\mathbf{k}), \tag{6.9}$$

substitution of which in Eq. (6.7) yields

$$\frac{d}{dt}u_z(\mathbf{k}) = \text{RaPr}\frac{k_\perp^2}{k^2}\theta(\mathbf{k}) - \text{Pr}k^2 u_z(\mathbf{k}). \tag{6.10}$$

Equations (6.8, 6.10) are rewritten in matrix form as

$$\frac{d}{dt}\begin{pmatrix} u_z(\mathbf{k}) \\ \theta(\mathbf{k}) \end{pmatrix} = \begin{pmatrix} -\text{Pr}k^2 & \text{RaPr}k_\perp^2/k^2 \\ 1 & -k^2 \end{pmatrix}\begin{pmatrix} u_z(\mathbf{k}) \\ \theta(\mathbf{k}) \end{pmatrix} = A\begin{pmatrix} u_z(\mathbf{k}) \\ \theta(\mathbf{k}) \end{pmatrix}. \tag{6.11}$$

We solve the above equations using matrix method. The eigenvalues of the stability matrix A are

$$\lambda_\pm = \frac{1}{2}\left(\text{Tr} \pm \sqrt{\text{Tr}^2 - 4\text{Det}}\right), \tag{6.12}$$

where $\text{Tr} = -(\text{Pr}+1)k^2$ is the trace of A, while

$$\text{Det} = \text{Pr}k^4 - \text{RaPr}k_\perp^2/k^2 \tag{6.13}$$

is the determinant of A. The corresponding eigenvectors are

$$\begin{pmatrix} \lambda_+ + k^2 \\ 1 \end{pmatrix}; \quad \begin{pmatrix} \lambda_- + k^2 \\ 1 \end{pmatrix}. \tag{6.14}$$

Hence, the general solution of the above equations is

$$u_z(\mathbf{k}, t) = a_1(\lambda_+ + k^2)\exp(\lambda_+ t) + a_2(\lambda_- + k^2)\exp(\lambda_- t), \tag{6.15}$$
$$\theta(\mathbf{k}, t) = a_1\exp(\lambda_+ t) + a_2\exp(\lambda_- t), \tag{6.16}$$

where $k^2 = (n\pi)^2 + k_\perp^2$, and a_1, a_2 are constants that need to be determined using initial conditions $[\theta(\mathbf{k}, 0), u_z(\mathbf{k}, 0)]$.

We can determine the properties of $\lambda_\pm$ from the matrix A. The eigenvalues $\lambda_\pm$ are always real since the argument of the square-root,

$$\text{Tr}^2 - 4\text{Det} = (\text{Pr}-1)^2 k^4 + 4\text{RaPr}\frac{k_\perp^2}{k^2}, \tag{6.17}$$

is always real and positive-definite. Note that $\text{Tr} < 0$. Therefore, when $\text{Det} > 0$, we obtain $\sqrt{\text{Tr}^2 - 4\text{Det}} < |\text{Tr}|$ that leads to $\lambda_\pm < 0$, and hence

$$u_z(\mathbf{k}, t), \theta(\mathbf{k}, t) \to 0. \tag{6.18}$$

This is the stable solution with asymptotic solution as $\mathbf{u} = \theta = 0$; for this case, heat is transported purely by conduction. However, for Det < 0, we obtain $\sqrt{\mathrm{Tr}^2 - 4\mathrm{Det}} > |\mathrm{Tr}|$. Therefore, $\lambda_+ > 0$ and $\lambda_- < 0$, hence the asymptotic solution

$$u_z(\mathbf{k}, t), \theta(\mathbf{k}, t) \to \exp(\lambda_+ t) \tag{6.19}$$

grows with time. Note that the component with $\exp(\lambda_- t)$ vanishes as $t \to \infty$. Hence Det > 0 yields unstable solution of the linearized equation. Since the eigenvalues are always real, the system admits non-oscillatory growing solution via *supercritical pitchfork bifurcation*.

The transition from the decaying solution to the growing solution occurs at $\lambda_+ = 0$, or when Det $= 0$. Using Eq. (6.13), we obtain the *critical Rayleigh number*

$$\mathrm{Ra}_c = \frac{k^6}{k_\perp^2} = \frac{(k_\perp^2 + n^2 \pi^2)^3}{k_\perp^2}. \tag{6.20}$$

This particular solution called *neutral solution* corresponds to *neutral stability* that neither decays nor grows with time.

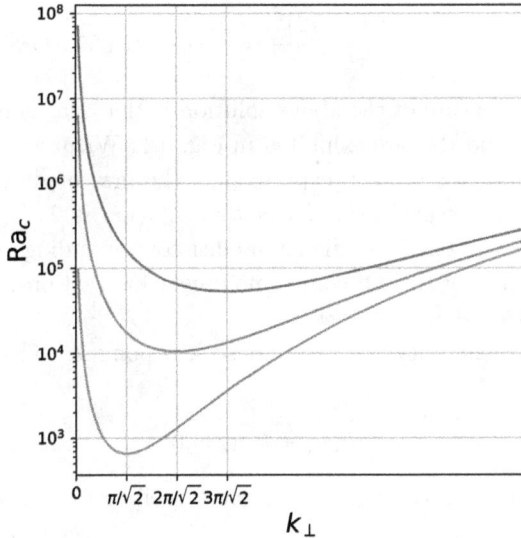

Fig. 6.1   Plots of Ra_c vs. $k_\perp$ for $n = 1, 2, 3$. See Eq. (6.20). In the figure we show that Ra_c is minimum at $k_c = n\pi/\sqrt{2}$.

The critical number Ra_c depends on n and $k_\perp$, as shown in Fig. 6.1 for $n = 1, 2, 3$. For a given n, the minimum value of Ra_c occurs at $k_\perp = k_c$, which is determined using

$$\partial_{k_\perp} \mathrm{Ra}_c |_{k_c} = 0. \tag{6.21}$$

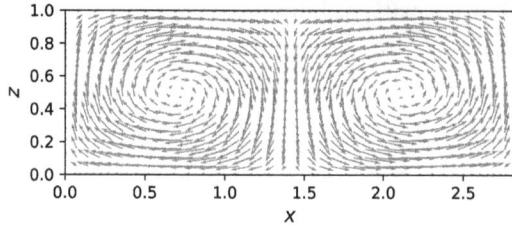

Fig. 6.2   Under the free-slip boundary condition, the primary convective rolls at the onset of RBC instability. Here $\mathrm{Ra} = 27\pi^4/4, k_c = \pi/\sqrt{2}$, and $L_x = 2\sqrt{2}$. The arrows represent the velocity vectors.

This particular condition yields

$$k_c = \frac{n\pi}{\sqrt{2}},\tag{6.22}$$

hence,

$$k^2 = k_c^2 + (n\pi)^2 = 3n^2\pi^2/2,\tag{6.23}$$

substitution of which in Eq. (6.20) yields

$$\mathrm{Ra}_{c,\min} = \frac{\left(\pi^2 + \pi^2/2\right)^3}{\pi^2/2} = \frac{27n^2\pi^4}{4} \approx 657.511n^2.\tag{6.24}$$

Another important feature of the above solution is that Ra_c is independent of Pr. The values of k_c and Ra_c are exhibited in Fig. 6.1. When we increase Ra, which corresponds to an increase of Δ in experiments, the first mode $\mathbf{k} = (\pi/\sqrt{2})\hat{x} + \pi\hat{z}$ becomes unstable. The wavelength L_x of this solution is $2\pi/k_c = 2\sqrt{2}$, and the height of the box is unity. The Higher modes corresponding to $n = 2, 3, ...$ get excited at larger values of Ra. Of course, nonlinear interactions, to be discussed in Chapters 7 and 8, excite more modes.

Without loss of generality, we can choose $\mathbf{k}_\perp$ along $\hat{x}$. The incompressibility condition yields

$$ik_x u_x + k_z u_z = 0\tag{6.25}$$

that helps us derive u_x given u_z. Using these recipes we construct the growing solution for $n = 1$, which is

$$u_z(x, z) = |u_z(\mathbf{k})|2\sin(\pi z)\cos(k_c x + \phi_k),\tag{6.26}$$

$$u_x(x, z) = -\frac{\pi}{k_c}|u_z(\mathbf{k})|2\cos(\pi z)\sin(k_c x + \phi_k),\tag{6.27}$$

where ϕ_k is the phase of the velocity Fourier mode $[u_z(\mathbf{k}) = |u_z(\mathbf{k})|\exp(i\phi_k)]$. The above solution, exhibited in Fig. 6.2, is the *primary convective roll*. The dimension of a pair of primary roll is $2\sqrt{2} \times 1$. Due to the translation symmetry, these rolls can be shifted along the x axis by an arbitrary distance.

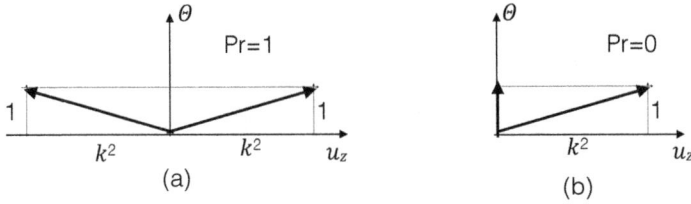

Fig. 6.3   The eigenvectors of free-slip RBC for (a) $\mathrm{Pr} = 1$; (b) $\mathrm{Pr} = 0$. See Eq. (6.28).

Let us study the eigenvectors of the matrix A of Eq. (6.11) at the neutral stability. In this case, the eigenvalues are 0 and $-(\mathrm{Pr}+1)k^2$ with the corresponding eigenvectors as

$$\begin{pmatrix} k^2 \\ 1 \end{pmatrix}; \quad \begin{pmatrix} -\mathrm{Pr}k^2 \\ 1 \end{pmatrix} \tag{6.28}$$

respectively, which are depicted in Fig. 6.3 for $\mathrm{Pr} = 1$ and 0. Interestingly, the eigenvector corresponding to $\lambda_+ = 0$ is independent of Pr. Since $\exp(-\lambda_- t) \to 0$, the asymptotic state of the neutral solution is

$$\theta(\mathbf{k}, t) = a_1, \tag{6.29}$$

$$u_z(\mathbf{k}, t) = a_1 k^2. \tag{6.30}$$

For Ra just above Ra_c, λ_+ becomes positive but remains small, and the eigenvectors are approximately in the same direction as shown in Fig. 6.3. The solution $u_z(\mathbf{k}, t), \theta(\mathbf{k}, t) \sim \exp(\lambda_+ t)$ grows with time. In Chapter 7 we show how these growing solution saturates due to nonlinearity.

The solution for the special case of $\mathrm{Pr} = 0$ is quite interesting. For this case, the equations for all Ra's are

$$\frac{d}{dt} u_z(\mathbf{k}) = 0, \tag{6.31}$$

$$\frac{d}{dt} \theta(\mathbf{k}) = u_z(\mathbf{k}) - k^2 \theta(\mathbf{k}), \tag{6.32}$$

whose solution is

$$u_z(\mathbf{k}, t) = c_1, \tag{6.33}$$

$$\theta(\mathbf{k}, t) = \frac{c_1}{k^2} + c_2 \exp(-k^2 t) \to \frac{c_1}{k^2}. \tag{6.34}$$

Thus, for $\mathrm{Pr} = 0$, the system admits constant solution that does not grow with time. The constant solution is also consistent with the fact that the eigenvalues of the matrix A of Eq. (6.11) are 0 and $-k^2$, and that $\lambda = 0$ corresponds to the constant solution for all t.

Using the eigenvectors we can estimate the relative strengths of buoyancy, viscous force, and pressure gradient near the onset of convection. First, using Eqs. (6.9, 6.30), we obtain

$$\frac{\text{buoyancy}}{\text{viscous force}} \approx \frac{\mathrm{RaPr}\theta(\mathbf{k})}{-k^2 \mathrm{Pr}u_z(\mathbf{k})} \approx \frac{\mathrm{Ra}a_1}{-k^4 a_1} \approx -\frac{k^2}{k_c^2} \approx -3, \tag{6.35}$$

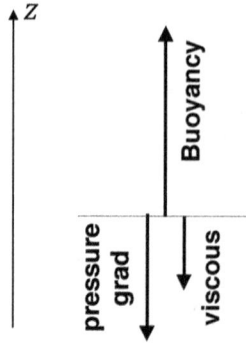

Fig. 6.4   A schematic of the force balance near the onset of thermal instability. Here, pressure gradient + viscous force balances buoyancy.

and

$$\frac{\text{pressure gradient}}{\text{viscous force}} \approx \frac{-ik_z\sigma}{-k^2 \text{Pr} u_z(\mathbf{k})} \approx \frac{k_z^2 \text{RaPr}\theta(\mathbf{k})}{k^4 \text{Pr} u_z(\mathbf{k})}$$

$$\approx \frac{k_z^2 \text{Ra} a_1}{k^6 a_1} \approx \frac{k_z^2}{k_c^2} \approx 2. \tag{6.36}$$

The force balance near the onset of convection is shown in Fig. 6.4. As expected, the viscous force opposes buoyancy. A surprising feature however is that the pressure gradient too opposes buoyancy; this is a unique feature of the free-slip boundary condition that persists even in turbulent regime (see Sec. 11.9). However, for RBC with the no-slip boundary condition, the pressure gradient is in the same direction as buoyancy (see Sec. 11.3).

In the next section we will rewrite the linearized RBC equations for deriving instability condition under the no-slip boundary condition.

6.2   Linearized version of Rayleigh-Bénard convection

Other than the free-slip boundary condition, in thermal convection we employ no-slip boundary conditions at top and bottom plates; and no-slip boundary condition at one plate and free-slip boundary condition at the other plate. Periodic boundary condition is employed along the horizontal directions. Dynamical analysis for these cases is more difficult than that for the free-slip case discussed in the previous section. However, for the no-slip boundary condition, it is relatively easy to derive the condition for neutral instability by setting $\partial_t = 0$; this exercise helps us determine the critical Rayleigh number.

We again start with Eqs. (6.2, 6.4). Since we are considering neutral stability, we set $\partial/\partial t = 0$ in these equations. First, we employ $\nabla \cdot \mathbf{u} = 0$ to Eq. (6.1) and obtain

$$-\nabla^2 \sigma + \text{Ra}_c \text{Pr} \partial_z \theta = 0. \tag{6.37}$$

By taking a derivative of Eq. (6.37) along z, we obtain

$$\nabla^2 \partial_z \sigma = \text{Ra}_c \text{Pr} \partial_{zz}^2 \theta. \qquad (6.38)$$

We employ Laplacian operator on Eq. (6.4) that yields

$$\nabla^2 \partial_z \sigma = \text{Ra}_c \text{Pr} \nabla^2 \theta + \text{Pr} \nabla^4 u_z. \qquad (6.39)$$

Substitution of Eq. (6.38) in the above equation yields the following equation for u_z:

$$\text{Ra}_c \nabla_\perp^2 \theta + \nabla^4 u_z = 0, \qquad (6.40)$$

where $\nabla_\perp^2 = \partial_{xx}^2 + \partial_{yy}^2$. Here $\perp$ is the direction perpendicular to that of buoyancy.

By employing Laplacian operator $\nabla_\perp^2$ to Eq. (6.2) and by setting $\partial_t = 0$, we obtain

$$\nabla_\perp^2 u_z + \nabla^2 \nabla_\perp^2 \theta = 0. \qquad (6.41)$$

Now we employ Laplacian operator to Eq. (6.40) and eliminate the θ term using Eq. (6.41) that yields

$$\nabla^6 u_z = \text{Ra}_c \nabla_\perp^2 u_z. \qquad (6.42)$$

This is a sixth-order differential equation that requires six boundary conditions for deriving its solution. We discuss these boundary conditions in the next section.

6.3 Boundary conditions for RBC instabilities

We employ no-slip or free-slip boundary conditions at the top and bottom plates, and periodic boundary conditions along the horizontal directions. For the same, we choose the following set of basis functions:

$$u_x(x, z) = U(z) \exp(ik_x x) + c.c., \qquad (6.43)$$
$$u_z(x, z) = W(z) \exp(ik_x x) + c.c., \qquad (6.44)$$
$$\theta(x, z) = \Theta(z) \exp(ik_x x) + c.c., \qquad (6.45)$$

where *c.c.* stands for complex conjugate. As discussed earlier, sin/cos functions are appropriate for $U(z), W(z), \Theta(z)$ under the free-slip boundary condition. Note that RBC's primary instability leads to two-dimensional pattern, hence near the onset of thermal instability, the velocity and temperature fields are functions only of x and z.

A common condition for both free-slip and no-slip boundary conditions is that $u_z(x, z) = 0$ at both the plates, hence

$$W(z)|_{\text{plates}} = 0, \qquad (6.46)$$

which are the first two boundary conditions at the two plates. Also, the condition for the conducting plates is $\theta(x, z) = 0$ at the plates, which yields

$$\Theta(z)|_{\text{plates}} = 0. \qquad (6.47)$$

Substitution of Eqs. (6.44, 6.45) in Eq. (6.40) yields

$$- \mathrm{Ra}_c k_x^2 \Theta(z) + \left(\frac{d^2}{dz^2} - k_x \right)^2 W(z) = 0. \tag{6.48}$$

Using Eq. (6.47) we obtain the third and fourth boundary conditions at the two plates:

$$\left(\frac{d^2}{dz^2} - k_x \right)^2 W(z)|_{\mathrm{plates}} = 0. \tag{6.49}$$

The conditions given by Eqs. (6.46, 6.49) apply to both free-slip and no-slip boundary conditions.

The fifth and sixth boundary conditions originate from the incompressibility condition $\nabla \cdot \mathbf{u} = 0$:

$$ik_x U(z) + \frac{d}{dz} W(z) = 0. \tag{6.50}$$

For the no-slip boundary condition, at the plates, $u_x(x, z) = 0$ or $U(z) = 0$, application of which to Eq. (6.50) yields

$$\frac{d}{dz} W(z)|_{\mathrm{plates}} = 0. \tag{6.51}$$

which are the fifth and sixth boundary conditions for the no-slip boundary condition. For the free-slip boundary condition,

$$\partial_z u_x(x, z) = 0, \tag{6.52}$$

hence the incompressibility condition yields

$$\partial_x \partial_z u_x + \partial_{zz}^2 u_z = 0. \tag{6.53}$$

The first term of the above equation vanishes for the free-slip boundary condition. Hence for the free-slip boundary condition, the conditions at the plates are

$$\frac{d^2}{dz^2} W(z)|_{\mathrm{plates}} = 0. \tag{6.54}$$

These are the fifth and sixth boundary conditions for the free-slip boundary condition.

Thus we have six boundary conditions required to solve Eq. (6.42). We summarise them below:

(1) No-slip boundary condition: The boundary conditions are

$$W(z)|_{\mathrm{plates}} = 0; \quad \left(\frac{d^2}{dz^2} - k_x \right)^2 W(z)|_{\mathrm{plates}} = 0; \quad \frac{d}{dz} W(z)|_{\mathrm{plates}} = 0. \tag{6.55}$$

(2) Free-slip boundary condition: The boundary conditions are

$$W(z)|_{\mathrm{plates}} = 0; \quad \left(\frac{d^2}{dz^2} - k_x \right)^2 W(z)|_{\mathrm{plates}} = 0; \quad \frac{d^2}{dz^2} W(z)|_{\mathrm{plates}} = 0. \tag{6.56}$$

The above conditions imply that

$$\frac{d^4}{dz^4}W(z)|_{\text{plates}} = 0. \tag{6.57}$$

Subsequent application of Laplacian operator to Eq. (6.42) and the boundary conditions of Eqs. (6.56, 6.57) yields [Chandrasekhar (2013)]

$$\frac{d^{(2m)}}{dz^{(2m)}}W(z)|_{\text{plate}} = 0, \tag{6.58}$$

where m is an integer. From the above, it follows that

$$W(z) = A\sin(m\pi z) \tag{6.59}$$

are the appropriate basis functions for the free-slip boundary condition at the top and bottom plates.

We can compute the solution of Eq. (6.42) using these boundary conditions. Also note that we may apply free-slip boundary condition at one plate and no-slip boundary condition at the other plate.

In the next section, we will solve for $W(z)$ and determine Ra_c for the no-slip boundary condition.

6.4 Thermal instability for the no-slip boundary condition

The derivation given below is very similar to that described in Chandrasekhar (2013). We consider a Boussinesq fluid confined between two infinite horizontal and conducting plates on which we employ no-slip boundary condition. It is convenient to consider the two infinite horizontal plates to be located at $z = \pm 1/2$. We employ periodic boundary condition along the horizontal direction, and conditions given by Eq. (6.55) at the two plates. Note that near the onset of convection, the velocity and temperature fields are given by Eqs. (6.43–6.45).

We solve Eq. (6.42) for the aforementioned boundary conditions. Since Eq. (6.42) is linear, its solution is of the form

$$W(z) = \exp(qz) \tag{6.60}$$

that yields the following sixth order polynomial for q:

$$(q^2 - k_x^2)^3 = -\text{Ra}_c k_x^2. \tag{6.61}$$

The solutions of the above equation are

$$\pm iq_0, \quad \pm q \tag{6.62}$$

with

$$q_0 = k_x\left[\tau - 1\right]^{1/2}, \tag{6.63}$$

$$q = k_x\left[1 + \frac{1}{2}\tau(1 \pm i\sqrt{3})\right]^{1/2} \tag{6.64}$$

$$= q_{\text{re}} \pm iq_{\text{im}}, \tag{6.65}$$

where

$$\tau = \left(\frac{\mathrm{Ra}_c}{k_x^4}\right)^{1/3}, \tag{6.66}$$

$$q_{re} = \frac{k_x}{\sqrt{2}}\left[\sqrt{1+\tau+\tau^2}+(1+\tau/2)\right]^{1/2}, \tag{6.67}$$

$$q_{im} = \frac{k_x}{\sqrt{2}}\left[\sqrt{1+\tau+\tau^2}-(1+\tau/2)\right]^{1/2}. \tag{6.68}$$

Therefore the desired even solution is of the form

$$W = c_1\cos q_0 z + c_2\cosh qz + c_3\cosh q^* z, \tag{6.69}$$

where q^* is complex conjugate of q. We choose the even solution because the first unstable mode is symmetric about the mid horizontal plane. Also note that the even solution guarantees simultaneous application of boundary conditions at the two plates.

An application of the boundary condition $W = dW/dz = (d^2/dz^2 - k_x^2)^2 W = 0$ at the walls yields

$$\begin{pmatrix} \cos\frac{q_0}{2} & \cosh\frac{q}{2} & \cosh\frac{q^*}{2} \\ -q_0\sin\frac{q_0}{2} & q\sinh\frac{q}{2} & q^*\sinh\frac{q^*}{2} \\ (q_0^2+k_x^2)^2\cos\frac{q_0}{2} & (q^2-k_x^2)^2\cosh\frac{q}{2} & (q^{*2}-k_x^2)^2\cosh\frac{q^*}{2} \end{pmatrix}\begin{pmatrix} c_1 \\ c_2 \\ c_3 \end{pmatrix} = 0. \tag{6.70}$$

For nontrivial c_1, c_2, c_3, the determinant of the above matrix must vanish. Using

$$(q_0^2+k_x^2)^2 = \tau^2 k_x^4, \tag{6.71}$$

$$(q^2-k_x^2)^2 = \frac{1}{2}\tau^2 k_x^4(-1\pm i\sqrt{3}), \tag{6.72}$$

several properties of determinants, and by setting the determinant of the matrix to zero, we obtain

$$\begin{aligned} 0 &= \begin{vmatrix} \cos\frac{q_0}{2} & \cosh\frac{q}{2} & \cosh\frac{q^*}{2} \\ -q_0\sin\frac{q_0}{2} & q\sinh\frac{q}{2} & q^*\sinh\frac{q^*}{2} \\ \cos\frac{q_0}{2} & \frac{1}{2}(-1+i\sqrt{3})\cosh\frac{q}{2} & -\frac{1}{2}(1+i\sqrt{3})\cosh\frac{q^*}{2} \end{vmatrix} \\ &= \begin{vmatrix} 1 & 1 & 1 \\ -q_0\tan\frac{q_0}{2} & q\tanh\frac{q}{2} & q^*\tanh\frac{q^*}{2} \\ 1 & \frac{1}{2}(-1+i\sqrt{3}) & -\frac{1}{2}(1+i\sqrt{3}) \end{vmatrix} \\ &= \begin{vmatrix} 1 & 1 & 1 \\ -q_0\tan\frac{q_0}{2} & q\tanh\frac{q}{2} & q^*\tanh\frac{q^*}{2} \\ 0 & \sqrt{3}-i & \sqrt{3}+i \end{vmatrix} \\ &\Longrightarrow \Im\left\{(\sqrt{3}+i)q\tanh(q/2)\right\} + q_0\tan\frac{q_0}{2} = 0. \tag{6.73} \end{aligned}$$

Given k_x, the above transcendental equation can be solved iteratively using midpoint method or any other appropriate method to determine τ. Thus determined τ can be used to compute Ra_c using Eq. (6.66). In Fig. 6.5 we plot the value of Ra_c

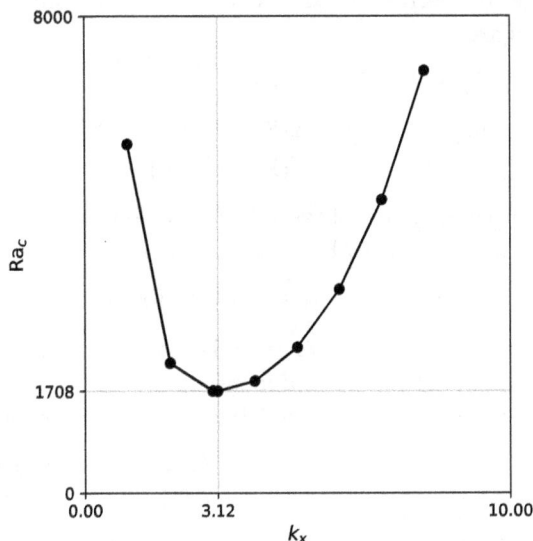

Fig. 6.5 Plot of Ra_c vs. k_x for the no-slip RBC. Ra is minimum for $k_c \approx 3.117$, and it takes value around 1707.762.

for various values of k_x; these values have been taken from Reid and Harris (1958). The minimum value of Ra_c occurs for $k_c \approx 3.117$ that yields

$$Ra_{c,\min} \approx 1707.762. \tag{6.74}$$

This is the minimum critical Rayleigh number for the no-slip boundary condition with periodic boundary condition along the horizontal direction. Higher modes are generated at larger Rayleigh numbers.

In the next section, we rederive the thermal instability for the free-slip boundary condition in the Craya-Herring basis.

6.5 Thermal instability in Craya-Herring basis

We can derive the instability condition for the free-slip boundary condition in Craya-Herring basis. This is a useful exercise, and it will be generalised later to rotating convection and magnetoconvection. We start with the linearized RBC equations:

$$\frac{d}{dt}\mathbf{u}(\mathbf{k}) = -i\mathbf{k}\sigma(\mathbf{k}) + \text{RaPr}\theta(\mathbf{k})\hat{z} + \text{Pr}k^2\mathbf{u}(\mathbf{k}), \tag{6.75}$$

$$\frac{d}{dt}\theta(\mathbf{k}) = u_z(\mathbf{k}) + k^2\theta(\mathbf{k}), \tag{6.76}$$

$$\mathbf{k} \cdot \mathbf{u}(\mathbf{k}) = 0. \tag{6.77}$$

We rewrite the above equation in the Craya-Herring basis in which the buoyancy term is

$$\mathbf{F}(\mathbf{k}) = \text{RaPr}\theta(\mathbf{k})\hat{z}$$
$$= -\hat{e}_2\text{RaPr}\theta(\mathbf{k})\sin\zeta + \hat{e}_3\text{RaPr}\theta(\mathbf{k})\cos\zeta, \tag{6.78}$$

where ζ is the angle between $\mathbf{k}$ and $\hat{z}$. Therefore, the dynamical equations in the Craya-Herring basis are

$$\dot{u}_1(\mathbf{k}) = -\mathrm{Pr}k^2 u_1(\mathbf{k}), \tag{6.79}$$

$$\dot{u}_2(\mathbf{k}) = -\mathrm{Pr}k^2 u_2(\mathbf{k}) - \mathrm{RaPr}\sin\zeta\,\theta(\mathbf{k}), \tag{6.80}$$

$$\dot{\theta}(\mathbf{k}) = -\sin\zeta\, u_2(\mathbf{k}) - k^2\theta(\mathbf{k}), \tag{6.81}$$

where $u_{1,2}(\mathbf{k})$ are the components of the velocity mode $\mathbf{u}(\mathbf{k})$ in Craya-Herring basis, and $\sin\zeta = k_\perp/k$. Equation (6.79) has a trivial solution:

$$u_1(\mathbf{k}, t) = u_1(\mathbf{k}, 0)\exp(-\mathrm{Pr}k^2 t) \tag{6.82}$$

that decays to zero asymptotically, hence we ignore Eq. (6.79). This is the reason why the unstable mode of RBC is two-dimensional.

The remaining two equations are rewritten in matrix form as

$$\frac{d}{dt}\begin{pmatrix} u_2(\mathbf{k}) \\ \theta(\mathbf{k}) \end{pmatrix} = \begin{pmatrix} -\mathrm{Pr}k^2 & -\mathrm{RaPr}k_\perp/k \\ -k_\perp/k & -k^2 \end{pmatrix}\begin{pmatrix} u_2(\mathbf{k}) \\ \theta(\mathbf{k}) \end{pmatrix} = A\begin{pmatrix} u_2(\mathbf{k}) \\ \theta(\mathbf{k}) \end{pmatrix}. \tag{6.83}$$

Matrix A of Eq. (6.83) has a slightly different form than that of Eq. (6.11). However, the eigenvalues and respective solutions of the two matrices are identical. Hence, we refer the reader to Sec. 6.1 for carrying forward the analysis of Eq. (6.83). We remark that the velocity field of the roll is two-dimensional because u_1 decays in time. It is important to note that we can convert the amplitudes of the modes in free-slip basis functions to those in Fourier basis using simple transformations described in Sec. 3.4.

Another important buoyancy-driven instability is the Rayleigh-Taylor instability, which is the topic of the next section.

6.6 Rayleigh-Taylor instability

In RBC, the density of the fluid varies linearly along the vertical. In contrast, *Rayleigh-Taylor instability* (RTI) has a sharp density variation near the interface between two fluids. In this section we will describe RTI following the same procedure as Drazin (2002).

We consider a fluid of constant density ρ_2 moving with velocity $U_2\hat{x}$ on top of another fluid of constant density ρ_1 moving with velocity of $U_1\hat{x}$. The surface at $z = 0$ separates the two flows. See Fig. 6.6 for an illustration. This configuration is stable for some parameters, and unstable for some others. For example, when $U_1 = U_2 = 0$ and $\rho_2 > \rho_1$, then the fluid is unstable since dense fluid sits over lighter one. On the contrary the system is stable when $\rho_1 > \rho_2$. In the following, we will provide a general derivation that includes Rayleigh-Taylor instability, shear instability, and surface and internal gravity waves.

Assuming the fluid to be inviscid and incompressible, the flow equation is

$$\frac{\partial \mathbf{u}}{\partial t} + (\mathbf{u}\cdot\nabla)\mathbf{u} = -\nabla\left(\frac{p}{\rho}\right) - g\hat{z}. \tag{6.84}$$

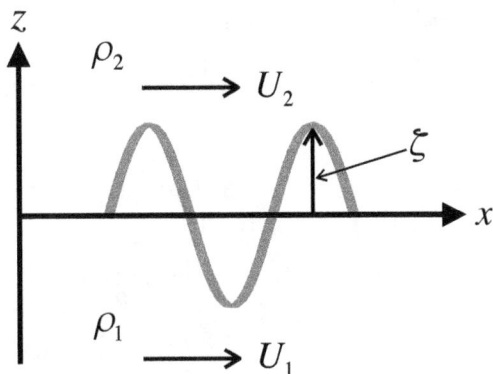

Fig. 6.6   Flow configuration for Sec. 6.6: In $z < 0$ region, fluid with density ρ_1 is moving with velocity $U_1\hat{x}$; in $z > 0$ region, fluid with density ρ_2 is moving with velocity $U_2\hat{x}$. We examine the stability of the flow under surface perturbation: $\zeta(x, y) = \zeta \exp(i\mathbf{k}_\perp \cdot \mathbf{r}_\perp)$ at $z = 0$.

Near the onset of instability, the flow is considered to be two-dimensional and irrotational everywhere except at the interface. Hence for $z \neq 0$,

$$\mathbf{u} = \nabla\phi, \tag{6.85}$$

where ϕ is the potential function. Using the incompressible condition, $\nabla \cdot \mathbf{u} = 0$, we obtain

$$\nabla^2\phi_1 = 0; \quad \nabla^2\phi_2 = 0, \tag{6.86}$$

where ϕ_1 and ϕ_2 are the potential functions for $z < 0$ and $z > 0$ respectively. We attempt linearised solution of the above with

$$\phi_1 = U_1 x + \phi_1', \quad \phi_2 = U_2 x + \phi_2' \tag{6.87}$$

where ϕ_1', ϕ_2' are perturbations for $z < 0$ and $z > 0$ respectively.

We also employ

$$-(\mathbf{u} \cdot \nabla)\mathbf{u} = \mathbf{u} \times (\nabla \times \mathbf{u}) - \nabla\frac{u^2}{2} = -\nabla\frac{u^2}{2} \tag{6.88}$$

since $\nabla \times \mathbf{u} = 0$. Substitution of the above in Eq. (6.84) yields

$$\frac{\partial \nabla\phi}{\partial t} = -\nabla\left(\frac{p}{\rho} + \frac{u^2}{2} + gz\right). \tag{6.89}$$

Hence

$$p = \rho\left[C - \frac{1}{2}(\nabla\phi)^2 - \frac{\partial\phi}{\partial t} - gz\right], \tag{6.90}$$

where C is a constant. The fluid densities above and below the interface are considered to be constant. The normal stress of the fluid is continuous at the fluid interface $z = \zeta(x, y)$, or

$$\rho_1\left[C_1 - \frac{1}{2}(\nabla\phi_1)^2 - \frac{\partial\phi_1}{\partial t} - g\zeta\right] = \rho_2\left[C_2 - \frac{1}{2}(\nabla\phi_2)^2 - \frac{\partial\phi_2}{\partial t} - g\zeta\right]. \tag{6.91}$$

Substitution of ϕ_1 and ϕ_2 of Eq. (6.87) in Eq. (6.91) yields

$$\rho_1 \left[C_1 - \frac{1}{2}U_1^2 - U_1\partial_x\phi_1' - \partial_t\phi_1' - g\zeta \right] = \rho_2 \left[C_2 - \frac{1}{2}U_2^2 - U_2\partial_x\phi_2' - \partial_t\phi_2' - g\zeta \right].$$

$$(6.92)$$

Here we ignore the quadratic terms. The basic flow profile satisfies the above equation when the perturbation is absent, thus

$$\rho_1 \left[C_1 - \frac{1}{2}U_1^2 \right] = \rho_2 \left[C_2 - \frac{1}{2}U_2^2 \right]. \tag{6.93}$$

Hence

$$\rho_1 \left[U_1\partial_x\phi_1' + \partial_t\phi_1' + g\zeta \right] = \rho_2 \left[U_2\partial_x\phi_2' + \partial_t\phi_2' + g\zeta \right]. \tag{6.94}$$

We attempt the following solution for the perturbations

$$\begin{pmatrix} \zeta(x,y) \\ \phi_1'(x,y,z) \\ \phi_2'(x,y,z) \end{pmatrix} = A \begin{pmatrix} \zeta \\ \Phi_1'(z) \\ \Phi_2'(z) \end{pmatrix} \exp(i\mathbf{k}_\perp \cdot \mathbf{r}_\perp + st) \tag{6.95}$$

where $\mathbf{k}_\perp = k_x\hat{x} + k_y\hat{y}$, $\mathbf{r}_\perp = x\hat{x} + y\hat{y}$, and s is the growth rate. Note that $\zeta \exp(st)$ is the amplitude of the interfacial fluctuations. Using Eq. (6.86, 6.87) we deduce that

$$\nabla^2\phi_1' = 0; \quad \nabla^2\phi_2' = 0. \tag{6.96}$$

Hence

$$\frac{d^2}{dz^2}\Phi_{1,2}' = -k_\perp^2\Phi_{1,2}', \tag{6.97}$$

whose solutions are

$$\Phi_2'(z) = c_1 \exp(k_\perp z) + c_2 \exp(-k_\perp z), \tag{6.98}$$
$$\Phi_1'(z) = d_1 \exp(k_\perp z) + d_2 \exp(-k_\perp z), \tag{6.99}$$

where $k_\perp = \sqrt{k_x^2 + k_y^2}$. Using the condition that $\mathbf{u} \to 0$ as $z \to \pm\infty$, we deduce that $c_1 = d_2 = 0$, and hence

$$\Phi_2'(z) = c_2 \exp(-k_\perp z), \tag{6.100}$$
$$\Phi_1'(z) = d_1 \exp(k_\perp z) \tag{6.101}$$

for $z > 0$ and $z < 0$ respectively.

We can relate $\phi_1'(x,y,z)$, $\phi_2'(x,y,z)$ to $\zeta(x,y)$ using the fact that the vertical velocity at the interface equals u_z, i.e.,

$$u_z = \frac{D\zeta}{Dt} = \partial_t\zeta + U_1\partial_x\zeta, \tag{6.102}$$

or

$$\partial_z\phi_1' = \partial_t\zeta + U_1\partial_x\zeta, \tag{6.103}$$
$$\partial_z\phi_2' = \partial_t\zeta + U_2\partial_x\zeta. \tag{6.104}$$

Substitution of $\phi'_{1,2}$ and ζ of Eqs. (6.95, 6.100, 6.101) in the above equations yields

$$d_1 = \frac{1}{k_\perp}(s + ik_x U_1)\zeta, \quad c_2 = -\frac{1}{k_\perp}(s + ik_x U_2)\zeta. \qquad (6.105)$$

Therefore

$$\phi'_1(x, y, z) = \frac{1}{k_\perp}(s + ik_x U_1)\exp(k_\perp z)\exp(i\mathbf{k}_\perp \cdot \mathbf{r}_\perp + st)\zeta, \qquad (6.106)$$

$$\phi'_2(x, y, z) = -\frac{1}{k_\perp}(s + ik_x U_2)\exp(-k_\perp z)\exp(i\mathbf{k}_\perp \cdot \mathbf{r}_\perp + st)\zeta. \qquad (6.107)$$

Substitution of Eq. (6.106, 6.107) in Eq. (6.94) yields

$$\rho_1\left[(ik_x U_1 + s)^2 + gk_\perp\right]\zeta = \rho_2\left[-(ik_x U_1 + s)^2 + gk_\perp\right]\zeta. \qquad (6.108)$$

Since $\zeta \neq 0$, we demand that

$$(\rho_1 + \rho_2)s^2 + 2isk_x(\rho_1 U_1 + \rho_2 U_2) + k_\perp g(\rho_1 - \rho_2) - k_x^2(\rho_1 U_1^2 + \rho_2 U_2^2) = 0, \quad (6.109)$$

whose solutions are

$$s = -ik_x\frac{\rho_1 U_1 + \rho_2 U_2}{\rho_1 + \rho_2} \pm \left[\frac{k_x^2 \rho_1 \rho_2 (U_1 - U_2)^2}{(\rho_1 + \rho_2)^2} - \frac{k_\perp g(\rho_1 - \rho_2)}{\rho_1 + \rho_2}\right]^{1/2}. \qquad (6.110)$$

The solution is oscillatory if the second part of s in the above equation is imaginary, or, if

$$k_\perp g(\rho_1^2 - \rho_2^2) \geq k_x^2 \rho_1 \rho_2 (U_1 - U_2)^2. \qquad (6.111)$$

This is called *neutrally stable solution*. However, when

$$k_\perp g(\rho_1^2 - \rho_2^2) < k_x^2 \rho_1 \rho_2 (U_1 - U_2)^2, \qquad (6.112)$$

s is complex with one of the root having positive real part that makes the amplitude of the perturbation grow in time. Thus, the flow is unstable when the parameters satisfy Eq. (6.112). Hence, Eq. (6.112) is called the *instability criterion*.

Equation (6.110) yields several interesting solutions that are given below:

(1) *Rayleigh-Taylor instability*: For static fluid ($U_1 = U_2 = 0$) with $\rho_2 > \rho_1$,

$$s = \pm\left[k_\perp g\frac{(\rho_2 - \rho_1)}{\rho_1 + \rho_2}\right]^{1/2}. \qquad (6.113)$$

The positive s is the growth rate of the interfacial fluctuations. This is called *Rayleigh-Taylor instability*.

(2) *Internal gravity waves*: When fluid is at rest ($U_1 = U_2 = 0$) and $\rho_1 > \rho_2$, we obtain

$$s = i\omega = \pm i\left[k_\perp g\frac{(\rho_1 - \rho_2)}{\rho_1 + \rho_2}\right]^{1/2}. \qquad (6.114)$$

Hence the wave travels along $\mathbf{k}_\perp$ with a frequency of ω. Recall that the internal gravity wave with linear $d\bar{\rho}/dz$ has $\omega = N$, which is different from the above wave. See Chapter 5 for details.

(3) *Surface gravity waves*: Consider $\rho_2 = 0$ and $U_1 = U_2 = 0$. In this case

$$s = i\omega = \pm i\sqrt{k_\perp/g}. \tag{6.115}$$

The wave travels along $\mathbf{k}_\perp$ with a phase speed

$$v_{ph} = \frac{\omega}{k_\perp} = \pm\sqrt{g/k_\perp}. \tag{6.116}$$

Surface ocean wave is an example of the above wave.

(4) *Shear instability*: When $\rho_1 = \rho_2$ but $U_1 \neq U_2$, we obtain

$$s = -\frac{1}{2}ik_x(U_1 + U_2) \pm \frac{1}{2}k_x(U_1 - U_2). \tag{6.117}$$

Clearly, s is complex with one of the solution having positive real part. This mode grows in time with the growth rate of $\mathrm{Re}(s)$. Note that shear instability is not covered in this book.

Thus, the flow configuration of Fig. 6.6 supports Rayleigh-Taylor and shear instabilities, as well as internal and surface gravity waves. The flow becomes even more interesting when surface tension is included. However, this discussion is beyond the scope of this book.

RBC and Rayleigh-Taylor instability satisfy Boussinesq approximation. However, atmospheric flows are typically non-Boussinesq involving thermodynamic processes. In the following section, we will discuss thermal instability that is based on adiabatic cooling and density gradient.

6.7 Schwarzschild criterion for convective instability

We consider a vertical movement of air in the atmosphere. The air is considered to be ideal, hence

$$p = \rho R T, \tag{6.118}$$

where p is the thermodynamic pressure, ρ, T are respectively the density and temperature of the gas, and R is the gas constant. Since air is a poor thermal conductor, the process is assumed to be adiabatic, hence

$$p = C\rho^\gamma \tag{6.119}$$

where C is a constant, and $\gamma = C_p/C_v$ is the ratio of the heat capacities at constant pressure and constant volume. Substitution of ρ from Eq. (6.118) into the above equation yields

$$p = C^{-1/(\gamma-1)} (RT)^{\frac{\gamma}{\gamma-1}} = C^{-1/(\gamma-1)} (RT)^{\frac{C_p}{R}}. \tag{6.120}$$

Now an application of the force balance under an approximate steady state yields

$$-g = \frac{1}{\rho}\frac{dp}{dz}. \tag{6.121}$$

Substitution of Eq. (6.120) in the above equation yields

$$-g = C_p \frac{dT}{dz}, \tag{6.122}$$

or

$$\frac{dT}{dz} = -\frac{g}{C_p}. \tag{6.123}$$

For the terrestrial atmosphere, $C_p = 1000$ J/(kgK), hence

$$\frac{dT}{dz} \approx -\frac{10}{1000} \approx -10 \text{ K/km}. \tag{6.124}$$

That is, the temperature falls by 10 degrees every km. This fall in temperature is called the *adiabatic lapse rate*[1]. The above condition, called *Schwarzschild criterion*, is a condition for neutral stability; the fluid (air) has zero velocity when $dT/dz = (dT/dz)_{ad} = -g/C_p$, where ad stands for adiabatic. The flow becomes unstable when $|dT/dz| > |dT/dz|_{ad}$, and becomes stable when $|dT/dz| < |dT/dz|_{ad}$.

Now let us explore the scenario when $dT/dz \neq (dT/dz)_{ad}$. First, we consider an atmosphere in which $|dT/dz| > |dT/dz|_{ad}$. Consider a rising fluid element from the bottom of a fluid layer (see Fig. 6.7(a)). When it reaches the top layer, its temperature, which is affected by adiabatic cooling, would be higher than the imposed background temperature. Since $\rho \propto p/T$, for a given p, the density of the rising parcel would be lower than its background. Therefore, the parcel of air will continue to rise. Similar arguments reveal that a descending air parcel would continue to descend. Thus, the fluid configuration with $|dT/dz| > |dT/dz|_{ad}$ is unstable.

Now we consider the other limiting case when $|dT/dz| < |dT/dz|_{ad}$. Under this scenario, a rising air parcel from the bottom would have lower temperature and higher density than the background, hence this element would descend and return to the original position (see Fig. 6.7(b)). Similarly, an air parcel descending downward also returns to its original position. Thus, the temperature profile $|dT/dz| < |dT/dz|_{ad}$ corresponds to a stable configuration. Note that the fluid is stable for this configuration even though denser fluid is above the lighter fluid. This stability is due to the adiabatic cooling. Also note that the above process is assumed to be slow or adiabatic, hence the fluid acceleration $\mathbf{u} \cdot \nabla \mathbf{u}$ is ignored. This is akin to the linear stability discussed in the earlier sections.

The instability analysis presented here appear to differ significantly from that for RBC. These two systems however have significant similarities. In both these systems, the adverse temperature gradient drives the instability, but the stabilising forces are different—it is the viscous force in RBC, but adiabatic cooling in Schwarzschild's calculation. A flow becomes unstable when the destabilising force dominates the stabilising force. Recall that RBC becomes unstable only when the temperature difference between the plates exceeds certain critical values so as to overcome the viscous effects. However, an atmospheric flow become unstable

---

[1]In reality, the temperature lapse rate in the Earth's atmosphere is approximately 6.5 K/km. The correction is due to the humidity of the atmosphere.

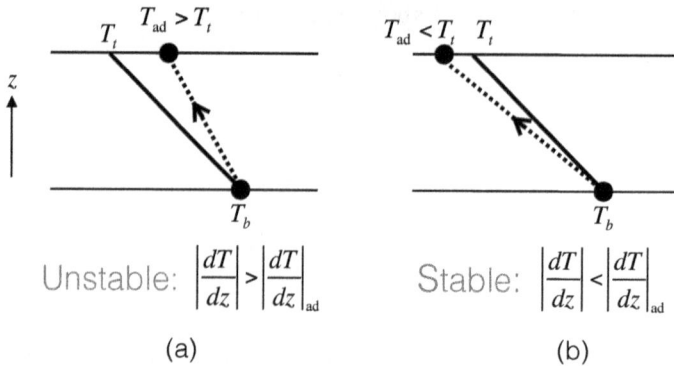

Fig. 6.7   An illustration of convective instability due to adiabatic cooling in an planetary atmosphere. The background temperature drops as dT/dz, while the temperature of an ascending fluid parcel decreases as $|dT/dz|_{ad}$. T_b, T_t are the temperatures of the bottom and top layers respectively. (a) When $|dT/dz| > |dT/dz|_{ad}$, an ascending fluid element at the top has temperature T_{ad}, which is higher than the background temperature, T_t. Hence it continues to rise, thus making the flow unstable. (b) When $|dT/dz| < |dT/dz|_{ad}$, an ascending fluid element is colder than its surrounding, and hence it returns to its original configuration, thus making the flow stable.

when it overcomes adiabatic cooling; here the viscous force is relatively negligible. The neutral stability condition for the two systems are—$|dT/dz| = |dT/dz|_{ad}$ and $\mathrm{Ra} = \mathrm{Ra}_c$. Given these analogies, the aforementioned two instabilities are quite similar.

With this, we conclude our preliminary discussion on Schwarzschild criterion. There are other buoyancy-driven instabilities, e.g., in rotating convection, magnetoconvection, Taylor-Couette, stellar convection, etc. In the third part of the book, we will cover instabilities in rotating convection and magnetoconvection. But, other instabilities are beyond the scope of this book.

In the next chapter we will show how nonlinearity saturates the growth of thermal instability.

Further reading

Chandrasekhar (2013), Kundu *et al.* (2015), Drazin (2002), and Bhattacharjee (1987) describe thermal and Rayleigh-Taylor instabilities in detail. We also refer to Gershuni and Znukhovitski (1976) and Gershuni *et al.* (1989) for discussion on thermal stability.

Exercises

(1) Consider an idealised experiment in which spherical ball of fluid of radius a is ascending vertically with a constant velocity v. The atmosphere has a temperature stratification of $d\bar{T}/dz$. Derive the condition for neutral instability by balancing the viscous force and buoyancy on the ball.

(2) Consider an isothermal atmosphere of ideal gas. Derive the density variation of the atmosphere as a function of z.

(3) Perform thermal instability analysis for the following systems:

    (a) In a unit box with periodic boundary condition in all three directions.

    (b) In a unit box with free-slip boundary condition in all three directions.

    (c) In a rectangular box with free-slip at the top surface and no-slip at the bottom surface. Assume periodic boundary condition along the horizontal direction.

    For the aforementioned systems, compute Ra_c. Compute and plot the velocity and temperature fields for the neutral mode.

(4) For RBC with no-slip boundary condition, compute $\mathbf{u}(\mathbf{r})$ and $\theta(\mathbf{r})$ for the neutral mode. Plot the velocity and temperature fields.

(5) For the RBC with no-slip boundary condition, in Sec. 6.4 we computed the neutral mode with the lowest wavenumber. Compute the next neutral mode of the flow.

(6) Contrast the Schwarzschild criterion for convective instability with the neutral instability condition in RBC.

(7) Contrast the Rayleigh-Taylor instability with RBC.

Chapter 7

Nonlinear Saturation

In Chapter 6, we performed linear instability analysis of RBC and showed that for Rayleigh number beyond Ra_c, the flow becomes unstable leading to primary convective rolls. The amplitude of these rolls grow in time. However, when the amplitude of the roll becomes significant, nonlinearity starts to dominate the flow and it generates higher Fourier modes. These new modes do many things—saturate the growth of the primary mode, create stationary and time-dependent patterns, and exhibit chaos and turbulence. We will cover these topics in the present and subsequent chapters.

In the next section, we will discuss how nonlinearity saturates the growth of the primary convective rolls. We will illustrate the saturation mechanism using the Lorenz model [Lorenz (1963); Strogatz (2014); Hilborn (2001)].

7.1 Lorenz equations

Lorenz (1963) (also see [Strogatz (2014); Hilborn (2001)]) constructed a simple but interesting nonlinear model of 2D RBC. Most features of the Lorenz model do not match with real RBC flows, yet, it captures the flow properties near the onset of convection, and the saturation mechanism of the growing primary mode. In the subsequent discussion we will reconstruct the Lorenz model. In literature of 2D RBC, gravity is considered to be along $-\hat{z}$, and the flow in the xz plane. This is in contrast to the discussion in Sec. 3.7 where xy plane was used for 2D flows. For consistency with the 2D RBC literature, we rename xyz coordinate system to xzy, as shown in Fig. 7.1. As discussed in Sec. 3.7, we choose the Craya-Herring basis vector $\hat{e}_1$ for representing the velocity field.

For the velocity field, Lorenz employed the free-slip boundary conditions at all the walls, and for the temperature field, he used conducting boundary condition at the horizontal plates, and insulating boundary condition at the vertical walls. RBC with free-slip boundaries along all the walls yield a single convection roll at the convection onset. Hence, $L_x = \sqrt{2}$, not $2\sqrt{2}$ as in Sec. 6.1 where the boundary condition was along $\hat{x}$ was periodic. Note that $L_z = 1$. Lorenz considered Fourier

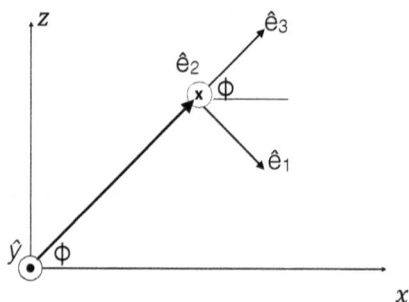

Fig. 7.1   For 2D RBC, an illustration of the wavenumber **k** and the basis vectors. The velocity vector is along $\hat{e}_1$.

modes whose wavenumbers are

$$\mathbf{k}_1 = (1,1) = k_c \hat{x} + \pi \hat{z},$$
$$\mathbf{k}_2 = (0,2) = 2\pi \hat{z}, \tag{7.1}$$

where $k_c = \pi/L_x = \pi/\sqrt{2}$. Note that $\mathbf{k}_1$ is the wavenumber of the first convective roll that gets excited first at the critical Rayleigh number

$$\mathrm{Ra}_c = \frac{k^6}{k_c^2} = \frac{27\pi^4}{4}, \tag{7.2}$$

where $k^2 = k_c^2 + \pi^2$ (see Sec. 6.1).

In Craya-Herring basis,

$$\mathbf{u}(\mathbf{k}_1) = u_1(\mathbf{k}_1)\hat{e}_1(\mathbf{k}_1) = \frac{U_{11}}{i}(\sqrt{2/3}, -1/\sqrt{3}, 0), \tag{7.3}$$

$$\theta(\mathbf{k}_1) = \frac{\theta_{11}}{i}, \tag{7.4}$$

$$\theta(\mathbf{k}_2) = \frac{\theta_{02}}{i}. \tag{7.5}$$

See Sec. 3.7 for details. In real space, temperature and velocity fields are

$$\theta(x,z) = \theta_{11}[4\cos(k_c x)\sin(\pi z)] + \theta_{02}[2\sin(2\pi z)], \tag{7.6}$$

$$\mathbf{u}(x,z) = \begin{pmatrix} 4U_{11}(\pi/k)\sin(k_c x)\cos(\pi z) \\ -4U_{11}(k_c/k)\cos(k_c x)\sin(\pi z) \end{pmatrix}. \tag{7.7}$$

It is important to note that $\mathbf{u}(x,z)$ and $\theta(x,z)$ contain Fourier modes $(1,1), (-1,1), (1,-1), (-1,-1)$ whose amplitudes are related to each other as described in Sec. 3.7:

$$\frac{U_{11}}{i} = u_1(1,1) = u_1(-1,-1) = -u_1(-1,1) = -u_1(1,-1), \tag{7.8}$$

$$\frac{\theta_{11}}{i} = \theta_1(1,1) = -\theta_1(-1,-1) = \theta_1(-1,1) = -\theta_1(1,-1), \tag{7.9}$$

$$\frac{\theta_{02}}{i} = \theta_1(0,2) = -\theta_1(0,-2), \tag{7.10}$$

Buoyancy in the Craya-Herring basis is

$$\mathbf{F}_g(\mathbf{k}_1) = \text{RaPr}\theta(\mathbf{k}_1)\hat{z} = \text{RaPr}\theta(\mathbf{k}_1)(\hat{e}_3 \sin\phi - \hat{e}_1 \cos\phi). \tag{7.11}$$

Therefore, the component of the Navier-Stokes equation along $\hat{e}_1$ is

$$\frac{d}{dt}u_1(\mathbf{k}_1) = -\text{Pr}k_1^2 u_1(\mathbf{k}_1) - \text{RaPr}\theta(\mathbf{k}_1)\cos\phi \tag{7.12}$$

or

$$\frac{d}{dt}U_{11} = -\frac{k_c}{k_1}\text{RaPr}\theta_{11} - \text{Pr}k_1^2 U_{11}, \tag{7.13}$$

Note that Lorenz ignored the nonlinear term in the velocity equation.

Second, we derive the equation for $\theta(\mathbf{k}_1)$, which is

$$\frac{d}{dt}\theta(\mathbf{k}_1) = \mathbf{u}(\mathbf{k}_1) \cdot \hat{z} - k_1^2\theta(\mathbf{k}_1) - N_\theta(\mathbf{k}_1)$$

$$= -u_1(\mathbf{k}_1)\cos\phi - k_1^2\theta(\mathbf{k}) - N_\theta(\mathbf{k}_1), \tag{7.14}$$

where $N_\theta(\mathbf{k}_1)$ is the nonlinear term. Given that we have have only wavenumbers $\mathbf{k}_1$ and $\mathbf{k}_2$ in the system, the condition $\mathbf{k} = \mathbf{p} + \mathbf{q}$ discussed in Chapter 4 yields the following interacting wavenumber triads for $N_\theta(\mathbf{k}_1)$:

$$\theta(1,1) = u_1(1,-1) \bigoplus \theta(0,2), \tag{7.15}$$

where $\bigoplus$ represents the nonlinear interaction. Hence

$$N_\theta(\mathbf{k}_1) = i\mathbf{k}_1 \cdot \mathbf{u}(1,-1)\theta(\mathbf{k}_2) = ik_1\hat{e}_3(\mathbf{k}_1) \cdot \hat{e}_1(1,-1)\frac{U_{11}}{-i}\frac{\theta_{02}}{i}. \tag{7.16}$$

Using $\hat{e}_1(1,-1) = (-\sqrt{2/3}, -1/\sqrt{3}, 0)$, $\hat{e}_3(\mathbf{k}_1) = (1/\sqrt{3}, \sqrt{2/3}, 0)$, $k_1 = \pi\sqrt{3/2}$, and Eqs. (7.8, 7.10) we obtain

$$N_\theta(\mathbf{k}_1) = -i\frac{2\pi}{\sqrt{3}}U_{11}\theta_{02} = -i\frac{2\pi k_c}{k_1}U_{11}\theta_{02}, \tag{7.17}$$

substitution of which in Eq. (7.14) yields

$$\frac{d}{dt}\theta_{11} = -2\pi k_c U_{11}\theta_{02} - \frac{k_c}{k_1}U_{11} - k_1^2\theta_{11}. \tag{7.18}$$

Similarly, the equation for $\theta(\mathbf{k}_2)$ is

$$\frac{d}{dt}\theta(\mathbf{k}_2) = -k_2^2\theta(\mathbf{k}_2) - N_\theta(\mathbf{k}_2), \tag{7.19}$$

where $N_\theta(\mathbf{k}_2)$ is generated by the following nonlinear interactions:

$$\theta(0,2) = u_1(1,1) \bigoplus \theta(-1,1) + u_1(-1,1) \bigoplus \theta(1,1). \tag{7.20}$$

Hence

$$N_\theta(\mathbf{k}_2) = i\mathbf{k}_2 \cdot [\mathbf{u}(1,1)\theta(-1,1) + \mathbf{u}(-1,1)\theta(1,1)]$$

$$= ik_3\hat{e}_3(\mathbf{k}_3) \cdot \left[\hat{e}_1(1,1)\frac{U_{11}}{i}\frac{\theta_{11}}{i} + \hat{e}_1(-1,1)\frac{U_{11}}{-i}\frac{\theta_{11}}{i}\right]$$

$$= i\frac{4\pi}{\sqrt{3}}U_{11}\theta_{02} = i\frac{4\pi k_c}{k_1}U_{11}\theta_{11} \tag{7.21}$$

Therefore, the dynamical equation for $\theta(\mathbf{k}_2)$ is

$$\frac{d}{dt}\theta_{02} = \frac{4\pi k_c}{k_1}U_{11}\theta_{11} - 4\pi^2\theta_{02}. \tag{7.22}$$

Thus we derive the dynamical equations, Eqs. (7.13, 7.18, 7.22), for the Lorenz model. In the linear limit, they are same as Eqs. (6.80, 6.81), and they exhibit thermal instability. We will show below how θ_{02} helps saturate the growth of θ_{11} and U_{11}.

To simplify the equations, we make the following change of variables: $t \to \tilde{t} = k_1^2 t$, $U_{11} \to \tilde{U}_{11} = (k_c/k_1^3)U_{11}$, and $r = \mathrm{Ra}/\mathrm{Ra}_c$ with $\mathrm{Ra}_c = k_1^6/k_c^2$. Note that r is the normalised Rayleigh number. The transformed equations in terms of new variables (with tilde sign dropped) are

$$\dot{U}_{11} = \mathrm{Pr}(-r\theta_{11} - U_{11}), \tag{7.23}$$
$$\dot{\theta}_{11} = -2\pi U_{11}\theta_{02} - U_{11} - \theta_{11}, \tag{7.24}$$
$$\dot{\theta}_{02} = 4\pi U_{11}\theta_{11} - b\theta_{02}, \tag{7.25}$$

where $b = 4\pi^2/k_1^2 = 8/3$. If we substitute

$$U_{11} = \frac{1}{4k_c}X, \tag{7.26}$$
$$\theta_{11} = -\frac{1}{4rk_c}Y, \tag{7.27}$$
$$\theta_{02} = -\frac{1}{2r\pi}Z, \tag{7.28}$$

then Eqs. (7.23–7.25) transform to

$$\dot{X} = \mathrm{Pr}(Y - X), \tag{7.29}$$
$$\dot{Y} = X(r - Z) - Y, \tag{7.30}$$
$$\dot{Z} = XY - bZ, \tag{7.31}$$

which are the equations derived by Lorenz. However, in this book we will work with Eqs. (7.23–7.25) since they are related to the spectral simulations.

We also remark that the above derivation, which is based on Craya-Herring decomposition, differs from that of Lorenz, which is based on stream function. These are two different methods to eliminate pressure from the equations.

In the next section we will discuss some of the properties of the Lorenz equations.

7.2 Stationary solution of Lorenz equations and nonlinear saturation

To study the Lorenz model, we take $b = 8/3$ and $\mathrm{Pr} = 10$, and vary r. For a set of r, the Lorenz equations admits solution that are constant in time ($d/dt = 0$). These solutions, called *stationary or fixed-point solutions*, are given below:

$$\begin{pmatrix} U_{11} \\ \theta_{11} \\ \theta_{02} \end{pmatrix} = \begin{pmatrix} 0 \\ 0 \\ 0 \end{pmatrix}; \quad \begin{pmatrix} -\frac{1}{2\pi}\sqrt{b(r-1)/2} \\ \frac{1}{2\pi r}\sqrt{b(r-1)/2} \\ -\frac{1}{2\pi}(1-\frac{1}{r}) \end{pmatrix}; \quad \begin{pmatrix} \frac{1}{2\pi}\sqrt{b(r-1)/2} \\ -\frac{1}{2\pi r}\sqrt{b(r-1)/2} \\ -\frac{1}{2\pi}(1-\frac{1}{r}) \end{pmatrix}. \tag{7.32}$$

In terms of X, Y, and Z the fixed points are

$$\begin{pmatrix} X \\ Y \\ Z \end{pmatrix} = \begin{pmatrix} 0 \\ 0 \\ 0 \end{pmatrix}; \quad \begin{pmatrix} -\sqrt{b(r-1)/2} \\ -\sqrt{b(r-1/2)} \\ r-1 \end{pmatrix}; \quad \begin{pmatrix} \sqrt{b(r-1)/2} \\ \sqrt{b(r-1)/2} \\ r-1 \end{pmatrix}. \quad (7.33)$$

The zero solution corresponds to the conduction state, while the other two solutions are the steady rolls with clockwise and anti-clockwise rotations. The former is stable for $r < 1$, while the later ones are stable for $r > 1$. Note that U_{11} and θ_{11} are anti-correlated because $u_z(1,1) \propto -U_{11} \propto \theta(1,1)$—hot plumes ascend, while the cold plumes descend.

We illustrate the time series of these modes in Fig. 7.2. Here we plot $U_{11}(t), r\theta_{11}(t), \theta_{02}(t)$ for $r = 0.5$, 5, and 30. We observe that all the modes are zeros for $r = 0.5$, but they take nonzero values for $r = 5$ and 30, consistent with Eq. (7.32). Thus, the Lorenz model predicts a transition from $\mathbf{u} = 0$, $\theta = 0$ to nonzero $\mathbf{u}$, θ that are proportional to $\pm\sqrt{r-1}$. Such transitions are classified as *supercritical pitchfork bifurcation*. The modes become chaotic for $r = 30$; this feature will be described subsequently.

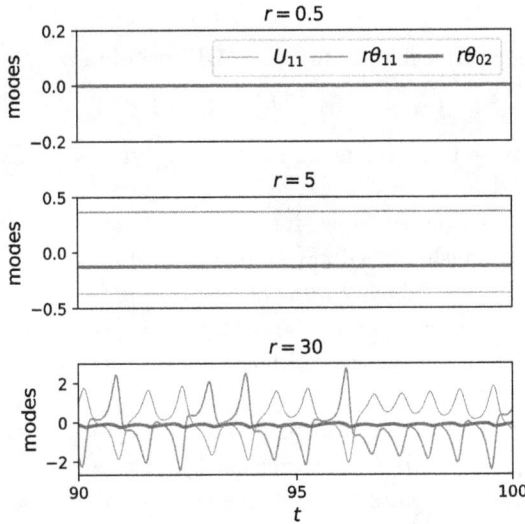

Fig. 7.2 For $Pr = 10$ and $b = 8/3$, time series of the Lorenz modes $U_{11}, r\theta_{11}$, and θ_{02} for $r = 0.5$, 5, and 30. We observe that $U_{11}, \theta_{11}, \theta_{02} = 0$ for $r = 0.5$, take fixed-point values consistent with Eq. (7.32) for $r = 5$, and become chaotic for $r = 30$.

Note that for $1 < r < 24.74$, the Lorenz modes saturate to constant values due to nonlinearity. To probe the dynamics further, we derive the following equation for the modal energy and entropies by multiplying Eqs. (7.23, 7.24, 7.25) with

$U_{11}, \theta_{11}, \theta_{02}$ respectively:

$$\frac{d}{dt} \frac{1}{2} U_{11}^2 = -\mathrm{Pr}(r U_{11}\theta_{11} + U_{11}^2), \tag{7.34}$$

$$\frac{d}{dt} \frac{1}{2} \theta_{11}^2 = -U_{11}\theta_{11} - \theta_{11}^2 - 2\pi\theta_{02} U_{11}\theta_{11}, \tag{7.35}$$

$$\frac{d}{dt} \frac{1}{2} \theta_{02}^2 = -b\theta_{02}^2 + 4\pi\theta_{11} U_{11}\theta_{02}. \tag{7.36}$$

For $1 < r < 24.74$, the flow is steady, hence $d/dt = 0$ in the above equations. Hence the energy gained by U_{11} from buoyancy, $-r\mathrm{Pr}U_{11}\theta_{11}$, is lost to viscous dissipation $\mathrm{Pr}U_{11}^2$. Note however that these two contributions are from the linear terms of Eq. (7.23).

Regarding the mode θ_{11}, the entropy gained from buoyancy by an amount $-U_{11}\theta_{11}$ is lost to thermal diffusion (amount $= \theta_{11}^2$) and to θ_{02} via nonlinear interactions (amount $= 2\pi\theta_{02}U_{11}\theta_{11}$). From Eq. (7.23), we obtain $|\theta_{11}/U_{11}| = 1/r$. Therefore, for moderate and large r, $|\theta_{11}| \ll |U_{11}|$, and hence from Eq. (7.35), we obtain

$$U_{11}\theta_{11} \approx -2\pi\theta_{02}U_{11}\theta_{11} \tag{7.37}$$

that yields

$$\theta_{02} \approx -\frac{1}{2\pi}. \tag{7.38}$$

Without the nonlinear transfers, from Eq. (7.24), we obtain

$$\dot{\theta}_{11} = -U_{11} - \theta_{11} \approx (r-1)\theta_{11}. \tag{7.39}$$

Hence θ_{11} would grow in time. Thus, nonlinearity $-2\pi U_{11}\theta_{02}$ of Eq. (7.24) saturates θ_{11}'s growth.

The mode θ_{02} receives entropy from $\theta(1,1)$ and $\theta(-1,1)$ by an amount $-4\pi\theta_{11}U_{11}\theta_{02}$ (each transfer $= -2\pi\theta_{11}U_{11}\theta_{02}$) via nonlinear transfers. This entropy is lost to diffusion. Thus, nonlinearity plays a very important role in the saturation of the linear growth of all the modes. Note however that the equations become more complex in the presence of large number of modes, some of which will be discussed in the next chapter.

The result that $\theta_{02} \approx -1/(2\pi)$ also bears out in analytic theory (see Sec. 9.2) and in large-scale simulations of RBC (see Sec. 13.3). The aforementioned arguments show that $\theta_{02} \approx -1/(2\pi)$ occurs essentially due to a balance between the entropy feed to θ_{11} by buoyancy and the nonlinear energy transfer from θ_{11} to θ_{02}. This feature persists even in the presence of large number of modes (to be discussed in the next chapter and in chapter 13).

Lorenz equation shows many interesting features. For $r \gtrsim 24.74$, the steady-state solution becomes unstable through another bifurcation, and the system becomes chaotic. We compute the chaotic time series of the modes for $r = 30$ and exhibit them in the bottom panel of Fig. 7.2. In Fig. 7.3 we draw the phase space plot that appears like a butterfly, hence it is called a *butterfly diagram*. These plots show that

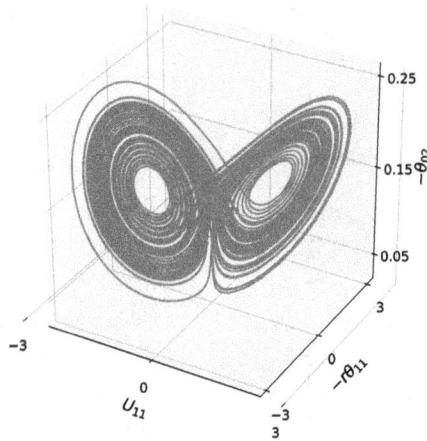

Fig. 7.3 Phase space plot of the Lorenz modes for $r = 30$. The system exhibits chaotic behaviour for this parameter.

the Lorenz system is chaotic for $r = 30$. Note however that the realistic chaotic and turbulent behaviour of RBC is not captured by the Lorenz model. This deficiency is due to the drastic truncation adopted in the Lorenz model in which only Fourier modes (1,1) and (0,2) are retained.

It is also important to note that Lorenz system is suitable for convection with large Prandtl numbers. This is because the nonlinear term of the momentum equation is weak for large-Pr convection, which is consistent with Eq. (7.23) of Lorenz equation in which the nonlinear term is absent. Note that the nonlinear term of the momentum equation dominates the convective flow with small Prandtl numbers.

In RBC, large number of Fourier modes are generated by nonlinear interactions. For example, $\theta(1,3)$ is generated due to nonlinear interactions between $\mathbf{u}(1,1)$ and $\theta(2,2)$. The mode $\theta(1,3)$ induces $\mathbf{u}(1,3)$. Subsequently, $\mathbf{u}(1,1)$ and $\mathbf{u}(1,3)$ generate $\mathbf{u}(2,2)$ by nonlinear interactions. Successively, more modes are born in a chain reaction. In the next chapter we will discuss how the secondary modes are generated in RBC.

In the next sections we will describe bifurcation analysis and amplitude equations that describe generic instabilities and nonlinear saturation in thermal convection.

7.3 Bifurcation analysis

Many dynamical systems are described by a set of ordinary differential equations (ODEs). A class of such system experience a dramatic change in their behaviour under variations of their control parameters. This phenomena is called *bifurcation* with the transition point named as the *bifurcation point*.

Mathematicians have shown that near a bifurcation point, systems exhibiting bifurcations with a single control parameter can be classified into four classes:

(a) Saddle-node   (b) Transcritical

(c) Supercritical Pitchfork   (d) Subcritical Pitchfork

(e) Supercritical Hopf   (f) Subcritical Hopf

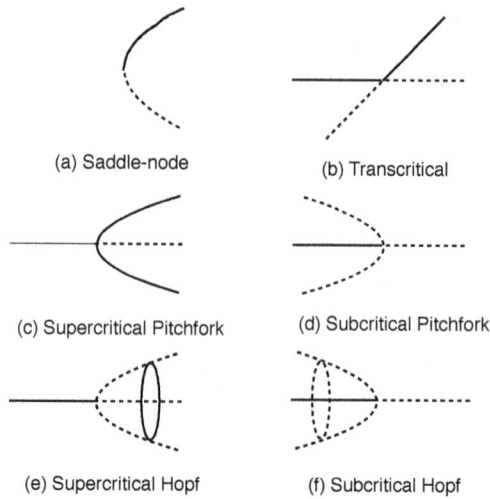

Fig. 7.4 Schematic diagrams illustrating various bifurcations. The x axis represents the control parameter c, while y axis the system variable X or Z. Here, solid lines represent the stable solutions, while the dashed lines represent the unstable ones.

(1) *Saddle-node:* $\dot{X} = c - X^2$.
(2) *Transcritical:* $\dot{X} = cX - X^2$.
(3) *Pitchfork::* $\dot{X} = cX - X^3$ *(supercritical)*, $\dot{X} = cX + X^3$ *(subcritical)*.
(4) *Hopf:* $\dot{Z} = (c + i\omega)Z - (1 + ib)|Z|^2 Z$ *(supercritical)*,
     $\dot{Z} = (c + i\omega)Z + (1 + ib)|Z|^2 Z$ *(subcritical)*.

where X, Z are the system variables, c is the control parameter, and ω, b are real constants. The dynamical variable for the Hopf bifurcation, $Z = X + iY$, is a complex variable, hence it describes two-dimensional systems whose variables are X and Y. Note that the aforementioned first three classes describe one-dimensional systems. We also remark that near the bifurcation point, all differential equations, however complex, could be reduced to one of the aforementioned *normal forms* [Guckenheimer and Holmes (2013)]. The bifurcation diagrams for the aforementioned classes are shown in Fig. 7.4. Figures 7.4(a–d) describe the stationary or fixed-point solutions ($\dot{X} = 0$)—stable solutions as solid curves, and unstable ones as dashed curves. But, Figs. 7.4(e,f) illustrate periodic solutions of the system. Due to the above nature of the fixed points, the first three bifurcations are also called *stationary*, and the fourth one is called *oscillatory*.

In the following discussion, we explain the pitchfork and Hopf bifurcations in some detail. The fixed point solutions for the supercritical pitchfork bifurcations are

$$X^* = 0; \quad X^* = \pm\sqrt{c} \text{ only for } c > 0. \tag{7.40}$$

For $c < 0$ and near $X \approx 0$, we deduce that $\dot{X} \approx cX < 0$, hence $X^* = 0$ is a stable solution of the system. However, for $c > 0$, $X^* = 0$ becomes unstable, while the new solutions $X^* = \pm\sqrt{c}$ are the stable solutions because

$$\dot{x} \approx -2cx \tag{7.41}$$

where $x = X - (\pm\sqrt{c})$.

We can perform similar analysis for the subcritical pitchfork bifurcation whose fixed points are

$$X^* = 0; \quad X^* = \pm\sqrt{c} \text{ only for } c < 0. \tag{7.42}$$

Following similar arguments as above, for $c < 0$, the only stable solution is $X^* = 0$. In literature, supercritical and subcritical bifurcations are also called *forward* and *backward* respectively. See Fig. 7.4 for illustrations.

When we compare the stationary solution of Eq. (7.32) with the fixed points of Eq. (7.40), we observe that the Lorenz system exhibits supercritical pitchfork bifurcation near $r = 1$, where r is the normalised Rayleigh number. Note that $c = r - 1$.

Now, we describe the Hopf bifurcation briefly. We rewrite the equation for the supercritical Hopf bifurcation in terms of the amplitude R and phase ϕ of Z. Substitution of $Z = R\exp(i\phi)$ in

$$\dot{Z} = (c + i\omega)Z - (1 + ib)|Z|^2 Z \tag{7.43}$$

yields

$$\dot{R} = R(c - R^2), \tag{7.44}$$
$$\dot{\phi} = \omega - bR^2. \tag{7.45}$$

Using fixed point analysis, we find the stationary solution of the above equations as

$$R = 0; \sqrt{c}, \tag{7.46}$$
$$\dot{\phi} = \omega - bc = \text{const.} \tag{7.47}$$

The solution $R = 0$ and $\sqrt{c}$ are stable solutions for $c < 0$ and $c > 0$ respectively. Thus, Z converges to $(0,0)$ for $c < 0$, but it asymptotically reaches a circle of radius $\sqrt{c}$ for $c > 0$. Note that Z rotates with a frequency of $\dot{\phi}$, as shown in Fig. 7.4(e). Following the similar arguments as above, we can show that the system

$$\dot{Z} = (c + i\omega)Z + (1 + ib)|Z|^2 Z \tag{7.48}$$

exhibits subcritical Hopf bifurcation.

In the following section, we will discuss amplitude equations related to RBC.

7.4 Amplitude equation

Many systems exhibiting spatio-temporal variations are described by partial differential equations (PDEs). Often, Galerkin method is adopted to convert the PDEs to a set of ODEs. A simple example of Galerkin truncation is the Lorenz system described in Sec. 7.1; here, from the RBC equations we derive three ODEs,

Eqs. (7.23, 7.24, 7.25), for the dominant Fourier modes U_{11}, θ_{11}, and θ_{02}. These equations exhibit onset of convective rolls and chaos.

There is an alternative approach to address bifurcations and pattern formation in nonlinear systems. One employs scale separation between the fast and slow time scales, and then derive appropriate equations for the slow variables. This is an extensive field of research, and it is beyond the scope of this book. We refer the readers to books by Manneville (2004), Cross and Greenside (2009), and Hoyle (2006), and review article by Cross and Hohenberg (1993). Here we sketch how an amplitude equation is derived for RBC.

The velocity field, Rayleigh number, and time scales are expanded perturbatively as

$$\mathbf{u} = \epsilon \mathbf{u}_1 + \epsilon^2 \mathbf{u}_2, \tag{7.49}$$

$$\mathrm{Ra} = \mathrm{Ra}_c + \epsilon^2, \tag{7.50}$$

$$\partial_t = \partial_{t_0} + \epsilon^2 \partial_{t_1}. \tag{7.51}$$

Perturbative solution of RBC equation at an order ϵ is

$$\mathbf{u}_1 = A_1 \exp(ik_c x) + c.c., \tag{7.52}$$

where A_1 is the amplitude of the mode. After several steps, one arrives at the amplitude equation [Manneville (2004)]

$$\tau_0 \partial_{t_1} A_1 = a A_1 - g|A_1|^2 A_1. \tag{7.53}$$

Using translational symmetry, one can argue that A_1, a, and g are real. Therefore, the aforementioned amplitude equation has the same form as the supercritical pitchfork bifurcation described in the earlier section. This is how we show using the amplitude equation that in RBC, the transition from the conduction state to the convection state is via supercritical pitchfork bifurcation, and that the nonlinearity saturates the growth of A_1.

The nonlinear term of Eq. (7.53), $-g|A_1|^2 A_1$, involves interactions of four Fourier modes, $k = k + (-k) + k$ with

$$\dot{A}_1(k) = -g A_1(k) A_1(-k) A_1(k) \tag{7.54}$$

This is unlike the Galerkin method where different Fourier modes participate in the nonlinear interactions. Fore example, in the Lorenz equations, the nonlinear interactions are of the form:

$$\dot{\theta}_{11} \sim u_{11} \theta_{02}, \tag{7.55}$$

with the participating modes being $(1, 1)$, $(-1, 1)$, and $(0, 2)$.

Thus, there are several differences between the amplitude equations and the Galerkin method. An amplitude equation provides a schematic description of the nonlinear interactions. However, a low-dimensional model derived using Galerkin method provides a somewhat more realistic representation of the interacting triads, and thus captures the instabilities, patterns, and onset of chaos more accurately.

For example, we can compute the critical Rayleigh number Ra_c using the truncated Galerkin equations, but Ra and Ra_c appear as free parameters in the amplitude equation. In the amplitude equation (7.53), the critical Rayleigh number is a function of the parameters a and g. An advantage of the amplitude equations however is that they are somewhat easier to derive using general symmetry arguments.

In the next section, we will discuss very briefly the nonlinear interactions in a stably stratified system.

7.5 Nonlinear interactions in stably stratified flows

In Chapter 5, we derived internal gravity waves from the linearised version of stably stratified flows. These waves interact nonlinearly and yield structures, such as planetary jets [Vallis (2006)]. For larger Reynolds numbers, the nonlinear interactions become strong and make the flow turbulent. In this book, we do not cover the patterns and chaos in stably stratified system, but we will cover stably stratified turbulence.

In the next chapter, we will describe how patterns and chaos emerge in RBC.

Further reading

The Lorenz model first appeared in Lorenz (1963). For a more detailed derivation of Lorenz equations and its subsequent bifurcations, refer to Hilborn (2001) and Strogatz (2014). The formalism of amplitude equation is covered in books by Manneville (2004), Cross and Greenside (2009), and Hoyle (2006), and its applications to RBC is in books by Getling (1998) and Lappa (2010).

Exercises

(1) Derive Lorenz equations using stream function formalism: replace $\mathbf{u} = \nabla \times (\psi \hat{z})$, and write down an equation for ψ. Fourier expand ψ and θ as done in this chapter. Then, derive ODEs for the Fourier coefficients. Contrast this method with one described in Sec. 7.1.

(2) Perform linear stability analysis of the Lorenz equations. Show that the equations exhibit

   (a) Supercritical pitchfork bifurcation at $r = 1$. Test the stability of the solutions before and after the bifurcation.

   (b) Show that nonzero solutions of Eq. (7.32) bifurcate at $r_H = \mathrm{Pr}(\mathrm{Pr} + b + 3)/(\mathrm{Pr} - b - 1)$ via subcritical Hopf bifurcation. For $\mathrm{Pr} = 10$, and $b = 8/3$, $r_H \approx 24.74$. Study the eigenvalues of the stability matrix.

(3) Study the Lorenz equations numerically for various parameter values.

Chapter 8

Patterns and Chaos in Buoyancy-Driven Flows

In Chapter 6 we described the thermal instability, and in Chapter 7 we showed how nonlinearity saturates the growth of the primary convective roll. In this chapter we will advance these discussions by showing how secondary modes are generated in RBC, and how patterns and chaos are created by these secondary modes. A further increase of Rayleigh number leads to a generation of turbulence that will be discussed in Part B of the book.

In the next section we will construct a seven-mode model of RBC that demonstrates generation of secondary modes and patterns.

8.1 Seven-mode model of RBC

The Lorenz model (see Chapter 7) consists of three interacting Fourier modes, U_{11}, θ_{11}, θ_{02}, whose wavenumbers are

$$\mathbf{k}_1 = (1, 1) = k_c \hat{x} + \pi \hat{z},$$

$$\mathbf{k}_2 = (0, 2) = \pi \hat{z}. \tag{8.1}$$

Lorenz constructed dynamical equations for U_{11}, θ_{11}, and θ_{02}, and showed stable solution for $r \lesssim 24.74$. In real RBC, we observe generation of many secondary modes much before $r = 24.74$. These modes are generated by various nonlinear interactions. Here we illustrate one such generation mechanism in a seven-mode model of RBC. This model also admits a three-dimensional square structure.

8.1.1 *Derivation of the seven-mode model*

We consider a three-dimensional rectangular geometry with free-slip boundary conditions at all the walls. The dimensions of the box is $\sqrt{2} \times \sqrt{2} \times 1$, and the critical Rayleigh number $\mathrm{Ra}_c = 27\pi^4/4$ (see Sec. 6.1). We take seven modes U_{101}, U_{011}, U_{112}, θ_{101}, θ_{011}, θ_{112}, and θ_{002}, whose wavenumbers are

$$\mathbf{k}_1 = (1, 0, 1) = k_c \hat{x} + \pi \hat{z},$$

$$\mathbf{k}_2 = (0, 1, 1) = k_c \hat{y} + \pi \hat{z},$$

$$\mathbf{k}_3 = (1, 1, 2) = k_c \hat{x} + k_c \hat{y} + 2\pi \hat{z},$$

$$\mathbf{k}_4 = (0, 0, 2) = 2\pi \hat{z} \tag{8.2}$$

with $k_c = \pi/L_x = \pi/L_y = \pi/\sqrt{2}$.

Note that the reality condition and the free-slip boundary condition require presence of Fourier modes $(\pm 1, 0, \pm 1)$, $(0, \pm 1, \pm 1)$, $(\pm 1, \pm 1, \pm 2)$, $(0, 0, \pm 2)$. We employ Craya-Herring basis to expand the velocity field as $\mathbf{u}(\mathbf{k}) = u_2(\mathbf{k})\hat{e}_2$. See Sec. 3.6 for details. Since $u_1(\mathbf{k}) = 0$, the flow has zero kinetic helicity. This is a good approximation for moderate and large Pr, but not for small Pr. The components of the velocity and the temperature fluctuation θ are presented in Table 8.1.

Table 8.1   The Fourier modes of the seven-mode model: $\hat{e}_2(\mathbf{k})$ and $\hat{e}_3(\mathbf{k})$ are the Craya-Herring basis vectors. Note that $u_1 = 0$ for all $\mathbf{k}$'s.

mode	$u_2(\mathbf{k})$	$\theta(\mathbf{k})$	$\hat{e}_2(\mathbf{k})$	$\hat{e}_3(\mathbf{k})$
$(1,0,1)$	U_{101}/i	θ_{101}/i	$(\sqrt{2/3}, 0, -1/\sqrt{3})$	$(1/\sqrt{3}, 0, \sqrt{2/3})$
$(0,1,1)$	U_{011}/i	θ_{011}/i	$(0, \sqrt{2/3}, -1/\sqrt{3})$	$(0, 1/\sqrt{3}, \sqrt{2/3})$
$(1,1,2)$	U_{112}/i	θ_{112}/i	$(\sqrt{2/5}, \sqrt{2/5}, -1/\sqrt{5})$	$(1/\sqrt{10}, 1/\sqrt{10}, 2/\sqrt{5})$
$(0,0,2)$	0	θ_{002}/i	-	$(0, 0, 1)$

The derivation of the dynamical equations for the aforementioned seven modes is similar to that adopted for the Lorenz equations. To illustrate, the momentum equation for $\mathbf{N}_u(1, 1, 2)$ involves the following nonlinear interaction:

$$u(1, 1, 2) = u(1, 0, 1) \bigoplus u(0, 1, 1). \tag{8.3}$$

We denote $\mathbf{k}_1 = (1, 0, 1)$, $\mathbf{k}_2 = (0, 1, 1)$, and $\mathbf{k}_3 = (1, 1, 2)$. Hence, component of $\mathbf{N}_u(1, 1, 2)$ along $\hat{e}_2(\mathbf{k}_3)$ is

$$
\begin{aligned}
N_u(\mathbf{k}_3) &= i k_3 \cdot \mathbf{u}(\mathbf{k}_2) u_2(\mathbf{k}_1)[\hat{e}_2(\mathbf{k}_1) \cdot \hat{e}_2(\mathbf{k}_3)] + i k_3 \cdot \mathbf{u}(\mathbf{k}_1) u_2(\mathbf{k}_2)[\hat{e}_2(\mathbf{k}_2) \cdot \hat{e}_2(\mathbf{k}_3)] \\
&= i k_3 [\{\hat{e}_3(\mathbf{k}_3) \cdot \hat{e}_2(\mathbf{k}_2)\}\{\hat{e}_2(\mathbf{k}_1) \cdot \hat{e}_2(\mathbf{k}_3)\} \\
&\quad + \{\hat{e}_3(\mathbf{k}_3) \cdot \hat{e}_2(\mathbf{k}_1)\}\{\hat{e}_2(\mathbf{k}_2) \cdot \hat{e}_2(\mathbf{k}_3)\}] u_2(\mathbf{k}_1) u_2(\mathbf{k}_2) \\
&= -i \frac{2\pi}{\sqrt{5}} U_{011} U_{101},
\end{aligned}
\tag{8.4}
$$

and $N_\theta(\mathbf{k}_3)$ is

$$
\begin{aligned}
N_\theta(\mathbf{k}_3) &= i k_3 \cdot [\mathbf{u}(\mathbf{k}_2)\theta(\mathbf{k}_1) + \mathbf{u}(\mathbf{k}_1)\theta(\mathbf{k}_2)] \\
&= i k_3 \left[\hat{e}_3(\mathbf{k}_3) \cdot \hat{e}_2(\mathbf{k}_2) \frac{U_{011}}{i} \frac{\theta_{101}}{i} + \hat{e}_3(\mathbf{k}_3) \cdot \hat{e}_2(\mathbf{k}_1) \frac{U_{101}}{i} \frac{\theta_{011}}{i} \right] \\
&= -i \frac{\pi}{\sqrt{3}} [U_{011}\theta_{101} + U_{101}\theta_{011}].
\end{aligned}
\tag{8.5}
$$

Buoyancy in the Craya-Herring basis is

$$\mathbf{F}_g(\mathbf{k}_1) = \text{RaPr}\theta(\mathbf{k}_1)\hat{z} = \text{RaPr}\theta(\mathbf{k}_1)(\hat{e}_3 \cos \zeta - \hat{e}_2 \sin \zeta). \tag{8.6}$$

Therefore, the component of the dynamical equation of $\mathbf{u}(\mathbf{k}_3)$ along $\hat{e}_2$ is

$$\frac{d}{dt} u_2(\mathbf{k}_3) = -N_u(\mathbf{k}_3) - \text{RaPr}\theta(\mathbf{k}_3) \sin \zeta - \text{Pr}k_3^2 u_2(\mathbf{k}_3). \tag{8.7}$$

Similarly, the dynamical equation for $\theta(\mathbf{k}_3)$ is

$$\frac{d}{dt}\theta(\mathbf{k}_3) = -N_\theta(\mathbf{k}_3) - u_2(\mathbf{k}_3) \sin \zeta - k_3^2 \theta(\mathbf{k}_3). \tag{8.8}$$

Substitution of nonlinear terms, and the following change of variables—$t \to \bar{t} = k_1^2 t$, $U_{11} \to \tilde{U}_{11} = (k_c/k_1^3)U_{11}$, $r = \text{Ra}/\text{Ra}_c = \text{Ra}/(k_1^6/k_c^2)$—yields

$$\dot{U}_{112} = 2\pi\sqrt{\frac{3}{5}}U_{101}U_{011} - \Pr\left[r\sqrt{\frac{3}{5}}\theta_{112} + (k_3/k_1)^2 U_{112}\right], \tag{8.9}$$

$$\dot{\theta}_{112} = \pi[U_{101}\theta_{011} + U_{011}\theta_{101}] - \sqrt{\frac{3}{5}}U_{112} - (k_3/k_1)^2\theta_{112}. \tag{8.10}$$

Following similar procedure, we obtain the following equations for other five modes:

$$\dot{U}_{101} = -2\pi\sqrt{\frac{3}{5}}U_{011}U_{112} - \Pr[r\theta_{101} + U_{101}], \tag{8.11}$$

$$\dot{U}_{011} = -2\pi\sqrt{\frac{3}{5}}U_{101}U_{112} - \Pr[r\theta_{011} + U_{011}], \tag{8.12}$$

$$\dot{\theta}_{101} = -2\pi[U_{011}\theta_{112} + U_{101}\theta_{002}] - U_{101} - \theta_{101}, \tag{8.13}$$

$$\dot{\theta}_{011} = -2\pi[U_{101}\theta_{112} + U_{011}\theta_{002}] - U_{011} - \theta_{011}, \tag{8.14}$$

$$\dot{\theta}_{002} = 4\pi[U_{101}\theta_{101} + U_{011}\theta_{011}] - b\theta_{002}, \tag{8.15}$$

where $b = 4\pi^2/k_1^2$. Some interesting observations on the 7-mode model are as follows:

(1) If we set $U_{011} = U_{112} = 0$ and $\theta_{011} = \theta_{112} = 0$, then we recover the Lorenz equations with rolls in the xz planes. The solutions of these equations are exactly same as the Lorenz equations.
(2) The complimentary solutions, rolls in the yz plane, are obtained by setting $U_{101} = U_{112} = 0$ and $\theta_{101} = \theta_{112} = 0$.

Now let us analyse the solution of the 7-mode model. When $r < 1$, all the modes vanish; this solution corresponds to the conduction state. For $r > 1$, we first obtain a roll either in xz plane or in yz plane (depending on the initial condition). For rolls in the xz plane, $U_{011} = U_{112} = 0$, and $\theta_{011} = \theta_{112} = 0$, and the asymptotic solutions are exactly same as that for Lorenz equations, namely

$$\begin{pmatrix} U_{101} \\ \theta_{101} \\ \theta_{002} \end{pmatrix} = \begin{pmatrix} 0 \\ 0 \\ 0 \end{pmatrix}; \quad \begin{pmatrix} -\frac{1}{2\pi}\sqrt{b(r-1)/2} \\ \frac{1}{2\pi r}\sqrt{b(r-1)/2} \\ -\frac{1}{2\pi}(1-\frac{1}{r}) \end{pmatrix}; \quad \begin{pmatrix} \frac{1}{2\pi}\sqrt{b(r-1)/2} \\ -\frac{1}{2\pi r}\sqrt{b(r-1)/2} \\ -\frac{1}{2\pi}(1-\frac{1}{r}) \end{pmatrix}, \tag{8.16}$$

where $b = 8/3$. We obtain zero solution for $r < 1$, and nonzero solution for $1 < r < r_{c2}$. Three-dimensional oscillatory solutions start to appear near $r = r_{c2}$. The above constant solutions are same as those of Lorenz model, which are illustrated in Fig. 7.2 (top and middle panels). Note that the transition of the amplitudes from zero to those proportional to $\propto \sqrt{r-1}$ is via a *supercritical pitchfork bifurcation*. Also note that U_{101} and θ_{101} are anti-correlated because $u_z(1,0,1) \propto -U_{101} \propto \theta(1,0,1)$—hot plumes ascend, while the cold plumes descend. Similar conclusions can be drawn for $(0,1,1)$ modes.

On a further increase of r, at $r \approx 12$, we start to get three-dimensional structures due to generation of secondary rolls. We will discuss this topic in the next section.

8.1.2 Generation of secondary modes

We study the generation of secondary modes by simulating Eqs. (8.9–8.15). For an illustration we focus our attention on $r = 13$. We start with the steady roll solution in the xz plane. Under a steady-sate, the mode U_{101} oscillates around a mean value of A_1:

$$U_{101} = A_1 \approx 0.55. \tag{8.17}$$

Now we substitute the above in Eqs. (8.12, 8.9) and focus only on the nonlinear terms. For shorthand, we denote $U_{011} = X$ and $U_{112} = Y$ that yields the following equations:

$$\dot{X} = -2\pi\sqrt{\frac{3}{5}}A_1 Y, \tag{8.18}$$

$$\dot{Y} = 2\pi\sqrt{\frac{3}{5}}A_1 X. \tag{8.19}$$

In terms of $Z = X + iY$, the above equation becomes

$$\dot{Z} = i2\pi\sqrt{\frac{3}{5}}A_1 Z \tag{8.20}$$

Hence Z (or X and Y) oscillates with the frequency of

$$\omega = 2\pi\sqrt{\frac{3}{5}}A_1 \approx 2.66 \tag{8.21}$$

or time period $T = 2\pi/\omega \approx 2.4$.

It is interesting to note that the nonlinear terms of Eqs. (8.9, 8.11, 8.12) have similar structures as those of the three-mode model of Example 3.3 of Chapter 3. There too, one mode oscillates around a mean value, while the other two modes oscillate around zero. Thus, such nonlinear behaviour is quite generic.

When we compare Eq. (8.20) with the equation for the Hopf bifurcation of Sec. 7.3, we can deduce that the new modes U_{011}, U_{112} are generated by the Hopf bifurcation. Considerations of all the terms, e.g., $rPr\theta_{011}$ and $rPr\theta_{112}$ of Eqs. (8.12, 8.9), would yield the nonlinear term $|Z|^2 Z$ of the Hopf bifurcation.

To verify the aforementioned deductions, we solve the equations of 7-mode model numerically for $Pr = 1$ and $r = (10, 15)$. At around $r \approx 12$, the oscillatory solution start to emerge. In Fig. 8.1, we plot the time series of U_{101}, U_{101}, U_{112} for $r = 13$ that exhibit such oscillations, and they can be approximated as

$$\begin{pmatrix} U_{101} \\ U_{011} \\ U_{112} \\ r\theta_{101} \end{pmatrix} \approx \begin{pmatrix} A_1 + B_1\cos(2\omega t) \\ B_2\cos(\omega t + \phi) \\ B_3\sin(\omega t) \\ -A_2 + B_4\sin(2\omega t) \end{pmatrix} \approx \begin{pmatrix} 0.55 + 0.066\cos(2\omega t) \\ 0.27\cos(\omega t + \phi) \\ 0.20\sin(\omega t) \\ -0.59 + 0.29\sin(2\omega t) \end{pmatrix} \tag{8.22}$$

with $\omega \approx 2.4$. The values of A_1 and ω are close to the theoretical estimates described earlier. Also, $A_2 \approx A_1$ as in Eq. (8.16). We remark that the Eq. (8.22) has minimal complexity without sub- and super harmonics; such forms have been been chosen

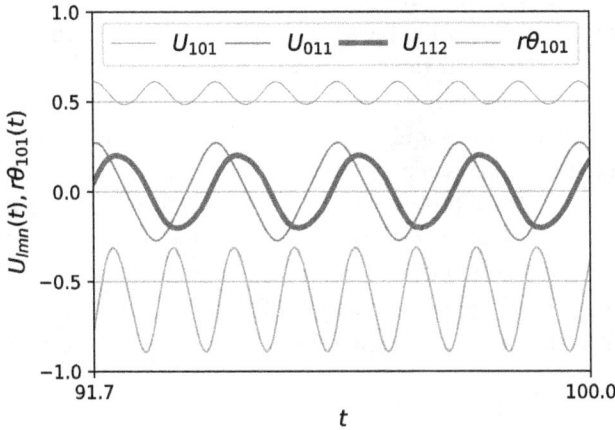

Fig. 8.1 For the 7-mode model with Pr $= 1$ and $r = 13$, the time series of U_{101}, U_{011}, U_{112}, and $r\theta_{101}$ (see legend). U_{101} oscillates around a constant mean of 0.55, while $r\theta_{101}$ oscillates around -0.59. The modes U_{011} and U_{112} oscillate around zero.

only to illustrate the generation mechanism of secondary modes. Also, the transients have been eliminated from the time series plots of Fig. 8.1.

For Pr $= 1$ and $r = 13$, U_{101} exhibits oscillatory solution around the mean value, $A_1 \approx 0.55$. When we substitute the approximate solution in Eq. (8.11), we obtain

$$-2\omega B_1 \sin(2\omega t) - \approx \pi\sqrt{\frac{3}{5}}B_2 B_3[\sin(2\omega t + \phi) - \sin\phi] + A_2 - A_1$$
$$-B_4 \sin(2\omega t) - B_1 \cos(2\omega t). \tag{8.23}$$

Time averaging of the above equation yields

$$\pi\sqrt{\frac{3}{5}}B_2 B_3 \sin\phi + A_2 - A_1 \approx 0, \tag{8.24}$$

that yields $\phi \approx 20$ degrees. Equation (8.23) also reveal why U_{101} and θ_{101} oscillate with frequencies twice that of U_{011} and U_{112}. These observations are consistent with the time series shown in Fig. 8.1. We remark that the amplitudes of the oscillations increase with r, and its frequency $\omega \propto A_1 \propto \sqrt{r-1}$.

In summary, constant A_1 generates oscillatory modes U_{011} and U_{112}. However, the new oscillatory modes in turn induce oscillations on U_{101} itself. This is how new secondary modes U_{011} and U_{112} are generated that fundamentally changes the behaviour of the primary mode itself. As described earlier, this is via *Hopf bifurcation* (see Sec. 7.3). We will show in the next that the aforementioned modes yield oscillating asymmetric square that have been observed in RBC experiments [Busse and Whitehead (1974)].

A critical missing piece in Eq. (8.20) is the derivation of the amplitudes of the oscillations; this is determined by the nonlinear term. Without all the terms, we cannot deduce the bifurcation point for the Hopf bifurcation, as we well as how the amplitudes and frequency of oscillations vary with the Prandtl and Rayleigh numbers. In fact, the bifurcation parameter r_c increases with the Prandtl number. Our preliminary computation shows that for $Pr = 100$, the Hopf bifurcation takes place beyond $r = 200$. In fact, we need to perform bifurcation analysis of the full 7-mode model for the above inquiry. In particular, we have to look for r at which the linearised matrix yields complex eigenvalues for the first time.

In Fig. 8.2 we illustrate the variations of the amplitudes of the modes with the Prandtl number—$Pr = 0.02, 1$, and 100 at $r = 13$. We observe that the amplitudes of oscillations of the modes decrease with the increase of Pr. The amplitudes of the secondary modes are zero for $Pr = 100$. The weakening of oscillations with the increase of Pr is essentially due to the dominance of the viscous force for large Pr.

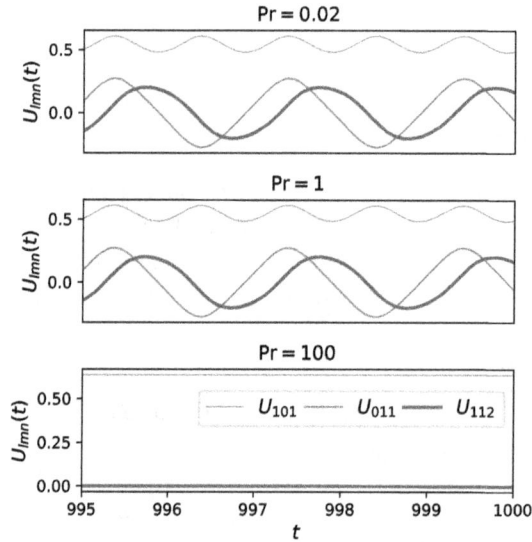

Fig. 8.2   For the 7-mode model with $r = 13$, the time series of U_{101}, U_{011}, and U_{112}. The time series for $Pr = 0.02, 1$, and 100 described in the three panels have similar behaviour as Fig. 8.1, except that the oscillation amplitudes decrease with the increase of Pr.

In 2D RBC too, new modes are generated via similar mechanism. Here, the primary convective mode is U_{11}, and the secondary modes are U_{22}, U_{31}, and U_{13}. In the following discussion, we sketch how modes U_{22}, U_{31} are generated in 2D RBC. Note that the Fourier modes $\{(1,1), (2,2), (3,1)\}$ form a triad since

$$(1,1) = (-2,2) + (3,-1). \tag{8.25}$$

Following the similar lines of arguments as the derivation of 7-mode model, we obtain the following set of equations for U_{11}, U_{22}, U_{31}:

$$\dot{U}_{11} = ()U_{22}U_{31} - \text{Pr}\{()r\theta_{11} + ()U_{11}\}, \tag{8.26}$$

$$\dot{U}_{22} = ()U_{11}U_{31} - \text{Pr}\{()r\theta_{22} + ()U_{22}\}, \tag{8.27}$$

$$\dot{U}_{31} = ()U_{22}U_{11} - \text{Pr}\{()r\theta_{31} + ()U_{31}\}, \tag{8.28}$$

where () stands for constant coefficients. The other Fourier modes θ_{11}, θ_{22}, θ_{31}, θ_{02} have similar equations. Given U_{11}, the nonlinear terms of the above equations generate U_{22} and U_{31} modes via a Hopf bifurcation. Following similar arguments as above, we can deduce that the modes U_{11}, U_{22}, U_{31} would exhibit temporal oscillations.

Thus, the aforementioned mode-generation mechanism via Hopf bifurcation appears to be generic, and it may be present in other fields as well. It appears to be similar to the mass generation in particle physics and condensed-matter physics through *Higgs mechanism*[1].

The 7-mode model described above has limited number of Fourier modes, which are obtained by a truncation of the Fourier series. This process is called *Galerkin truncation*, and the models thus constructed are called *Galerkin-based low-dimensional models*.

In the next section we will provide further description of secondary bifurcations and chaos in RBC.

8.2   Patterns and chaos in moderate and large-Pr RBC

In this section we provide a brief description of patterns and chaos in large-Pr RBC. First, we provide a physical interpretation to the aforementioned secondary modes generated via Hopf bifurcation.

The secondary mode, U_{011}, discussed in the previous section is a roll solution perpendicular to the primary roll, U_{101}. Hence, we expect square patterns when observed from the top. The secondary mode U_{011} however is weaker than the primary mode U_{011}, hence the square pattern is asymmetric, as illustrated in Fig. 8.3(a). Such structures, called *cross roll patterns*, have been observed in experiments [Busse and Whitehead (1974)]

The modes U_{101} and U_{011} oscillate in time, hence the strengths of the primary and secondary rolls vary in time in a periodic fashion. When we work out the energetics of such oscillations [Verma *et al.* (2006a)], we observe that U_{101} transfers energy to U_{112}, who in turn transfers energy to U_{011}. The mode U_{011} transfers energy back to U_{112}, who transfers energy to U_{101}. This process continues cyclically.

---

[1]In RBC with walls along the $\hat{z}$ and periodic boundary condition along $\hat{x}$, the phase of the primary Fourier mode U_{101} is arbitrary. That is, the convective rolls could be shifted horizontally by any amount. However, the system chooses a particular phase at the onset of convection. This symmetry breaking is akin to generation of a *Goldstone mode*. At a later r, secondary modes, e.g. U_{011}, U_{112}, are generated by nonlinear interactions, to be more precise via Hopf bifurcation. Generation of the secondary modes by this process has similarities with the Higgs mechanism.

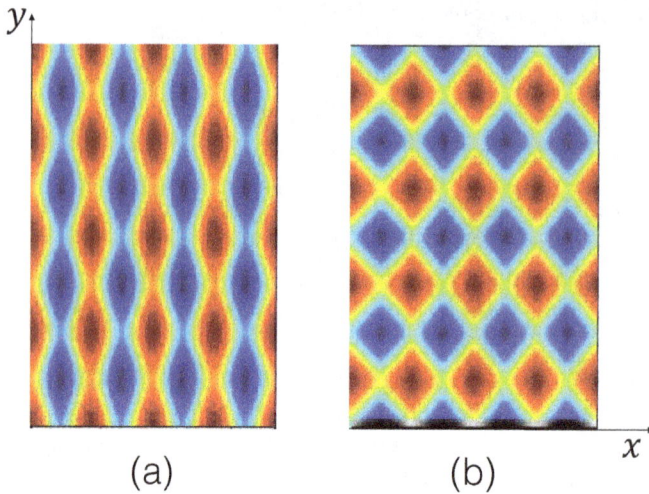

Fig. 8.3   A horizontal cross sections at the mid-plane exhibits ascending warm fluids (red color) and descending cold fluids (blue color). (a) The flow has primary rolls U_{101} aligned along y, and weak rolls U_{011} aligned along x. The resulting pattern is a set of *asymmetric squares*. (b) In the flow, the modes U_{101} and U_{011} have equal amplitudes, which leads to *symmetric squares*.

This feature is similar to the oscillatory behaviour observed in Example 4.2 of Chapter 4. As a result, the asymmetric squares of Fig. 8.3 flicker periodically in time. For illustration, see videos turbulencehub.org (2017).

For larger r's, RBC with Pr $\gtrsim$ 1 becomes chaotic. One such chaotic state of the 7-mode model for $r = 50$ and Pr $= 1$ is illustrated in Fig. 8.4. Here, we plot the phase space projections on $(U_{101}, \theta_{011}, -\theta_{002})$ and (U_{101}, U_{011}) subspaces. The trajectories are visually chaotic, and Fig. 8.4(a) is similar to the butterfly diagram (see Fig. 7.3). Here, we do not discuss characterisation of the chaotic orbits, as well as route to chaos, of the 7-mode model.

When we consider low-dimensional models with a larger number of modes, the flow becomes more complex. Paul *et al.* (2011) constructed a low-dimensional model with 14 complex and 2 real modes and analysed various states of RBC. Similar to the 7-mode model, they observed a supercritical pitchfork bifurcation at the transition $(r = 1)$, and then a supercritical Hopf bifurcation at $r \approx 27.6$. At $r \approx 40.3$, another Hopf bifurcation occurs yielding another frequency of oscillation[2]. These two frequencies generated by the Hopf bifurcations are incommensurate, and they lead to quasi-periodic motion in the phase space. The system becomes *chaotic* subsequently via *quasiperiodicity and phase locking* [Ecke *et al.* (1991); Ecke (1991); Paul *et al.* (2011)]. Figure 8.5 illustrates the bifurcation diagram in which we plot $|W_{101}|$, the magnitude of $[\mathbf{u}_{101}]_z$, vs. r. The figure clearly illustrates the sequence of bifurcations—supercritical pitchfork, supercritical Hopf, Neimark Sacker, and then

---

[2]The bifurcation, also called Nemark-Sacker bifurcation, is for 2D maps.

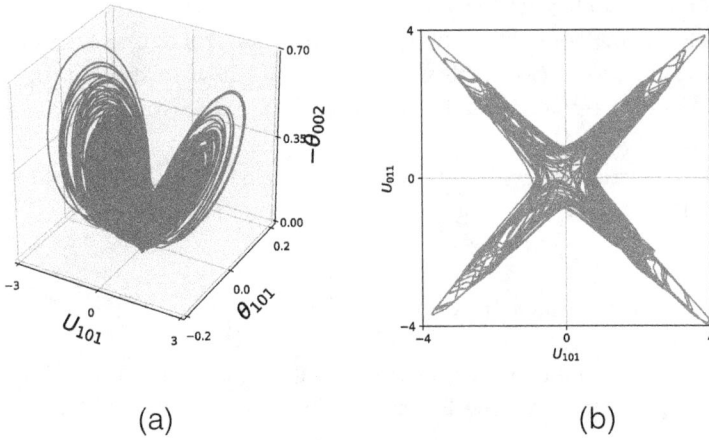

Fig. 8.4   For Pr = 1 and $r = 50$: (a) A phase space projection of the chaotic attractor on the $(U_{101}, \theta_{101}, -\theta_{002})$ subspace. (b) Phase space projection on the (U_{101}, U_{011}) subspace.

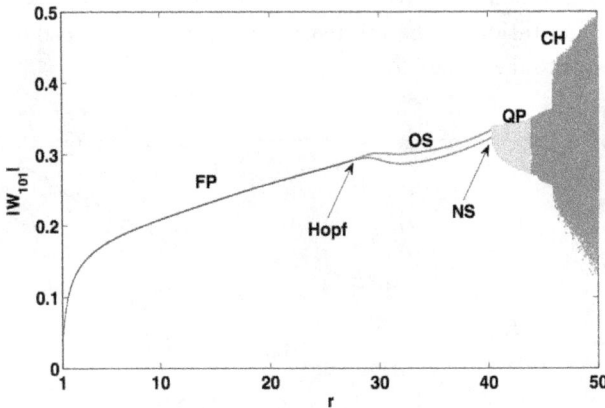

Fig. 8.5   Bifurcation diagram of Paul *et al.*'s low-dimensional model for large Pr-RBC. In the figure, FP, OS, QP, CH represent steady rolls, time-periodic rolls, quasiperiodic rolls, and chaotic rolls respectively, while NS indicates the NeimarkSacker bifurcation. From Paul *et al.* (2011). Reprinted with permission from Elsevier.

chaos. The flow patterns exhibited by the model are *primary steady rolls (FP)*, *time-periodic or oscillatory rolls (OS)*, *quasiperiodic rolls (QP)*, and *chaotic rolls (CH)*, which are labelled in Fig. 8.5. In this book, we will not detail the route to chaos in such systems, for which we refer the reader to Ecke *et al.* (1991); Ecke (1991); Paul *et al.* (2011).

For very large r and Pr ≈ 1, numerical simulations reveal that the modes U_{101} and U_{011} dominate the other modes of the flow. The rms values of the amplitudes of these modes are expected to be the approximately the same due to the xy symmetry of the system. See Fig. 8.4(b) as illustration. In a turbulent flow, the system jumps from U_{101}-dominant configuration to U_{101}-dominant configuration randomly. We will revisit this issue in Sec. 17.3 when we discuss flow reversals in a rectangular geometry. We also remark that for very large Pr, the system become quasi2D as shown in Fig. 8.2(c).

In the next section we will discuss some of the patterns and chaos of RBC with small Prandtl numbers.

8.3   Patterns and chaos in small-Pr RBC

Small-Pr convection tends to be more unstable than large-Pr convection due to the presence of vortical modes. The Fourier modes in small-Pr RBC have components along both $\hat{e}_1$ and $\hat{e}_2$ basis[3] [Pal *et al.* (2009)], unlike large-Pr convection in which the Fourier modes have components only along $\hat{e}_2$. In the following discussion, we will present the results based on low-dimensional models of Pal *et al.* (2009), Mishra *et al.* (2010), and Nandukumar and Pal (2016). They report very rich and complex behaviour for low-Pr RBC in the parameter range $1 < r < 1.35$. See Fig. 8.6 for an illustration of the bifurcation diagram for Pr $= 0.025$ and $1 \leq r \leq 1.25$ [Nandukumar and Pal (2016)].

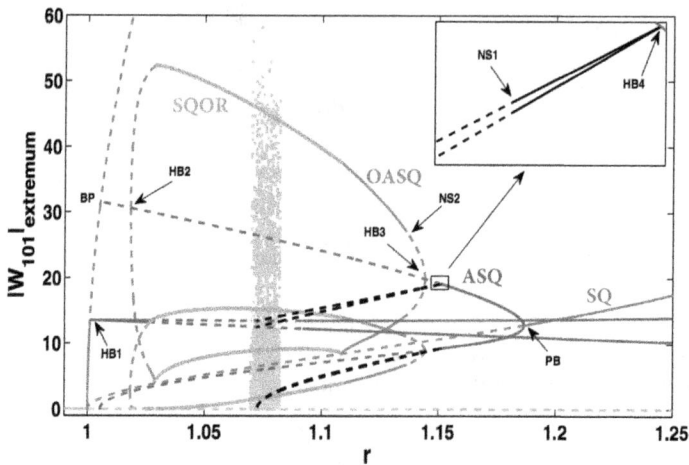

Fig. 8.6   For RBC with Pr $= 0.025$ and $1 \leq r \leq 1.25$: the bifurcation diagram of Nandukumar and Pal (2016)'s low-dimensional model. In the figure, PB, HBx, and NSx indicate the pitchfork, Hopf, and NeimarkSacker bifurcations respectively. The solid and dashed lines represent the stable and unstable fixed points respectively. The patterns are square (SQ), asymmetric square (ASQ), oscillating asymmetric square (OASQ), and relaxation oscillations of squares (SQOR). From Nandukumar and Pal (2016). Reprinted with permission from Elsevier.

---

[3]For $\mathbf{u} = u_2\hat{e}_2$ of the Craya-Herring decomposition, $\omega_z^{(2)} = u_2[i\mathbf{k} \times \hat{e}_2] \cdot \hat{z} = iku_2\hat{e}_1 \cdot \hat{z} = 0$. However, for $\mathbf{u} = u_1\hat{e}_1$, $\omega_z^{(1)} = u_1[i\mathbf{k} \times \hat{e}_1] \cdot \hat{z} = -iku_1\hat{e}_2 \cdot \hat{z} \neq 0$.

Some of the salient features of small-Pr RBC are:

(1) The generation of secondary modes via the Hopf bifurcation occurs at a much lower Rayleigh number (shown as HB's in Fig. 8.6). This is consistent with the fact that flows with small Pr are more dynamic than those with larger Pr.

(2) Due to the unstable nature, small-Pr convection exhibits a zoo of secondary bifurcations and patterns near the onset of convection itself. We illustrate some of these bifurcations using the low-dimensional model of Nandukumar and Pal (2016) for $1.02 \lessapprox r \lessapprox 1.2$ (see Fig. 8.6). These bifurcations are best understood when we study the bifurcation diagram backward (from large r to small r). Near $r = 1.2$, $U_{101} = U_{011}$ that yields a symmetric square pattern [see Fig. 8.3(b)]. This branch bifurcates to steady asymmetric square (ASQ) or stationary cross roll (CR) with $U_{101} \neq U_{011}$ via a pitchfork bifurcation (PB); here the Fourier modes are constant in time. Later, the ASQ branch is subjected to a Hopf bifurcation (HB4), and two Neimark-Sacker bifurcations, NS1, and NS2. After this, an oscillating asymmetric square (OASQ) and relaxation oscillation of squares (SQOR) appear. In Fig. 8.7 we illustrate the phase space projections of OASQ and SQOR states on the W_{101}-W_{101} plane. Note that $W_{101} = [\mathbf{u}_{101}]_z$. Also see videos of these states in turbulencehub.org (2017).

(3) Pal *et al.* (2009), Mishra *et al.* (2010), and Nandukumar and Pal (2016) argue that after the emergence of SQOR, the system becomes *chaotic via homoclinic gluing*. Refer to Pal *et al.* (2009) for further details on the chaotic states.

(4) RBC with low Prandtl number also exhibits chaos for large r's (e.g., $r > 1.2$ in Fig. 8.6) via another set of secondary bifurcations. These regimes have not been studied in detail. Fauve *et al.* (1981) performed convection experiments with mercury to study chaos in low-Pr RBC. They reported period-doubling route to chaos for such systems. Note however that Fauve *et al.* (1981) varied r in the neighbourhood of 3.5 to 3.65, which is far beyond those studied by Pal *et al.* (2009), Mishra *et al.* (2010), and Nandukumar and Pal (2016).

In Fig. 8.8, we summarise the results on patterns and chaos in RBC. For a moderate and large Prandtl numbers, a sequence of bifurcations lead to steady rolls → oscillating rolls → double frequency → chaos [Krishnamurti (1970a,b)]. However, for RBC with small Prandtl number, the system exhibits steady rolls → oscillating rolls, and then to chaos with patterns interspersed in between. Due to unstable nature of flow, instability and turbulence appear for small-Pr convection at lower Ra compared to large-Pr convection.

In the next section we will discuss pattern formation using amplitude equation.

8.4 Understanding patterns using amplitude equation

Thermal convection exhibits a large number of patterns—square, asymmetric square or cross-roll, hexagons, zig-zag knot, oscillatory blob, spiral defect, etc. [Krishnamurti (1970a,b); Getling (1998); Lappa (2010)]. They arise due to nonlinear

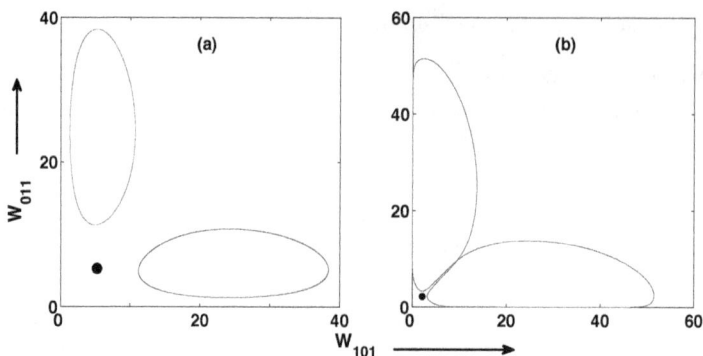

Fig. 8.7   For Pr $= 0$, the phase space projections of limit cycles on the W_{101}-W_{011} planes for (a) r $= 1.0494$ and (b) r $= 1.0099$. The limit cycles in (a) exhibit oscillatory asymmetric squares (OASQ). They merge to form a single limit cycle in (b). Black dots indicate the symmetric square saddle. From Pal *et al.* (2009). Reprinted with permission from Institute of Physics.

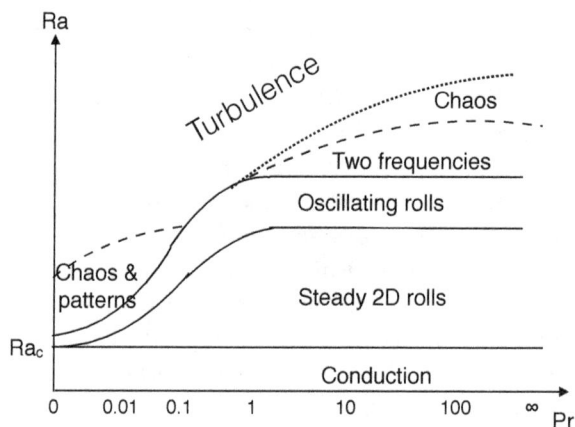

Fig. 8.8   A schematic diagram of RBC states as a function of Ra and Pr.

interactions among the modes. For example, hexagonal patterns involve interactions among Fourier modes $(1/2, \sqrt{3}/2)$, $(1, 0)$, $(-1/2, \sqrt{3}/2)$. In this book we will not discuss these patterns.

In Sec. 7.4 we introduced amplitude equation and showed how it captures bifurcations and nonlinear saturation in RBC. In this section, we briefly describe how amplitude equations help us understand formation of various patterns in RBC. For details refer to Cross and Greenside (2009), Fauve (1998), and Hoyle (2006).

The amplitude equations take advantage of the symmetry properties of the system. Hoyle (2006) showed that a square pattern is obtained by the following amplitude equations:

$$\frac{dz_1}{dt} = \mu z_1 - a_1|z_1|^2 z_1 - a_2|z_2|^2 z_1, \tag{8.29}$$

$$\frac{dz_2}{dt} = \mu z_2 - a_1|z_2|^2 z_2 - a_2|z_1|^2 z_2, \tag{8.30}$$

where z_1, z_2 (complex variables) are the amplitudes of the perpendicular rolls that form square patterns, and μ, a_1, a_2 are real coefficients. Though, the above equations yield square patterns, they cannot describe bimodal solution or asymmetric squares [Hoyle (2006)]. Using symmetry arguments, Hoyle (2006) also showed that the amplitudes equations for generating hexagonal patterns are

$$\frac{dz_1}{dt} = \mu z_1 + a\bar{z}_2\bar{z}_3 - b|z_1|^2 z_1 - c(|z_2|^2 + |z_3|^2)z_1, \tag{8.31}$$

$$\frac{dz_2}{dt} = \mu z_2 + a\bar{z}_3\bar{z}_1 - b|z_2|^2 z_2 - c(|z_3|^2 + |z_1|^2)z_2, \tag{8.32}$$

$$\frac{dz_3}{dt} = \mu z_3 + a\bar{z}_1\bar{z}_2 - b|z_3|^2 z_3 - c(|z_1|^2 + |z_2|^2)z_3, \tag{8.33}$$

where a, b, c, d are constants.

Amplitude equations have been used to derive host of patterns, e.g., Eckhaus, zig-zag, spiral, defects, etc. These patterns are schematically represented using Busse balloon [Cross and Greenside (2009); Hoyle (2006); Lappa (2010)]. Using amplitude equation, Knobloch (1992) showed that the zero-Pr convection and finite-Pr convection have very different behaviour, consistent with the discussion of the previous section. Due to lack of space, we will not discuss pattern formation using amplitude equations.

Amplitude equations are primarily based on symmetry arguments, hence they are somewhat simpler to derive than Galerkin-based low-dimensional models. However, computation of transitional parameters such as critical Rayleigh number, as well as definitive predictions on the emergence of patterns and chaos, are impractical using amplitude equation. Galerkin models have advantages in this front, and they can predict when the patterns actually emerge. For example, the nature of patterns and their emergence predicted by the Galerkin models of Pal *et al.* (2009) match quite closely with simulation results.

With this we close our discussion on the patterns and chaos in RBC near the onset of convection. In Chapter 17 we will revisit them in the turbulent regime. In the next chapter, we will discuss the boundary layers and exact relations of RBC.

Further reading

For derivation of Galerkin truncation of fluid equations, refer to textbooks by Lesieur (2008) and Pope (2000). We refer the reader to Paul *et al.* (2011) for low-dimensional model of large-Pr convection, and Pal *et al.* (2009) and Nandukumar and Pal (2016) for models of zero- and small-Pr convection. Strogatz (2014) provides introduction to bifurcation analysis. Applications of amplitude equations to RBC are covered in Manneville (2004), Fauve (1998), Getling (1998), and Lappa (2010). For discussion on various patterns in RBC, refer to Krishnamurti (1970a,b), Getling (1998), Lappa (2010), and references therein.

Exercises

(1) Derive all the equations of the seven-mode model.
(2) Compute the critical r_c where the first Hopf bifurcation in the seven-mode model takes place.
(3) Study behaviour of the seven-mode model for various r's. How does the system become chaotic?
(4) Verify that the nonlinear terms of the 7-mode model satisfy appropriate conservation laws. For example, show that $\sum |\mathbf{u}(\mathbf{k})|^2$ is conserved in the absence of buoyancy and viscosity.
(5) Take the limiting case of $\mathrm{Pr} = \infty$ in the seven-mode model. Study the properties of the new model that has only 4 temperature modes.
(6) Construct low-dimensional model(s) for small-Pr RBC.
(7) What are the qualitative differences between the small-Pr and large-Pr convection?
(8) What are the advantages and disadvantages of amplitude equations and Galerkin-based low-dimensional models? Contrast the two approaches.

Chapter 9

Boundary Layer, Bulk, and Exact Relations in Thermal Convection

In RBC, a fluid is confined between two thermal plates, around which viscous and thermal boundary layers are formed. The region between the top and bottom boundary layers contains the *bulk flow*. In this chapter we will describe the properties of the boundary layers and temperature profile of RBC. We will also state several exact relations of RBC.

9.1 Viscous and thermal boundary layers

If the temperature of the top and bottom plates of RBC are T_t and T_b respectively, then near the bottom plate, the temperature drops from T_b to $T(z) = (T_b + T_t)/2$, which is the temperature of most of the bulk fluid. The temperature drops further near the top plate to T_t. Here $T(z) = \langle T \rangle_{xy}$ is the planar-averaged temperature at the height z. The regions near the plates where the temperature drops steeply are called *thermal boundary layers*. In Fig. 9.1(a), we illustrate $T(z)$ vs. z for a no-slip RBC. The thickness of the thermal boundary layer is denoted by δ_T.

The velocity field at the thermal plates is zero due to the no-slip boundary condition at the plates. The flow in the bulk however is dominated by the large-scale circulation (LSC, detailed in Chapter 17). Thus, the velocity field rises from zero at the plates to that of the LSC in a short distance called *viscous boundary layer*, which is denoted by δ_u. In Fig. 9.1(b) we illustrate plane-averaged rms value of $u_x^{\mathrm{rms}}(z)$, $u_y^{\mathrm{rms}}(z)$, $u_z^{\mathrm{rms}}(z)$ for a no-slip RBC simulation. Note that $u_i^{\mathrm{rms}}(z = 0) = 0$. The thickness of the viscous boundary layer, δ_u, is typically given by [Landau and Lifshitz (1987)]

$$\delta_u = \frac{d}{\sqrt{\mathrm{Re}}}. \tag{9.1}$$

This is the prediction of Prandtl-Blasius theory. Several experiments and numerical simulations reveal that δ_u in RBC differs marginally from that of Eq. (9.1). We will discuss these issues briefly in Sec. 11.6.

For the free-slip boundary condition, the profiles of temperature $T(z)$ and $u_z^{\mathrm{rms}}(z)$ are similar to those for no-slip boundary condition. However, $u_x^{\mathrm{rms}}(z)$, $u_y^{\mathrm{rms}}(z)$ are nonzero at the plates due to the free-slip boundary condition.

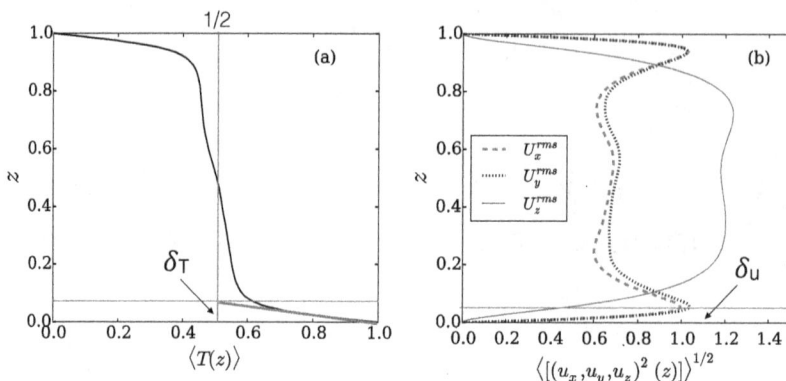

Fig. 9.1   For RBC with Pr $= 1$, Ra $= 10^7$ and no-slip boundary condition, plots of planar-averaged (a) $T(z)$, and (b) $U_i^{\mathrm{rms}}(z)$ vs. z, where $i = x, y, z$. At the walls $\mathbf{u} = 0$ and $T = 1, 0$. The regions where the velocity and temperate fields change abruptly are called the viscous and thermal boundary layers respectively. In the figure we depict the thickness of the thermal boundary layer, δ_T, and that of viscous boundary layer, δ_u.

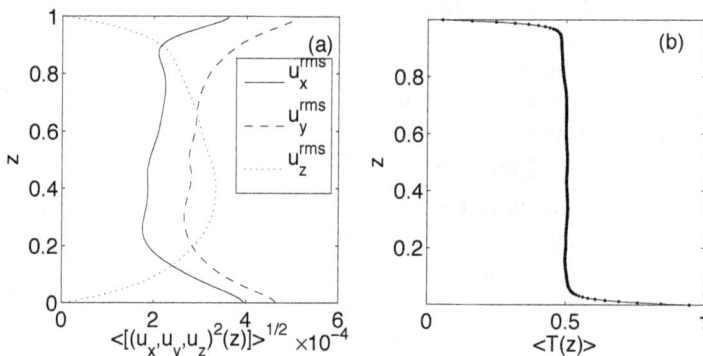

Fig. 9.2   For RBC with Pr $= 6.8$, Ra $= 6.6 \times 10^6$, and free-slip boundary condition, plots of planar-averaged (a) $U_i^{\mathrm{rms}}(z)$, and (b) $T(z)$. At the walls $u_z^{\mathrm{rms}} = 0$ and $T = 1, 0$. Note that u_x, u_y are nonzero at the plates due to the free-slip boundary condition. From Mishra and Verma (2010). Reprinted with permission from APS.

The velocity field in the boundary layers is small, hence the heat transport in the boundary layer is primarily due to conduction. In each of the thermal boundary layers, the temperature drops by $\approx \Delta/2$. Therefore, in the thermal boundary layer,

$$H_{\text{total}} \approx H_{\text{cond}} \approx \kappa \frac{\Delta/2}{\delta_T}. \tag{9.2}$$

Hence the Nusselt number is

$$\text{Nu} \approx \frac{H_{\text{total}}}{\kappa \Delta / d} = \frac{d}{2\delta_T}. \tag{9.3}$$

Therefore we can estimate the thickness of the thermal boundary layer as

$$\delta_T = \frac{d}{2\mathrm{Nu}}. \tag{9.4}$$

Note that the relations of Eqs. (9.1, 9.4) work well for large Ra for which the boundary layers are thin. In the turbulent regime, i.e., for large Ra or Re, $\delta_T \ll d$ and $\delta_u \ll d$. Note that in general, $\delta_T \neq \delta_u$, specially when Pr is either too large or too small.

The physics of viscous and thermal boundary layers in RBC is quite complex, and it remains primarily unresolved. In the present book, we will not cover the boundary layers of RBC in detail. We refer the reader to recent articles by Shi *et al.* (2012), Li *et al.* (2012), van der Poel *et al.* (2015), Schumacher *et al.* (2016), and references therein.

In the next section we will study the mean temperature profile in turbulent convection.

9.2 Temperature profile in RBC

In this section we derive the properties of temperature fluctuations in RBC. For convenience we work with nondimensional variables.

For analytic treatment, we model the temperature profile in the boundary layer to be linear, and the temperature in the bulk to be constant, $T_m = 1/2$, as shown in Fig. 9.3. Therefore, $T_m(z) = \langle T \rangle_{xy}$ is approximated as

$$T_m(z) = \begin{cases} 1 - \frac{z}{2\delta_T} & \text{for } 0 < z < \delta_T \\ 1/2 & \text{for } \delta_T < z < 1 - \delta_T \\ \frac{1-z}{2\delta_T} & \text{for } 1 - \delta_T < z < 1. \end{cases} \tag{9.5}$$

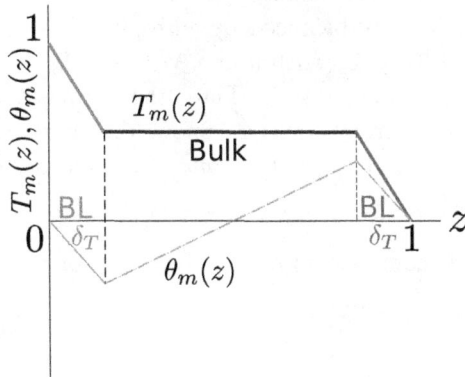

Fig. 9.3   A schematic diagram of the planar-averaged temperature $T_m(z)$ and $\theta_m(z)$. From Pandey and Verma (2016). Reprinted with permission from AIP.

A horizontal averaging of

$$T(x, y, z) = T_c(z) + \theta(x, y, z) = 1 - z + \theta(x, y, z) \tag{9.6}$$

yields

$$\theta_m(z) = \langle \theta(x, y, z) \rangle_{xy} = T_m(z) + z - 1$$

$$= \begin{cases} z \left(1 - \frac{1}{2\delta_T}\right) & \text{for } 0 < z < \delta_T \\ z - 1/2 & \text{for } \delta_T < z < 1 - \delta_T \\ (z-1)\left(1 - \frac{1}{2\delta_T}\right) & \text{for } 1 - \delta_T < z < 1, \end{cases} \tag{9.7}$$

as exhibited in Fig. 9.3. Note that $T_c(z)$ is the temperature profile for the pure conduction state. The Fourier transform of $\theta_m(z)$ is

$$\theta_m(0, 0, k_z) = \int_0^1 \theta_m(z) \sin(k_z \pi z) dz. \tag{9.8}$$

For thin boundary layers, we can ignore the contributions from the boundary layers, and hence

$$\theta_m(0, 0, k_z) \approx \int_0^1 (z - 1/2) \sin(k_z \pi z) dz$$

$$\approx \begin{cases} -\frac{1}{\pi k_z} & \text{for even } k_z \\ 0 & \text{otherwise.} \end{cases} \tag{9.9}$$

The Fourier modes $\theta_m(0, 0, k_z)$ with amplitudes of $-1/(\pi k_z)$ play an important role in making the temperature profile shown in Fig. 9.3. For $k_z = 2$, the above equation yields

$$\theta_m(0, 0, 2) \approx -\frac{1}{2\pi}, \tag{9.10}$$

which is same as that derived for the Lorenz model and the 7-mode model discussed in Chapters 7 and 8. We showed that the entropy feed to the mode $\theta(1, 0, 1)$ by buoyancy is approximately balanced by entropy loss to $\theta(0, 0, 2)$; this process leads to $\theta_m(0, 0, 2) \approx -1/(2\pi)$. Mishra and Verma (2010) generalised the above arguments to $\theta(n, 0, n)$ and $\theta(0, 0, 2n)$. Thus, the temperature profile of RBC is intricately related to the entropy transfers among the large-scale modes.

The velocity mode, $u_z(0, 0, k_z) = 0$ because of the incompressibility condition:

$$\mathbf{k} \cdot \mathbf{u}(0, 0, k_z) = k_z u_z(0, 0, k_z) = 0. \tag{9.11}$$

Regarding the x and y components of $\mathbf{u}(0, 0, k_z)$, a nonzero value of $u_x(0, 0, k_z)$ would yield

$$u_x(0, 0, z) \sim \cos k_z z \tag{9.12}$$

that implies a mean horizontal flow along $\hat{x}$ at a given z. RBC flows do not have such horizontal velocity, hence

$$u_x(0, 0, k_z) = 0. \tag{9.13}$$

Similar arguments yields

$$u_y(0, 0, k_z) = 0. \tag{9.14}$$

Thus, $\mathbf{u}(0, 0, k_z) = 0$. Therefore, for wavenumber $\mathbf{k} = (0, 0, k_z)$, the RBC equation for the velocity field, Eq. (3.8), yields

$$\frac{\partial}{\partial t} u_z(\mathbf{k}) = 0 = -\frac{ik_z \sigma(\mathbf{k})}{\rho_m} + \alpha g \theta(\mathbf{k}) \tag{9.15}$$

that translates to the following equation in real space:

$$\frac{d}{dz} \sigma_m(z) = \rho_m \alpha g \theta_m(z), \tag{9.16}$$

which is essentially the force balance between the static pressure and the gravitation force. The dynamics of the remaining set of Fourier modes is governed by the following equation:

$$\frac{\partial \mathbf{u}(\mathbf{k})}{\partial t} + i \sum_{p+q=k} [\mathbf{k} \cdot \mathbf{u}(\mathbf{q})]\mathbf{u}(\mathbf{p}) = -\frac{ik \sigma_{\mathrm{res}}(\mathbf{k})}{\rho_m} + \alpha g \theta_{\mathrm{res}}(\mathbf{k})\mathbf{z} - \nu k^2 \hat{\mathbf{u}}(\mathbf{k}), \tag{9.17}$$

where the *residual temperature*, $\theta_{\mathrm{res}}(\mathbf{k})$, and the *residual pressure*, $\sigma_{\mathrm{res}}(\mathbf{k})$, are the fluctuations excluding the modes $(0, 0, k_z)$. Hence, the modes $\theta_m(0, 0, k_z)$ and $\sigma_m(0, 0, k_z)$ do not couple with the velocity modes in the momentum equation, but $\theta_{\mathrm{res}}(\mathbf{k})$ and $\sigma_{\mathrm{res}}(\mathbf{k})$ do.

Inverse transforms of $\theta_{\mathrm{res}}(\mathbf{k})$ and $\sigma_{\mathrm{res}}(\mathbf{k})$ yield $\theta_{\mathrm{res}}(\mathbf{r})$ and $\sigma_{\mathrm{res}}(\mathbf{r})$ respectively. Note that

$$\theta(\mathbf{r}) = \theta_{\mathrm{res}}(\mathbf{r}) + \theta_m(z); \quad \sigma = \sigma_{\mathrm{res}}(\mathbf{r}) + \sigma_m(z). \tag{9.18}$$

When we take a planar average of Eq. (9.6)

$$T(x, y, z) = T_c(z) + \theta(x, y, z), \tag{9.19}$$

we obtain the mean temperature as

$$T_m(z) = T_c(z) + \theta_m(z) = 1 - z + \theta_m(z)$$

$$= 1 - z + \sum_{l=1}^{N} \theta(0, 0, 2l) 2 \sin(2l\pi z) \tag{9.20}$$

with $N \to \infty$, and

$$T(x, y, z) = T_m(z) + \theta_{\mathrm{res}}(x, y, z). \tag{9.21}$$

Hence, $\theta_{\mathrm{res}}(x, y, z)$ is the fluctuation over $T_m(z)$. The experiments measure the temperature fluctuations as $T(x, y, z) - T_m(z)$, which is same as $\theta_{\mathrm{res}}(x, y, z)$ (see Niemela *et al.* (2000)). Thus, $\theta_{\mathrm{res}}(\mathbf{r})$ is a realistic measure of temperature fluctuations in RBC.

In Fig. 9.4 we depict approximation $T_m(z)$ of Eq. (9.20) for various $N = 1, 2, 4,$ 10. For $N = 10$, $T_m(z)$ resembles the real $T_m(z)$ profile, as shown in Fig. 9.4. As shown in the figure, $T_m(z) \approx 1/2$ only in the bulk, but it exhibits sharp variations in the boundary layers, as in realistic RBC. The modes $\theta_m(0, 0, k_z)$ thus construct

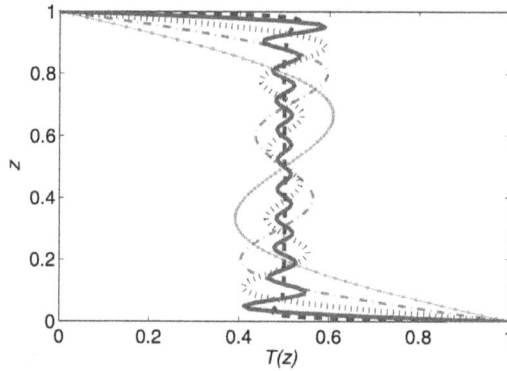

Fig. 9.4 Plot of $T_m(z) = T_c(z) + \theta_m(z) = 1 - z + \theta_m(z)$ with $\theta_m(z) = \sum_{l=1}^{N} \theta(0,0,2l)2\sin(2l\pi z)$ for $N = 1$ (red small-dots), $N = 2$ (sky-blue dash-dot), $N = 4$ (black bigger-dots), and $N = 10$ (brown solid curve) respectively. The curve for $N = 10$ appears as realistic $T_m(z)$. From Pandey *et al.* (2014). Reprinted with permission from APS.

the complex $T_m(z)$. Note that $\theta_m(0,0,2n)$ also couple nonlinearly with θ_{res}, e.g., in 7-mode model, $\theta(1,0,1)$ couples with $\theta(0,0,2)$. Hence, $\theta_m(0,0,2n)$ is not like a mean constant temperature.

From Eq. (9.17), we derive the following equation for the velocity field in the real space:

$$\frac{\partial \mathbf{u}}{\partial t} + (\mathbf{u} \cdot \nabla)\mathbf{u} = -\nabla \sigma_{\mathrm{res}} + \alpha g \theta_{\mathrm{res}}\hat{z} + \nu \nabla^2 \mathbf{u}, \qquad (9.22)$$

that is an equation for the temperature and velocity fluctuations. Equation (9.22) has major consequences on the scaling of the Reynolds and Nusselt numbers, which will be discussed in Chapter 11. In addition, the relationship $\theta(0,0,k_z) \sim -1/(\pi k_z)$ of Eq. (9.9) yields $E_\theta(k) \sim k^{-2}$, which will be discussed in Sec. 13.3.

The aforementioned results indicates that the walls affect the bulk flow in a significant way. In RBC, all the modes in the flow work together so as to maintain $\theta(0,0,2n) \approx -1/(2n\pi)$. The above feature appears to be generic, and similar feature should also hold for channel, Couette, and Taylor-Couette flows where the mean velocity has nontrivial profile. In channel flow, for example, the fluctuations and the mean flow work together to maintain the D shape of the mean velocity.

In the next section, we will discuss Shraiman and Siggia (1990)'s exact relations for RBC.

9.3   Exact relations of RBC

RBC equations are nonlinear, and there is no general analytic solution for them. However, Shraiman and Siggia (1990) derived the following exact relations that relate the Nusselt number, Nu, with the viscous dissipation rates, ϵ_u, and entropy dissipation rate, ϵ_θ:

$$\epsilon_u = \frac{\nu^3}{d^4} \frac{(\text{Nu} - 1)\text{Ra}}{\text{Pr}^2}, \tag{9.23}$$

$$\epsilon_\theta = \kappa \frac{\Delta^2}{d^2}(\text{Nu} - 1), \tag{9.24}$$

where ν, κ are respectively the kinematic viscosity and thermal diffusivity of the fluid, Δ is the temperature difference between the two horizontal plates that are placed at a distant d apart. At the plates, we assume no-slip boundary condition for the velocity field, and conducting boundary condition for the temperature field. Here we reproduce the derivation of Shraiman and Siggia (1990).

We start with Eq. (2.87) for the kinetic energy of RBC:

$$\frac{\partial}{\partial t} \frac{u^2}{2} + \nabla \cdot \left[\frac{u^2}{2} \mathbf{u} \right] = -\nabla \cdot (\sigma \mathbf{u} - \nu \mathbf{u} \times \boldsymbol{\omega}) + \alpha g \theta u_z - \nu \omega^2. \tag{9.25}$$

We perform a volume integration of the above equation over the box. We also use the following identities. Under a statistical steady state,

$$\partial_t \int d\mathbf{r} \frac{\langle u^2 \rangle}{2} = 0, \tag{9.26}$$

where $\langle . \rangle$ denotes the volume average. For the periodic or no-slip boundary conditions,

$$\int d\mathbf{r} \nabla \cdot \left[\frac{u^2}{2} \mathbf{u} \right] = \int \frac{u^2}{2} \mathbf{u}.d\mathbf{S} = 0, \tag{9.27}$$

$$\int d\mathbf{r} \nabla \cdot (\sigma \mathbf{u} - \nu \mathbf{u} \times \boldsymbol{\omega}) = \int [\sigma \mathbf{u} - \nu \mathbf{u} \times \boldsymbol{\omega}].d\mathbf{S} = 0, \tag{9.28}$$

where $d\mathbf{S}$ is the elemental surface area. Using these relations, we connect the average viscous dissipation rate with the energy feed by buoyancy as

$$\langle \epsilon_u \rangle = \frac{1}{\text{Vol}} \int d\mathbf{r} \nu \omega^2 = \frac{1}{\text{Vol}} \int d\mathbf{r} \alpha g u_z \theta = \alpha g \langle u_z \theta \rangle, \tag{9.29}$$

where Vol is the volume of the box.

Recall that the volume-averaged Nusselt number is defined as (see Sec. 2.9)

$$\text{Nu} = 1 + \frac{\langle u_z \theta \rangle}{\kappa(\Delta/d)}, \tag{9.30}$$

substitution of which in Eq. (9.29) yields

$$\begin{aligned} \langle \epsilon_u \rangle &= \alpha g (\text{Nu} - 1)\kappa \frac{\Delta}{d} \\ &= \frac{\nu^3}{d^4} \frac{(\text{Nu} - 1)\text{Ra}}{\text{Pr}^2}. \end{aligned} \tag{9.31}$$

This is the first exact relation of Shraiman and Siggia (1990). In our future discussion, we drop the angular brackets of $\langle \epsilon_u \rangle$ for convenience.

To derive the second exact relation of Shraiman and Siggia, we start with Eq. (2.92) for the entropy:

$$\frac{\partial}{\partial t}\frac{\theta^2}{2} + \nabla \cdot \left[\frac{\theta^2}{2}\mathbf{u} - \kappa\theta\nabla\theta \right] = \frac{\Delta}{d}\theta u_z - \kappa(\nabla\theta)^2. \tag{9.32}$$

We integrate the above equation over the volume. Again, under a steady state

$$\partial_t \int d\mathbf{r}\frac{\theta^2}{2} = 0, \tag{9.33}$$

and for the periodic or no-slip boundary condition for the velocity field, and for the conducting or insulating boundary condition for the temperature field,

$$\int d\mathbf{r}\nabla \cdot \left[\frac{\theta^2}{2}\mathbf{u} - \kappa\theta\nabla\theta \right] = \int \left[\frac{\theta^2}{2}\mathbf{u} - \kappa\theta\nabla\theta \right].d\mathbf{S} = 0. \tag{9.34}$$

Using these relations, we relate the averaged entropy dissipation rate and the entropy supply rate by buoyancy:

$$\langle \epsilon_\theta \rangle = \frac{1}{\text{Vol}} \int d\mathbf{r}\kappa(\nabla\theta)^2 = \frac{1}{\text{Vol}} \int d\mathbf{r}\frac{\Delta}{d}u_z\theta = \frac{\Delta}{d}\langle u_z\theta \rangle. \tag{9.35}$$

Using Eq. (9.30) we obtain

$$\langle \epsilon_\theta \rangle = \frac{\kappa\Delta^2}{d^2}(\text{Nu} - 1). \tag{9.36}$$

In terms of T, using

$$\nabla T = \nabla\theta - \frac{\Delta}{d}\hat{z}, \tag{9.37}$$

we obtain

$$\epsilon_T = \kappa\langle(\nabla T)^2\rangle = \frac{\kappa\Delta^2}{d^2}\text{Nu}. \tag{9.38}$$

This is the second exact relation of Shraiman and Siggia.

In the next section, we present the conservation laws of RBC.

9.4   Conservation laws

We derived the conservation laws in Secs. 2.8 and 4.1. Here we state these results because they belong to exact relations. The total kinetic energy, $\int d\mathbf{r} u^2/2$, and total entropy, $\int d\mathbf{r}\theta^2/2$, are not conserved for RBC due to the exchange of kinetic energy and entropy via buoyancy. However, as shown in Sec. 2.8,

$$E_u + \frac{\alpha g}{d\bar{T}/dz}E_\theta, \quad \text{or} \quad E_u - \frac{\alpha g}{|d\bar{T}/dz|}E_\theta \tag{9.39}$$

is conserved for periodic and vanishing boundary conditions.

The corresponding conservation law for the stably stratified flow is that the total energy

$$E = \frac{1}{2}\int \left(\frac{u^2}{2} + \frac{b^2}{2} \right) d\mathbf{r} \tag{9.40}$$

is conserved for periodic and vanishing boundary conditions. Here $b^2/2$ is interpreted as the potential energy. See Sec. 2.8 for the derivation.

With this, we close our discussion on part A of this book that covers basic formulation, waves and instabilities, patterns and chaos, and some exact relations. In part B, we will cover the properties of turbulent buoyant flows.

Further reading

Refer to Shraiman and Siggia (1990) for the exact relations of RBC. For a more detailed and recent account of boundary layers in RBC, refer to Li *et al.* (2012) and van der Poel *et al.* (2015).

Exercises

(1) Contrast the conservation laws of stably stratified flows and RBC.
(2) How does the flow profile of Fig. 9.3 change under the variation of Prandtl number?
(3) How does the nature of thermal and viscous boundary layers change under the variation of Prandtl number?

PART 2

Turbulent Buoyant Flows

This part covers the properties of turbulent buoyant flows, primarily

(1) Scaling of large-scale quantities in RBC
(2) Turbulence phenomenologies of stably stratified flows
(3) Turbulence phenomenologies of RBC and related systems
(4) Anisotropy in RBC
(5) Large-scale structures and flow-reversals in RBC

Note however that Chapter 11 has discussions on the scaling of large-scale quantities in laminar flows.

Chapter 10

A Survey of Hydrodynamic Turbulence

In this book we focus on the present understanding of buoyancy-driven turbulence. The present models of buoyancy-driven turbulence borrow concepts from hydrodynamic turbulence, for example the energy flux. Buoyancy-driven turbulence involve the density or temperature, hence, it has similarities with passive scalar turbulence. Due to these reasons, in this chapter we review some of the important ideas of hydrodynamic and passive scalar turbulence.

10.1 Kolmogorov's theory for 3D hydrodynamic turbulence

Incompressible fluid flows are descried by Navier-Stokes equation and the incompressibility condition:

$$\frac{\partial \mathbf{u}}{\partial t} + (\mathbf{u} \cdot \nabla)\mathbf{u} = -\nabla(p/\rho) + \nu\nabla^2\mathbf{u} + \mathbf{f}_u, \qquad (10.1)$$

$$\nabla \cdot \mathbf{u} = 0, \qquad (10.2)$$

where $\mathbf{u}$, ρ, p are the velocity, density, and pressure fields, $\mathbf{f}_u$ is the external force field, and ν is the viscosity. The flow becomes turbulent when $\mathrm{Re} = UL/\nu \gg 1$. For convenience, we take $\rho = 1$. The corresponding equations in Fourier space are same as Eqs. (3.8, 3.10) but without the buoyancy term.

Turbulence is a complex problem. However, notable simplification is achieved when we focus on hydrodynamic flows away from the walls, and with no forces other than pressure gradient and viscous drag. Kolmogorov (1941a) and Kolmogorov (1941b) assumed that such flow are statistically homogeneous and isotropic, an assumption that has been validated in experiments and numerical simulations away from the walls, and then derived various properties of turbulent flow.

Kolmogorov's theory of turbulence has been presented in different ways; here we describe this phenomenology using energetics arguments. In Sec. 4.3 we derived the time-evolution of the energy spectrum $E(k)$ as

$$\frac{d}{dt}E_u(k) = -\frac{d}{dk}\Pi_u(k) + \mathcal{F}(k) - D_\nu(k), \qquad (10.3)$$

where $\Pi_u(k)$ is the energy flux, and

$$\mathcal{F}(k) = \sum_{k-1<k'\leq k} \Re[\mathbf{f}_u(\mathbf{k'}) \cdot \mathbf{u}^*(\mathbf{k'})], \tag{10.4}$$

$$D_u(k) = 2\nu \sum_{k-1<k'\leq k} k'^2 E_u(\mathbf{k'}), \tag{10.5}$$

are respectively the energy supply rate by external force $\mathbf{f}_u$, and the viscous dissipation rate.

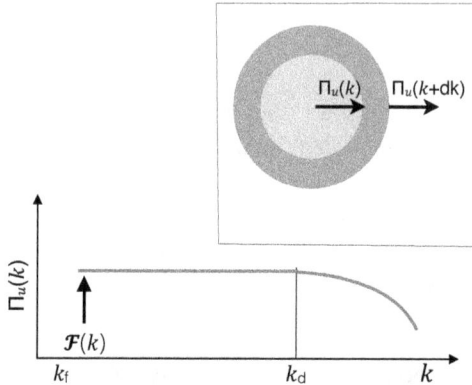

Fig. 10.1   Schematic diagrams of the kinetic energy flux $\Pi_u(k)$ for hydrodynamic turbulence. The energy supplied $\mathcal{F}(k)$ at small k cascades to intermediate wavenumbers, and then to the dissipative wavenumbers. In the intermediate or inertial range, $\Pi_u(k) = \text{const}$.

The key ingredients of Kolmogorov's theory are as follows:

(1) The external force $\mathbf{f}_u$ is active at large length scales (of the order of system size), i.e. for $k = k_f \sim 1/L$. Hence $\mathcal{F}(k)$ is nonzero only at small k.
(2) The kinematic viscosity ν is very small, or $\mathrm{Re} = UL/\nu \to \infty$. Hence, $D_u(k)$ is significant only at large k due to the k^2 factor.
(3) The energy supplied by $\mathbf{f}_u$ cascades to intermediate scales and then to small scales where the energy is dissipated. The wavenumber range of forcing and dissipation are called *forcing range* $(0 < k < 1/L)$ and *dissipation range* $(k > k_d = 1/l_d)$ respectively, where k_d is the dissipation wavenumber, and l_d is the dissipation length. The intermediate wavenumber band between the forcing range and dissipation range is called *inertial range*, where the nonlinear term and the pressure gradient play a dominant role.
(4) In the early stages, $E_u(k) \approx 0$ at the intermediate and small scales. Nonlinear interactions quickly transfer energy to these scales from the forcing range. After these transients, the flow reaches a quasi steady state in which the net energy supply rate by the external force balances the total dissipation rate (ϵ_u).

Under a steady state, or $dE_u(k)/dt \approx 0$. In the inertial range, $\mathcal{F}(k) = 0$ and $D_u(k) \to 0$, hence using Eq. (10.3) we deduce that

$$\frac{d}{dk}\Pi_u(k) \approx 0 \qquad (10.6)$$

or, as illustrated in Fig. 10.1,

$$\Pi_u(k) = \Pi_u = \text{const}, \qquad (10.7)$$

and it equals the total dissipation rate ϵ_u:

$$\epsilon_u = 2\nu \int_0^\infty k^2 E_u(k) dk. \qquad (10.8)$$

Now, using dimensional analysis, one can derive the one-dimensional kinetic energy spectrum as

$$E_u(k) = K_{\text{Ko}}\Pi_u^{2/3}k^{-5/3}, \qquad (10.9)$$

where K_{Ko} is the *Kolmogorov's constant*. Numerical simulations, experiments, and analytical computations report that $K_{\text{Ko}} \approx 1.6$. Note that Eqs. (10.7, 10.9) provide a *universal* description of hydrodynamic turbulence, and they are applicable irrespective of the boundary condition, initial condition, dissipation and forcing mechanisms. The aforementioned energy spectrum and energy flux have been observed in experiments and numerical simulations. See for example, Figs. 10.2 and 10.3 where we exhibit $E_u(k)$ and $\Pi_u(k)$ computed using numerical simulation on 4096^3 grid [Verma *et al.* (2017a)].

It is important to contrast the above derivation with that by Kolmogorov (1941a,b). Kolmogorov showed that a homogeneous and isotropic flow, under a steady state and in the limit of $\nu \to 0$ (or Re $\to \infty$) exhibits the following property:

$$\langle [\{\mathbf{u}(\mathbf{x}+\mathbf{r}) - \mathbf{u}(\mathbf{x})\} \cdot \hat{\mathbf{r}}]^3 \rangle = -\frac{4}{5}\Pi_u r, \qquad (10.10)$$

where $l_d \ll r \ll L$. Using the above equation and by making certain assumptions, we can derive the energy spectrum of Eq. (10.9). When we compare the above theory with the energetics arguments described earlier, we observe that both the theories assume $\nu \to 0$ and steady nature of the flow, but the energetics arguments do not explicitly assume homogeneity and isotropy of the flow. In fact, Eq. (10.3) is applicable to anisotropic flows as well. Hence, the energetics arguments are more general. In fact, buoyancy-driven flows are anisotropic, hence Eq. (10.3) becomes very handy for analysing such flows; we will employ this equation in Chapters 12 and 13. Note however that anistropic energy spectrum depends on the polar angle. We will discuss these issues in Chapter 15.

Kolmogorov's theory of turbulence provides many important predictions for a turbulent flow. Here we list some of them:

(1) At length scale l in the inertial range, the magnitude of the velocity fluctuation is given by

$$u_l \approx \epsilon_u^{1/3}l^{1/3}, \qquad (10.11)$$

and the effective interaction time-scale is

$$\tau_l \approx \frac{l}{u_l} \approx \epsilon_u^{-1/3}l^{2/3}. \qquad (10.12)$$

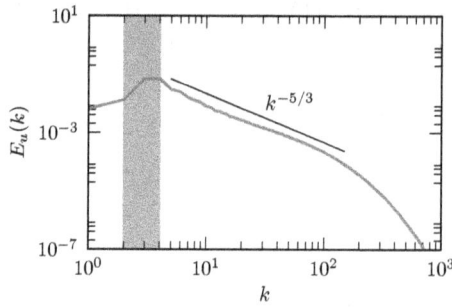

Fig. 10.2 The energy spectrum $E_u(k)$ computed using 4096^3-grid spectral simulation. In the inertial range $E_u(k) \sim k^{-5/3}$. From Verma *et al.* (2017a).

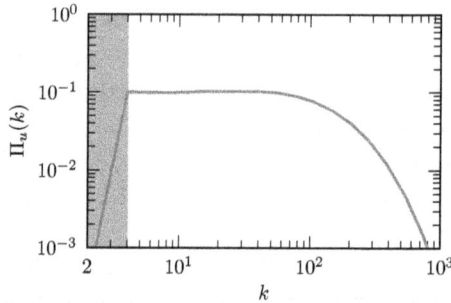

Fig. 10.3 The energy flux $\Pi_u(k)$, computed using 4096^3-grid spectral simulation. In the inertial range, $\Pi_u(k) \approx$ constant. From Verma *et al.* (2017a).

(2) When we extend Kolmogorov's scaling to the forcing and the dissipative scales, we obtain

$$\epsilon_u \approx \frac{U^3}{L} \approx \frac{u_{l_d}^3}{l_d},$$
(10.13)

where $l_d = 1/k_d$ is the dissipation scale, referred to as *Kolmogorov's length*. Using $\nu \approx u_{l_d} l_d$, we obtain[1]

$$l_d \approx \left(\frac{\nu^3}{\epsilon_u} \right)^{1/4}.$$
(10.14)

Note that the dissipation scale l_d is function of ν and, surprisingly, of the energy supply rate ϵ_u.

(3) We can also derive Eq. (10.14) from Eq. (10.11). We multiply Eq. (10.11) with l_d and set $l = l_d$. Equation (10.14) follows from here when we substitute $\nu \approx u_{l_d} l_d$.

---

[1] From kinetic theory of dilute gases, $\nu = c_s \lambda$, where c_s is the sound speed, and λ is the mean free path length. Thus, $\nu \approx u_{l_d} l_d$ and $\nu = c_s \lambda$ connect the nonequilibrium processes of hydrodynamic turbulence to the equilibrium processes at the kinetic scales. Note that $\lambda \neq l_d$, in fact, $\lambda \ll l_d$. The relation $\nu = c_s \lambda$ also extends to liquids, at least approximately.

(4) We can relate the Reynolds number to L/l_d using

$$\text{Re} = \frac{UL}{\nu} \approx \frac{UL}{u_{l_d} l_d} \approx \left(\frac{L}{l_d}\right)^{4/3}, \tag{10.15}$$

or

$$\frac{L}{l_d} \approx \text{Re}^{3/4}. \tag{10.16}$$

The aforementioned relation provides the range of length scales in a turbulent flow. Note that a realistic numerical simulation of turbulent flow requires minimum grid resolution of L/l_d along each direction. To illustrate, a numerical simulation of a turbulent flow with $\text{Re} = 10^4$ requires

$$\frac{L}{l_d} \approx \text{Re}^{3/4} \approx 10^3. \tag{10.17}$$

Hence, we need $(10^3)^3 = 10^9$ grid points to simulate a turbulent flow with $\text{Re} = 10^4$. Similarly, a numerical simulation of $\text{Re} = 10^8$ would require 10^{18} grid points, which is beyond the memory capacity of the best available supercomputer. Turbulence simulations are very challenging due to these reasons.

(5) An important property of 3D hydrodynamic turbulence is forward and local shell-to-shell energy transfers in the inertial range. We illustrate the shell-to-shell energy transfers $T_{u,n}^{u,m}$ in Fig. 10.4. See Eq. (4.55) for the definition. In the figure, x, y axes represent the receiver and giver shells respectively. Since the maximum energy transfer is from shell n to shell $(n+1)$, we conclude that the energy transfer in 3D hydrodynamic turbulence is local and forward.

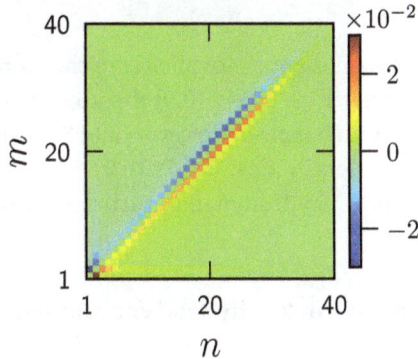

Fig. 10.4   The shell-to-shell energy transfers computed using a spectral simulation. See Eq. (4.55) for definition. The x, y axes represent the receiver and giver shells respectively. The figure demonstrates forward and local shell-to-shell energy transfer.

(6) Kolmogorov's theory for turbulence is applicable to the incompressible flows. The phenomenology of compressible turbulence is quite different from the Kolmogorov's theory of turbulence [Lesieur (2008)]. Some compressible features appear in the non-Boussinesq buoyancy-driven flows. These discussions however are beyond the scope of this book.

(7) Kolmogorov's theory assumes that $\nu \to 0$. An inviscid flow ($\nu = 0$) however has a very different behaviour; here, in the asymptotic limit, all the modes of the system have the same energy, thus the energy spectrum $E(k) \sim k^{d-1}$. This phenomena is related to the equilibrium behaviour for which $\Pi_u(k) \approx 0$. Contrast this with the Kolmogorov's $k^{-5/3}$ spectrum where the energy cascades from large scales to small scales. The directional energy cascade makes turbulence a nonequilibrium phenomena.

(8) Kolmogorov (1941a) and Kolmogorov (1941b) showed that

$$\langle [\{\mathbf{u}(\mathbf{x}+\mathbf{r}) - \mathbf{u}(\mathbf{x})\} \cdot \hat{\mathbf{r}}]^3 \rangle = -\frac{4}{5}\Pi_u r. \tag{10.18}$$

The above quantity is called *third-order structure function*. A simple extrapolation of the above scaling yields

$$\langle [\{\mathbf{u}(\mathbf{x}+\mathbf{r}) - \mathbf{u}(\mathbf{x})\} \cdot \hat{\mathbf{r}}]^q \rangle \sim (\Pi_u r)^{\zeta_q} \tag{10.19}$$

with $\zeta_q = q/3$. For example, $\zeta_2 = 2/3$, which leads to the Kolmogorov's spectrum of Eq. (10.9). Experiments and numerical simulations however reveal that $\zeta_q \neq q/3$, and the correction $\zeta_q - q/3$ is called the *intermittency correction* [Meneveau and Sreenivasan (1987); She and Leveque (1994); Sreenivasan (1991); Lesieur (2008)]. For example, numerical simulations and experiments show that $\zeta_2 \approx 0.71 \neq 2/3$ and $\zeta_{10} \approx 2.6 \neq 10/3$ [Meneveau and Sreenivasan (1987); She and Leveque (1994); Sreenivasan (1991)]. Note that $\zeta_2 \approx 0.71$ yields $E(k) \sim k^{-1.71}$, thus the real spectral exponent is closer to -1.71, not $-5/3$. This topic is quite interesting but quite complex, and it is beyond the scope of this book. Throughout the book, we assume that the hydrodynamic turbulence follows $E(k) \sim k^{-5/3}$.

The space dimension does not appear explicitly in the aforementioned derivation. Therefore one may expect that Eqs. (10.7, 10.9) describe both two-dimensional (2D) and three-dimensional (3D) turbulence. However only 3D hydrodynamic turbulence exhibits $\Pi_u(k)$ and $E_u(k)$ given by Eqs. (10.7, 10.9) respectively. Equations (10.7, 10.9) are not applicable to 2D hydrodynamic turbulence, which will be described in the next section.

10.2 Kraichnan's theory for 2D hydrodynamic turbulence

In the inviscid 2D hydrodynamics (with $\nu = 0$), the total kinetic energy, $\int d\mathbf{r} u^2/2$, and the total enstrophy, $\int d\mathbf{r} \omega^2/2$, are conserved. This differs from 3D hydrodynamics in which the total energy, $\int d\mathbf{r} u^2/2$, and the total kinetic helicity, $\int d\mathbf{r}(\mathbf{u} \cdot \boldsymbol{\omega})$, are conserved. Using the aforementioned conservation laws for 2D hydrodynamics, Kraichnan (1967) derived the following turbulence phenomenology for 2D hydrodynamics.

Kraichnan assumed that the fluid is forced at $k_f \gg 1/L$. Using field-theoretic calculation, Kraichnan (1967) showed that for $k < k_f$,

$$E_u(k) = K_{2D}\Pi_u^{2/3}k^{-5/3}, \quad \Pi_u = \text{const} < 0, \tag{10.20}$$

where Π_u is the energy flux, and K_{2D} is a constant whose numerical value is approximately 5.5–7.0. However for $k > k_f$,

$$E_u(k) = K'_{2D}\Pi_\omega^{2/3}k^{-3}, \quad \Pi_\omega = \text{const} > 0, \tag{10.21}$$

where Π_ω is the enstrophy flux, and K'_{2D} is another constant whose numerical value is 1.3–1.7. Note that 2D hydrodynamic turbulence exhibits inverse cascade of kinetic energy in contrast to forward cascade of kinetic energy in 3D hydrodynamic turbulence.

The scaling laws, Eqs. (10.7, 10.9) hold in the inertial range of hydrodynamic turbulence. Pao (1965) extended the aforementioned phenomenology to include dissipation range. We describe Pao's formalism in the next section. We also describe formulas for the energy spectrum and flux of laminar flows.

10.3 Energy spectrum in the dissipation wavenumber regime

Pao (1965) assumed that in the inertial and dissipative ranges, $\Pi_u(k)/E_u(k)$ is independent of ν, and it is only function of k and ϵ_u (also see Leslie (1973)). Under this assumption, dimensional analysis yields

$$\frac{E_u(k)}{\Pi_u(k)} = K_{\text{Ko}}\epsilon_u^{-1/3}k^{-5/3}. \tag{10.22}$$

Note that under a steady state, the flux equation (10.3) with $\mathcal{F}(k) = 0$ is

$$\frac{d}{dk}\Pi_u(k) = -D_u(k) = -2\nu k^2 E(k). \tag{10.23}$$

Substitution of Eq. (10.22) in the above equation yields

$$\Pi_u(k) = \epsilon_u \exp\left(-\frac{3}{2}K_{\text{Ko}}(k/k_d)^{4/3}\right), \tag{10.24}$$

$$E(k) = K_{\text{Ko}}\epsilon_u^{2/3}k^{-5/3}\exp\left(-\frac{3}{2}K_{\text{Ko}}(k/k_d)^{4/3}\right), \tag{10.25}$$

where

$$k_d = \left(\frac{\epsilon_u}{\nu^3}\right)^{1/4} \tag{10.26}$$

is the Kolmogorov's wavenumber. Thus, the energy spectrum in the dissipation range has an exponential behaviour $(\exp(-k^{4/3}))$. Verma *et al.* (2017a) have verified the above scaling using numerical simulations.

Equations (10.24, 10.25) describe the turbulent regime quite well, but they fail to model the laminar flows. Using similar arguments as above, Verma *et al.* (2017a) showed that in the laminar regime,

$$E(k) = \frac{U^2}{\text{Re}^3}\frac{1}{k}\exp(-k/\bar{k}_d), \tag{10.27}$$

$$\Pi(k) = \frac{1}{\text{Re}^3}2\nu U^2\bar{k}_d^2(1 + (k/\bar{k}_d))\exp(-k/\bar{k}_d), \tag{10.28}$$

where Re is the Reynolds number, and

$$\bar{k}_d \approx \left(\frac{\epsilon_u}{\nu^3}\right)^{1/4} \approx \left(\frac{\nu U^2}{\nu^3}\right)^{1/4} \approx \frac{\sqrt{\text{Re}}}{L} \tag{10.29}$$

is the dissipative wavenumber for laminar flows, and L is the box size.

In the next section we will describe important models of passive scalar turbulence that act as starting point for several turbulence phenomenologies of buoyancy-driven turbulence.

10.4   Passive scalar turbulence

Turbulent flows often carry dust particles and pollution along with it. Massless or light particles are advected by the velocity field, but these particles do not affect the velocity field. For this reason, such light particles are called *passive scalar*. On the contrary, heavy particles or energy-releasing particles/fields often affect the flow, hence such particles or fields are called *active scalars*. In buoyancy-driven flows, Eqs. (2.10, 2.11) reveal that the mass density field ρ is advected by the velocity field, and in a gravitational field, ρ affects the flow via buoyancy. Hence ρ is an active scalar.

We postpone the discussion on ρ as a active scalar to later chapters. Here we focus on flows with a passive scalar. Since a passive scalar is advected by the velocity field, it is described by the incompressible Navier Stokes equation and by an advection equation for the passive scalar ζ:

$$\frac{\partial \mathbf{u}}{\partial t} + (\mathbf{u} \cdot \nabla)\mathbf{u} = -\nabla(p/\rho) + \nu\nabla^2\mathbf{u} + \mathbf{f}_u, \tag{10.30}$$

$$\frac{\partial \zeta}{\partial t} + (\mathbf{u} \cdot \nabla)\zeta = \kappa\nabla^2\zeta + \mathbf{f}_\zeta, \tag{10.31}$$

$$\nabla \cdot \mathbf{u} = 0, \tag{10.32}$$

where κ is the diffusion coefficient for the passive scalar, and $\mathbf{f}_\zeta$ is the force field for ζ. The ratio of the viscosity and the diffusivity coefficient is called *Schmidt number*:

$$\text{Sc} = \frac{\nu}{\kappa}. \tag{10.33}$$

This is equivalent to the Prandtl number of buoyancy-driven flows. In the following discussion we describe the formalism of the passive scalar turbulence.

10.4.1   *Formalism of passive scalar turbulence*

Since the equation for the velocity field $\mathbf{u}$ in the above system is same as that for hydrodynamics, the properties of $\mathbf{u}$ field for the passive scalar are same as that for hydrodynamic flows. Hence, for turbulent passive scalar, the kinetic energy spectrum and flux covering both inertial and dissipation regime would be described by Eqs. (10.24–10.26). However, for the laminar flows, the corresponding equations are Eqs. (10.27–10.29).

Using scaling arguments, it can be shown that the scalar flux [Lesieur (2008); Verma (2001)]

$$\Pi_\zeta(k) \propto E_\zeta(k). \tag{10.34}$$

Hence, a dimensional analysis yields the following formula for the scalar spectrum:

$$E_\zeta(k) = K_{OC}\Pi_\zeta(\Pi_u)^{-1/3}k^{-5/3}, \tag{10.35}$$

where K_{OC} is the *Obukhov-Corrsin constant*. Note that the scalar flux $\Pi_\zeta(k)$ can be computed using the formula

$$\Pi_\zeta(k) = \sum_{k'>k} \sum_{p \leq k} S^{\zeta\zeta}(k' \,|\mathbf{p}|\, \mathbf{q}), \tag{10.36}$$

where

$$S^{\zeta\zeta}(\mathbf{k}|\mathbf{p}|\mathbf{q}) = \Im\left[\{\mathbf{k}\cdot\mathbf{u}(\mathbf{q})\}\{\zeta(\mathbf{p})\zeta^*(\mathbf{k})\}\right] \tag{10.37}$$

is the mode-to-mode transfer of $\zeta^2/2$ from mode $\mathbf{p}$ to mode $\mathbf{k}$ with mode $\mathbf{q}$ as a mediator.

Following the arguments of Sec. 10.1, we deduce the evolution equation for the scalar spectrum $E_\zeta(k)$ as

$$\frac{d}{dt}E_\zeta(k) = -\frac{d}{dk}\Pi_\zeta(k) + \mathcal{F}_\zeta(k) - D_\zeta(k), \tag{10.38}$$

where $\mathcal{F}_\zeta(k)$ is the supply rate of ζ^2 by $\mathbf{f}_\zeta$ of Eq. (10.31), and $D_\zeta(k)$ is the diffusion rate of ζ^2 in shell k. Under a steady state, $dE_\zeta(k)/dt \approx 0$, and under the assumption that $\mathcal{F}_\zeta(k)$ is active only at small k's, we obtain

$$\frac{d}{dk}\Pi_\zeta(k) = -D_\zeta(k) = -2\kappa k^2 E_\zeta(k). \tag{10.39}$$

Now we extend Pao (1965)'s arguments to passive scalar turbulence: we assume that in the inertial and dissipation range, $E_\zeta(k)/\Pi_\zeta(k)$ is independent of ν and κ. Under this assumption and using Eq. (10.35), we obtain

$$\frac{E_\zeta(k)}{\Pi_\zeta(k)} = K_{OC}\epsilon_u^{-1/3}k^{-5/3}, \tag{10.40}$$

substitution of which in Eq. (10.39) yields the following solution based on Pao (1965)'s extended conjecture:

$$\Pi_\zeta(k) = \epsilon_\zeta \exp\left(-\frac{3}{2}K_{OC}(k/k_c)^{4/3}\right), \tag{10.41}$$

$$E_\zeta(k) = K_{OC}\epsilon_\zeta\epsilon_u^{-1/3}k^{-5/3}\exp\left(-\frac{3}{2}K_{OC}(k/k_c)^{4/3}\right), \tag{10.42}$$

where ϵ_ζ is the diffusion rate of the scalar, $\Pi_\zeta = \epsilon_\zeta$, and

$$k_c = \left(\frac{\epsilon_u}{\kappa^3}\right)^{1/4} \tag{10.43}$$

is the diffusion wavenumber. Therefore,

$$\frac{k_c}{k_d} = \left(\frac{\nu}{\kappa}\right)^{3/4} = \text{Sc}^{3/4}. \tag{10.44}$$

Thus, turbulent scalar spectrum and flux are given by Eqs. (10.41, 10.42). Note that Péclet number

$$\text{Pe} = \frac{Ud}{\kappa} \gg 1 \tag{10.45}$$

in this regime.

When the nonlinear term of the scalar equation is much smaller than the diffusion term, or when $\text{Pe} \ll 1$, the scalar flux is quite small. However, $\Pi_\zeta \neq 0$ as long as Pe is finite. Following arguments similar to those in Sec. 10.3, we deduce the scalar spectrum and flux for $\text{Pe} \ll 1$ regime as:

$$E_\zeta(k) = \zeta_{\text{rms}}^2 \frac{1}{k} \exp(-k/\bar{k}_c), \tag{10.46}$$

$$\Pi_\zeta(k) = \epsilon_\zeta(1 + k/\bar{k}_c) \exp(-k/\bar{k}_c). \tag{10.47}$$

Using

$$\epsilon_\zeta = \frac{1}{L} \zeta_{\text{rms}}^2 U = 2\zeta_{\text{rms}}^2 \kappa \bar{k}_c^2, \tag{10.48}$$

we deduce that

$$\bar{k}_c = \frac{1}{L}\sqrt{\frac{UL}{L}} = \frac{\sqrt{\text{Pe}}}{L}. \tag{10.49}$$

Using Eqs. (10.29, 10.49) we deduce that for the laminar regime,

$$\frac{\bar{k}_c}{\bar{k}_d} \approx \sqrt{\frac{\text{Pe}}{\text{Re}}} = \sqrt{\text{Sc}}. \tag{10.50}$$

With this background, now we work out the spectra and fluxes for various regimes passive scalar flows.

10.4.2   *Passive-scalar turbulence for various regime*

For 3D hydrodynamic flows, in Sections 10.1 and 10.3 we worked out the kinetic energy spectrum and flux for the turbulent regime ($\text{Re} \gg 1$) and laminar regime ($\text{Re} \lesssim 1$). In Sec. 10.4.1, we derived $E_\zeta(k)$ for $\text{Pe} \gg 1$ and $\text{Pe} \ll 1$. Using these results we can derive the spectra and fluxes for $\mathbf{u}$ and ζ for $\text{Sc} \approx 1$, $\text{Sc} \ll 1$, and $\text{Sc} \gg 1$. These results are stated below:

(1) $\text{Sc} \approx 1$: Using the fact that $\text{Sc} = \text{Pe}/\text{Re}$, we deduce that $\text{Pe} \approx \text{Re}$. Hence, in the turbulent regime, both $\text{Re}, \text{Pe} \gg 1$. Therefore, $E_u(k)$ and $E_\zeta(k)$ are given by Eq. (10.25) and Eq. (10.42) respectively. In the laminar regime, the corresponding spectra would be described by Eq. (10.27) and Eq. (10.46). See Fig. 10.5(a,b) for illustrations. Note that $k_c \approx k_d$ for this case.

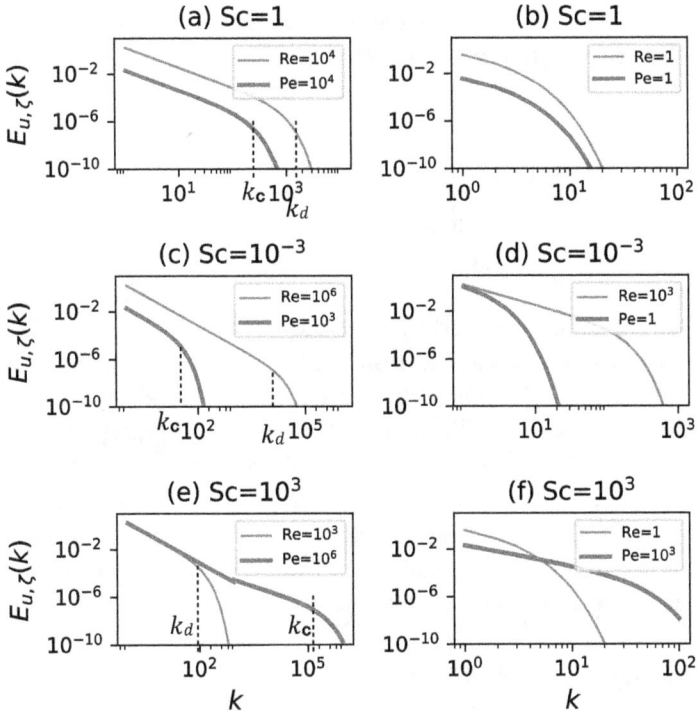

Fig. 10.5   The kinetic energy spectrum $E_u(k)$ (thin red curve) and scalar spectrum $E_\zeta(k)$ (thick green curve) for (a,b) Sc = 1, (c,d) Sc = 10^{-3}, (e,f) Sc = 10^3. $E_u(k)$ is described by Eq. (10.25) for large Re, but by Eq. (10.27) for small Re. Regarding $E_\zeta(k)$, for Sc $\lesssim$ 1(a,b,c,d), it is described by Eq. (10.42) for large Pe, but by Eq. (10.46) for small Pe; however, for Sc $\gg$ 1, $E_\zeta(k)$ is described by Eq. (10.42) for $k < k_d$, but by Eq. (10.46) for $k_d < k < \bar{k}_c$.

(2) Sc $\ll$ 1: For this case Re $\gg$ Pe and $k_c \ll k_d$. The spectral properties are very similar to case (1), except that $k_c \ll k_d$. We can choose the appropriate formulas depending on the magnitudes of Re and Pe. See Fig. 10.5(c,d) for illustrations.

(3) Sc $\gg$ 1: This case is more complex. Here Re $\ll$ Pe and $k_c \gg k_d$. Hence the scalar spectrum extends much beyond k_d, and the spectra in the wavenumber regime $k_d < k < k_c$ is somewhat complex. Since $u_k \to 0$ for $k > k_d$, we can argue that the effective Pe $\approx u_{k_d} L/\kappa \ll 1$ in the wavenumber regime $k_d < k < k_c$, and we expect Eq. (10.46) to describe the scalar turbulence in this range. Regarding the nonlinear transfers of ζ^2, in the range $k_d < k < k_c$, we expect $\mathbf{u}(k_d)$ to act as a mediator. Hence, ζ^2 transfers involve triads with $k, p \gg q \approx k_d$ that differs from typical local and forward transfers that involves $k \approx p \approx q$.

Therefore, for Sc $\gg$ 1, $E_u(k)$ is determined by Eq. (10.25) when Re $\gg$ 1, and by Eq. (10.27) when Re $\lesssim$ 1. Regarding the scalar spectrum, for $k < k_d$, it is given either by Eq. (10.42) or Eq. (10.46) depending on Pe $\gg$ 1 or Pe $\lesssim$ 1. However, the scalar spectrum for $k_d < k < k_c$ is given by Eq. (10.46). See Fig. 10.5(e,f) for illustrations.

For earlier work on passive scalar spectrum and flux for $Sc \gg 1$, refer to Batchelor (1959); Kraichnan (1968); Lesieur (2008); Gotoh and Yeung (2013) for detailed discussion. Here we briefly describe two popular models of $E_\zeta(k)$. Batchelor (1959) proposed that

$$E_\zeta(k) = K_{\mathrm{Ba}}\epsilon_\zeta (\nu/\epsilon_u)^{1/2}k^{-1}\exp(-K_{\mathrm{Ba}}\kappa k^2(\nu/\epsilon_u)^{1/2}), \qquad (10.51)$$

but Kraichnan (1968) argued that

$$E_\zeta(k) = K_{\mathrm{Ba}}\epsilon_\zeta(\nu/\epsilon_u)^{1/2}(1+\sqrt{6K_{\mathrm{Ba}}})k^{-1}\exp(-\sqrt{6K_{\mathrm{Ba}}}(k/\bar{k}_c)). \qquad (10.52)$$

Note that Eq. (10.46) derived using flux equation has the same form as Eq. (10.52), proposed by Kraichnan. Numerical results of Gotoh *et al.* (2014) and Yeung *et al.* (2004) tend to support the predictions of Kraichnan (1968) and Eq. (10.46). We refer the reader to Gotoh and Yeung (2013) for a detailed discussion.

As discussed earlier, in buoyancy-driven flows, the density field ρ affects the velocity field, hence it acts as an active scalar. The analysis of this active scalar however borrows concepts of passive scalar turbulence. In the subsequent chapters we will discuss the phenomenology of the buoyancy-driven turbulence in the framework of active scalar, and contrast them in relation to passive scalar.

Further reading

The textbooks by Leslie (1973), McComb (1990), Pope (2000), and Lesieur (2008) describe hydrodynamic and passive-scalar turbulence.

Exercises

(1) Derive the energy spectrum for 2D hydrodynamic turbulence using dimensional analysis.
(2) Estimate the Kolmogorov's length l_d in
  (a) A glass of turbulent water whose $Re \sim 10^5$.
  (b) Turbulent atmospheric flow with $Re \sim 10^6$.
  (c) In solar convection

  Compare these results with the mean free path length in such fluids.
(3) What grid resolutions are required to simulate the aforementioned flows?
(4) Is the passive scalar formalism discussed in this chapter appropriate for the dust particles in the atmosphere?

Chapter 11

Scaling of Large-Scale Quantities in RBC

In this chapter we will quantify the large-scale quantities of RBC, namely the Nusselt and Reynolds numbers, and the viscous and thermal dissipation rates. We also describe the relative strengths of various terms of RBC equations, for example, the pressure gradient and buoyancy. In addition, we will review the phenomenologies of Kraichnan (1962), Shraiman and Siggia (1990), Grossmann and Lohse (2000), Grossmann and Lohse (2001), Doering *et al.* (2006), Pandey and Verma (2016), and others.

In this chapter we will study the steady-state properties of RBC, which are derived by setting $\partial/\partial t = 0$ in the dynamical equations. We start with a preliminary discussion on the scaling of the Péclet number, Pe, and Nusselt number, Nu. We remark that many researchers report scaling of Reynolds number, Re, which is related to the Péclet number as $\mathrm{Re} = \mathrm{Pe}/\mathrm{Pr}$.

11.1 Preliminary models for scaling of Péclet and Nusselt numbers

11.1.1 *Scaling of Péclet number*

We start with the equations (2.17, 2.18) for the velocity and temperature fields described in Chapter 2:

$$\frac{\partial \mathbf{u}}{\partial t} + (\mathbf{u} \cdot \nabla)\mathbf{u} = -\nabla \sigma + \alpha g \theta \hat{z} + \nu \nabla^2 \mathbf{u}, \tag{11.1}$$

$$\frac{\partial \theta}{\partial t} + (\mathbf{u} \cdot \nabla)\theta = \frac{\Delta}{d} u_z + \kappa \nabla^2 \theta, \tag{11.2}$$

where Δ is the temperature difference the two thermal plates that are separated by a vertical distance d. We denote the large-scale velocity and temperature fluctuations by U and Θ respectively. Note that $\mathrm{Pe} = \frac{Ud}{k}$ and $\mathrm{Re} = \frac{Ud}{\nu}$. A naive analysis of the above equations yield the following scaling laws for thermal convection. We will show later that these scaling relations need to be modified so as to match with the numerical and experimental findings.

The rudimentary scaling analysis of the RBC equations yields the following two regimes [Verma *et al.* (2012, 2014)]:

(1) Pe ≫ 1: This case arises when the nonlinear term $(\mathbf{u} \cdot \nabla)\theta$ is much larger than the thermal diffusion term $\kappa \nabla^2 \theta$. Hence, in Eq. (11.2), the nonlinear term will balance $(\Delta/d)u_z$, or

$$(\mathbf{u} \cdot \nabla)\theta \approx \frac{\Delta}{d} u_z \implies \Theta \approx \Delta. \tag{11.3}$$

When Re ≪ 1, in Eq. (11.1), the viscous term dominates the nonlinear and pressure gradient terms, and it is balanced by buoyancy. Hence,

$$\nu \nabla^2 \mathbf{u} \approx \alpha g \theta \implies \frac{\nu U}{d^2} \approx \alpha g \Theta. \tag{11.4}$$

Therefore, using $\Theta = \Delta$ (from Eq. (11.3)) we obtain

$$\mathrm{Pe} = \frac{Ud}{\kappa} \approx \frac{\alpha g \Delta d^3}{\nu \kappa} = \mathrm{Ra}. \tag{11.5}$$

For the other extreme, Re ≫ 1, the nonlinear and pressure gradient terms dominate the viscous term. Hence, we may expect

$$(\mathbf{u} \cdot \nabla)\mathbf{u} \approx \alpha g \theta \implies \frac{U^2}{d} \approx \alpha g \Delta. \tag{11.6}$$

Therefore,

$$\mathrm{Pe} = \frac{Ud}{\kappa} \approx \frac{d}{\kappa}\sqrt{\alpha g \Delta d} = \sqrt{\mathrm{RaPr}}. \tag{11.7}$$

(2) Pe ≪ 1: An important point to note that Pe ≪ 1 is inconsistent with the Oberbeck-Boussinesq (OB) approximation. In Chapter 2 we showed that

$$\frac{\nabla \cdot \mathbf{u}}{UL} \approx \frac{1}{\mathrm{Pe}}. \tag{11.8}$$

For the Boussinesq approximation to hold, we require that $\nabla \cdot \mathbf{u} = 0$ or Pe ≫ 1. Thus the limiting case Pe ≪ 1 is not admissible under OB approximation. Still we perform scaling analysis for this case for completeness and with a hope that it may be useful for Non-Boussinesq flows.

For Pe ≪ 1, in Eq. (11.2), the nonlinear term is much smaller than the thermal diffusion term. Therefore,

$$-\frac{\Delta}{d}u_z = \kappa \nabla^2 \theta \implies \Theta \approx \frac{Ud\Delta}{\kappa}. \tag{11.9}$$

For Re ≫ 1,

$$(\mathbf{u} \cdot \nabla)\mathbf{u} \approx \alpha g \Theta \implies \frac{U^2}{d} \approx \alpha g \frac{Ud\Delta}{\kappa}, \tag{11.10}$$

from which we deduce that

$$\mathrm{Pe} = \frac{Ud}{\kappa} \approx \mathrm{RaPr} \implies \mathrm{Re} = \mathrm{Pe}/\mathrm{Pr} = \mathrm{Ra}. \tag{11.11}$$

Verma *et al.* (2014) verified the above scaling for zero-Prandtl RBC. Using numerical simulations, they showed that Re ∼ Ra for zero-Pr RBC. Also, from Eq. (11.9) we obtain

$$\frac{\Theta}{\Delta} = \frac{Ud}{\kappa} = \mathrm{RaPr}. \tag{11.12}$$

However, when $\mathrm{Re} \ll 1$,

$$\nu\nabla^2\mathbf{u} \approx \alpha g\Theta \implies \nu\frac{U}{d^2} \approx \alpha g\frac{Ud\Delta}{\kappa} \tag{11.13}$$

that yields

$$\frac{\alpha g\Delta d^3}{\nu\kappa} = \mathrm{Ra} \approx 1, \tag{11.14}$$

which is much less than Ra_c. Therefore, the velocity and temperature fluctuations will die out and convection will not take place. Hence, flows with $\mathrm{Pe} \ll 1$ and $\mathrm{Re} \ll 1$ do not exhibit thermal convection.

Note that most terrestrial experiments have $\mathrm{Pe} > 1$. Terrestrial experiments typically employ oil, water, air, and liquid metals as fluids. The Prandtl number of oil, water, and air are $\gtrsim 1$, while that of liquid metals lies in the range of 0.003 to 0.02 (see Table A.1 and A.2). Also, the critical Rayleigh number is $\mathrm{Ra}_c \gtrsim 1000$. Hence, Eqs. (11.5, 11.7) indicate that the Péclet number for terrestrial experiments is always greater than unity.

Numerical simulations and experiments of RBC reveal that Eq. (11.7) holds only for moderate Ra. Equations (11.5) is violated when $\mathrm{Re} \ll 1$. Similarly, Eq. (11.3) is not observed in experiments and simulations because the velocity field couples with θ_{res} (see Sec. 9.2). Thus, though the aforementioned scaling arguments appear to be quite robust, they are incorrect due to certain subtle reasons, which will be described in the following discussion.

11.1.2 *Scaling of Nusselt number*

Using the aforementioned scaling for U and Θ, Kraichnan (1962) derived scaling relation for the Nusselt number as a function of Ra and Pr when $\mathrm{Re} \gg 1$ and $\mathrm{Pe} \gg 1$. As discussed in Sec. 2.9, Nusselt number is defined as

$$\mathrm{Nu} = 1 + \frac{\langle u_z\theta\rangle_{xyz}}{\kappa\Delta/d} \approx \left\langle \left(\frac{u_z d}{\kappa}\right)\left(\frac{\theta}{\Delta}\right)\right\rangle_{xyz}, \tag{11.15}$$

where $\langle.\rangle_{xyz}$ stands for volume average. For large Ra, $\mathrm{Nu} \gg 1$, hence unity of Eq. (11.15) is ignored. For $\mathrm{Re}, \mathrm{Pe} \gg 1$, the above averaging process is simplified to

$$\mathrm{Nu} = \left\langle \left(\frac{Ud}{\kappa}\right)\right\rangle_{xyz} \left\langle \left(\frac{\theta}{\Delta}\right)\right\rangle_{xyz} \approx \mathrm{Pe}\frac{\Theta}{\Delta}. \tag{11.16}$$

Now using Eqs. (11.3, 11.7), Kraichnan deduced that

$$\mathrm{Nu} \sim \sqrt{\mathrm{RaPr}}. \tag{11.17}$$

The aforementioned arguments however have several serious flaws. First, in general,

$$\langle fg\rangle \neq \langle f\rangle\langle g\rangle. \tag{11.18}$$

The equality holds only when f and g are constants, or they have zero average, which is not the case for RBC. In Sec. 11.4.2 we will revisit the above correlations.

Another error in the above computation is $\Theta/\Delta \approx 1$, which is not the case because the velocity field couples with θ_{res}. We will correct these errors in the subsequent discussion.

For very small Pr and Pe $\ll 1$, by substitutions of Eqs. (11.11, 11.12) in Eq. (11.15), Kraichnan (1962) deduced that

$$\text{Nu} \approx 1 + (\text{RaPr})^2. \tag{11.19}$$

However, this scaling has not been observed in any experiment or numerical simulation due to stringent condition of Pe $\ll 1$. Note that Pe $\ll 1$ is not admissible under OB approximation.

In the next section, we correct some of the flaws of the aforementioned scaling arguments. First, we will work out the scaling of the Péclet number.

11.2   Modified scaling arguments for the Péclet number and Θ

In the following discussion we present the scaling arguments of Pandey *et al.* (2016a) and Pandey and Verma (2016). These arguments take into account subtle features of RBC. As discussed in the earlier section, we take Pe $\gg 1$.

In Sec. 9.2, we showed that the mean temperature profile $\theta_m(z)$ does not couple with the velocity field. In fact, the equation for the velocity field is

$$\frac{\partial \mathbf{u}}{\partial t} + (\mathbf{u} \cdot \nabla)\mathbf{u} = -\nabla \sigma_{\text{res}} + \alpha g \theta_{\text{res}} \hat{z} + \nu \nabla^2 \mathbf{u}, \tag{11.20}$$

where $\theta_{\text{res}}(\mathbf{r}) = \theta(\mathbf{r}) - \theta_m(z)$, and it differs from Eq. (11.1). Thus, $\mathbf{u}$ is coupled with θ_{res} that represented temperature fluctuations over $T_m(z)$. Note however that $\theta_m(z)$ couples nonlinearly with θ_{res}. For example, in 7-mode model, $\theta(1, 0, 1)$ couples with $\theta(0, 0, 2)$. Since $\theta_m(z)$ is significant, $\Theta_{\text{res}} < \Delta$, where Θ_{res} is the rms value of θ_{res}. We will also show in the following discussion that the pressure gradient $-\nabla \sigma_{\text{res}}$ too plays an important role in RBC.

A dimensional analysis of Eq. (11.20) yields the following equation for RBC:

$$c_1 \frac{U^2}{d} = c_2 \frac{U^2}{d} + c_3 \alpha g \Theta_{\text{res}} - c_4 \nu \frac{U}{d^2}, \tag{11.21}$$

where c_i's are dimensionless coefficients defined as

$$c_1 = \frac{\langle |\mathbf{u} \cdot \nabla \mathbf{u}| \rangle}{U^2/d}; \quad c_2 = \frac{\langle |\nabla \sigma|_{\text{res}} \rangle / \rho_0}{U^2/d}; \quad c_3 = \Theta_{\text{res}}/\Delta; \quad c_4 = \frac{\langle |\nabla^2 \mathbf{u}| \rangle}{U/d^2}, \tag{11.22}$$

where $\langle . \rangle$ could be the volume average. Multiplication of Eq. (11.21) with d^3/κ^2 yields

$$c_1 \text{Pe}^2 = c_2 \text{Pe}^2 + c_3 \text{RaPr} - c_4 \text{PePr}, \tag{11.23}$$

where $\text{Pe} = Ud/\kappa$ is the Péclet number. The solution of the above equation is

$$\text{Pe} = \frac{-c_4 \text{Pr} + \sqrt{c_4^2 \text{Pr}^2 + 4(c_1 - c_2)c_3 \text{RaPr}}}{2(c_1 - c_2)}. \tag{11.24}$$

We can now compute Pe as a function of Ra and Pr.

For unbounded or free turbulence, as in Kolmogorov's theory of 3D hydrodynamic turbulence, we expect the c_i's to be constants. Surprisingly, for RBC, c_i's are functions of Ra and Pr, as reported by Pandey *et al.* (2016a):

$$c_1 = 1.5 \mathrm{Ra}^{0.10} \mathrm{Pr}^{-0.06}, \tag{11.25}$$

$$c_2 = 1.6 \mathrm{Ra}^{0.09} \mathrm{Pr}^{-0.08}, \tag{11.26}$$

$$c_3 = 0.75 \mathrm{Ra}^{-0.15} \mathrm{Pr}^{-0.05}, \tag{11.27}$$

$$c_4 = 20 \mathrm{Ra}^{0.24} \mathrm{Pr}^{-0.08}. \tag{11.28}$$

The error bars in the above exponents are $\lesssim 0.01$, except for the Ra exponent of c_4 that has an error of the order of 0.10. In Figs. 11.1 and 11.2 we illustrate the dependence of c_i's on Ra and Pr. The above relations were derived using RBC simulation data for $\mathrm{Pr} = 1, 6.8, 10^2, 10^3$ and Ra from 10^6 to 5×10^8. Hence we expect these relations to hold for $\mathrm{Pr} \gtrsim 1$, but they may not work for small Pr. The box geometry (cylinder or rectangular) or aspect ratio could also affect the above relations, but we expect the correction to be small.

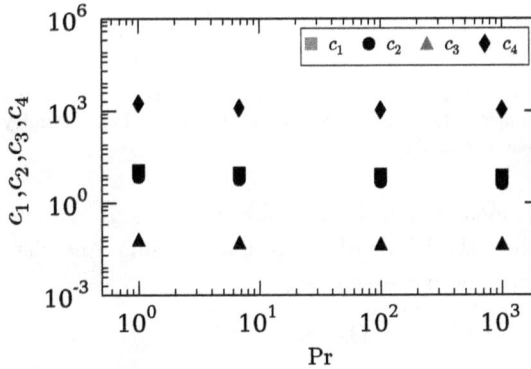

Fig. 11.1 Plot of c_i's vs. Pr for $\mathrm{Ra} = 2 \times 10^7$. c_i's decrease very slowly with the increase of Pr. From Pandey and Verma (2016). Reprinted with permission from AIP.

The aforementioned unexpected dependence of c_i's on Ra and Pr is due to the boundary walls. First, using the definition of c_3 from Eq. (11.22) and Eq. (11.27), we deduce that

$$\frac{\Theta_{\mathrm{res}}}{\Delta} \approx c_3 = 0.75 \mathrm{Ra}^{-0.15}, \tag{11.29}$$

i.e. $\Theta_{\mathrm{res}} \neq \Delta$. The above relation affects Pe, Re, and Nusselt scaling significantly. Niemela *et al.* (2000) observed similar scaling in an RBC experiment with Helium gas confined in a cylinder. They varied Ra of their experiment from 10^7 to 10^{15}, and measured temperature near the lateral wall in the mid plane of the cylinder. In Fig. 11.3 we illustrate the the probability distribution function (PDF) of the

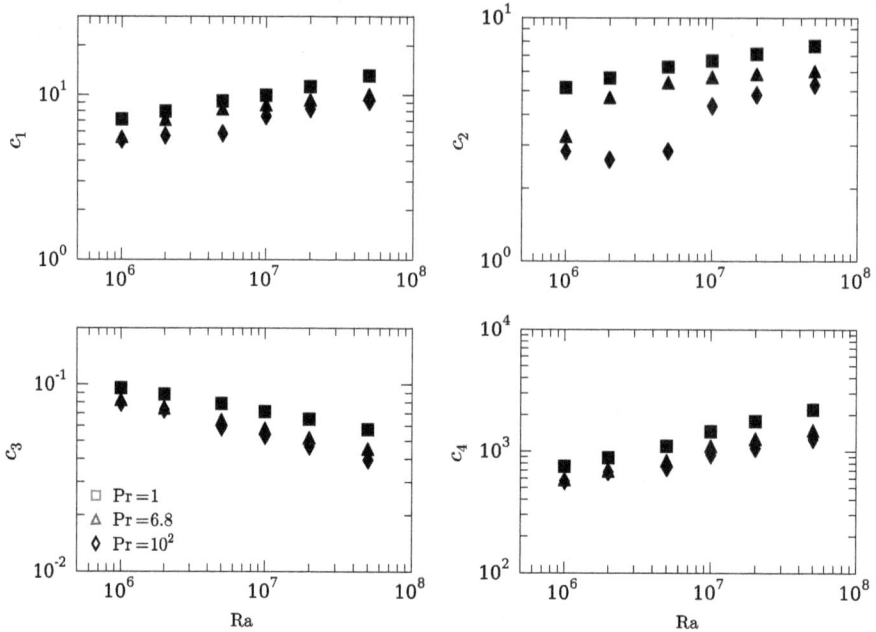

Fig. 11.2 Plot of c_i's vs. Ra for Pr $= 1$, 6.8, and 100. c_1, c_2, and c_4 increase with the increase of Ra, whereas c_3 decreases. Legend applies to all the plots. From Pandey and Verma (2016). Reprinted with permission from AIP.

temperature fluctuations, $\theta_{\rm res}$, as well as plot of $\theta_{\rm res}$ vs. Ra [Niemela *et al.* (2000)][1]. As shown in the figure, the PDFs of $\theta_{\rm res}$ is non-gaussian, and the width of the PDF or $\Theta_{\rm res}$ decreases with the increase of Ra:

$$\Theta_{\rm res} \sim {\rm Ra}^{-0.145}, \qquad (11.30)$$

which is close to the $\Theta_{\rm res} \sim {\rm Ra}^{-0.15}$ scaling of c_3. Castaing *et al.* (1989) observed similar scaling for $\Theta_{\rm res}$.

Thus, the temperature fluctuations relevant for Péclet and Nusselt number scaling is not of the order of Δ, but it is suppressed by a factor of ${\rm Ra}^{-0.15}$. These observations, based on modelling and numerical simulations, have very important consequences on the scaling of large-scale quantities.

For an unbounded or hydrodynamic turbulence, the ratio of the nonlinear term, $\mathbf{u} \cdot \nabla \mathbf{u}$, and the viscous term is the Reynolds number Ud/ν. But this is not the case for RBC for which the ratio is

$$\frac{\text{Nonlinear term}}{\text{Viscous term}} = \frac{\langle |\mathbf{u} \cdot \nabla \mathbf{u}| \rangle}{\langle |\nu \nabla^2 \mathbf{u}| \rangle} = \frac{Ud}{\nu} \frac{c_1}{c_4} \sim {\rm ReRa}^{-0.14}. \qquad (11.31)$$

---

[1] Niemela *et al.* (2000) reported PDF and fluctuations of $T(\mathbf{r}) - T_m(z)$. As shown in Sec. 9.2, $T(\mathbf{r}) - T_m = \theta_{\rm res}(\mathbf{r})$.

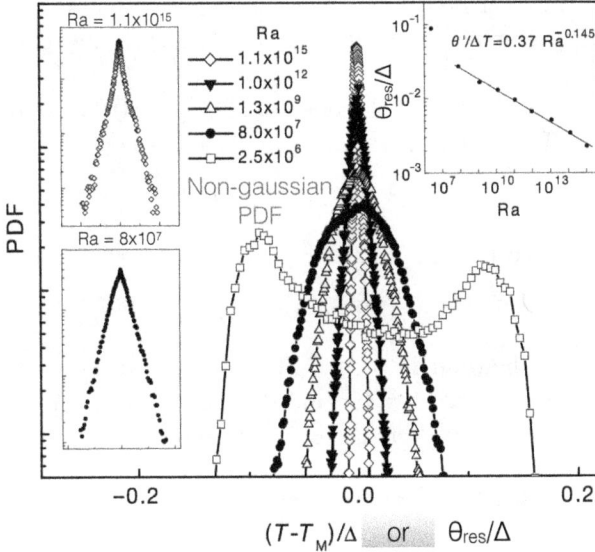

Fig. 11.3   For the temperature data measured in an RBC experiment with cryogenic helium gas: the PDFs of the temperature fluctuation θ_{res} for large Ra's are non-gaussian. The width of the PDF shrinks with the increase of Ra, and $\Theta_{res} \sim Ra^{-0.145}$. Adopted from a figure of Niemela *et al.* (2000).

Thus, in the turbulent regime, for the same U, L, and ν, RBC in a bounded domain has a weaker nonlinearity compared to the free or unbounded turbulence. It leads to a weaker energy flux and, hence, $\epsilon_u < U^3/d$. This phenomenon could be treated as a drag reduction in RBC, and it may be related to the drag reduction in turbulence flow with polymers and bubbles; we will revisit this issue in Sec. 11.6. These issues need further experimental and numerical inputs.

Now we derive the Pe scaling for the turbulent and viscous regimes. In the turbulent regime, the viscous term of Eq. (11.23) can be ignored, and hence

$$\text{Pe} \approx \sqrt{\frac{c_3}{|c_1 - c_2|}}\text{RaPr} \approx \sqrt{7.5\text{Pr}}\text{Ra}^{0.38}. \tag{11.32}$$

In the above analysis, pressure term is as important as the nonlinear term. Also, c_i's are functions of Ra. These features differ from the preliminary dimensional analysis performed in Sec. 11.1.

On the contrary, in the viscous regime, the nonlinear term and the pressure gradient are negligible. Therefore, we equate the the buoyancy and viscous terms of Eq. (11.23) that yields

$$\text{Pe} \approx \frac{c_3}{c_4}\text{Ra} \approx 0.038\text{Ra}^{0.60} \tag{11.33}$$

that matches with earlier numerical and experimental results. The above relation deviates from the preliminary scaling of Eq. (11.5) due to the nontrivial Ra dependence of c_3 and c_4.

We can also derive the above relations from Eq. (11.24) by comparing $c_4^2 Pr^2$ and $4(c_1 - c_2)c_3 RaPr$. The turbulent regime is applicable when

$$c_4^2 Pr^2 \ll 4|c_1 - c_2|c_3 RaPr. \tag{11.34}$$

Substitution of c_4 from Eq. (11.28) reduces the above condition to

$$Ra^{0.47} \gg 10^3 Pr, \quad \text{or} \quad Ra \gg 10^6 Pr^2. \tag{11.35}$$

A word of caution is however in order. The above condition is derived based on numerical data for $Pr = 1, 6.8, 10^2, 10^3$ and Ra from 10^6 to 5×10^8. Simulations for lower Pr and higher Ra may alter the condition somewhat.

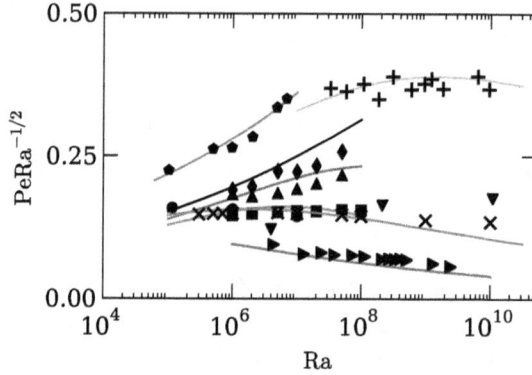

Fig. 11.4   Plots of normalized Péclet number ($PeRa^{-1/2}$) vs. Ra for various numerical simulations [Pandey *et al.* (2016a); Silano *et al.* (2010); van Reeuwijk *et al.* (2008); Scheel and Schumacher (2014)] and for the experiments [Xin and Xia (1997); Cioni *et al.* (1997); Niemela and Sreenivasan (2003)]. Here, Pr = 0.022 (brown right triangles), 0.7 (green crosses and green down-triangles), 1 (red squares), 6.8 (blue triangles, orange plusses), 100 (black diamonds), 1000 (magenta pentagons). The continuous curves representing the model predictions using Eq. (11.24) match with numerical and experimental results quite well. From Pandey and Verma (2016). Reprinted with permission from AIP.

In Fig. 11.4, we plot the normalized Péclet number, $PeRa^{-1/2}$, for various experiments and numerical simulations as various symbols, and compare them with the predictions of Pandey *et al.* (2016a) [Eq. (11.24)]. We observe that the model predictions match with the numerical and experimental results quite well. It is remarkable that the Pe of Eq. (11.24), which is functions only of Ra and Pr and is independent of box size and box geometry, describes a large number of experimental and numerical results very well. We make a cautionary remark that the predictions for $Pr = 0.022$ and $Pr = 6.8$ have been multiplied with 2.5 and 1.2 respectively for the best fit with the experimental results. These prefactors possibly arise due to differences in box geometry and boundary conditions. These results provide strong validation for the model.

The figure indicates that

(1) For large Pr and moderate Ra (blue triangle, black diamond, and magenta pentagons), we observe Pe $\sim$ Ra$^{0.60}$. This is consistent with the scaling for viscous regime.
(2) For small Pr and moderate Ra (brown right triangles), we observe Pe $\sim$ Ra$^{0.38}$, consistent with the scaling for turbulent regime.
(3) For moderate Pr and Ra, we obtain Pe $\sim$ Ra$^{1/2}$. For Pr $= 6.8$, this scaling is observed for large Ra as well. This is an intermediate scaling between the turbulent and laminar regimes. Here, Ra $\sim 10^6$Pr^2.

Thus, our model predictions capture the numerical and experimental data quite well.

In the following discussion we compare the model predictions of Eq. (11.24) with Cioni *et al.* (1997)'s experimental results on mercury (Pr $= 0.025$). Cioni *et al.* (1997) reported that for $5 \times 10^6 \leq$ Ra $\leq 5 \times 10^9$, Re $\sim$ Ra$^{0.424}$, which is close to the model's exponent of 0.38 for the turbulent regime. Note that according to Eq. (11.35), the experimental Ra of Cioni *et al.* is in the turbulent regime. The aforementioned results are in general agreement with those of Grossmann and Lohse (2000) and Grossmann and Lohse (2001).

For the viscous regime, let us apply the model predictions of Eq. (11.33) to the Earth's mantle for which the parameters are $d \approx 2900$ km, $\kappa \approx 10^{-6}$, Pr $\approx 10^{23}$-10^{24}, Ra $\approx 5 \times 10^7$. The observational value of the rms speed of the mantle $U \approx 2$ cm/yr yields an estimate of the Péclet number as

$$\text{Pe}_{\text{est.}} = \frac{Ud}{\kappa} \approx \frac{0.02 \times 2.9 \times 10^6}{86400 \times 365 \times 10^{-6}} \approx 1840. \tag{11.36}$$

The model predictions by Eq. (11.24) is

$$\text{Pe}_{\text{model}} \approx 0.038 \text{Ra}^{0.60} \approx 1582, \tag{11.37}$$

which is quite close to the estimated value.

Strictly speaking, the Péclet number scaling described above is valid for Pr $\geq$ 1 since c_i's have been derived for this range. Note however that the formula of Eq. (11.32) is in good agreement with the experimental results on mercury whose Pr $= 0.02$. Thus, low-Pr regime needs to be investigated in detail.

In Fig. 11.5 we summarize the scaling discussed in this section. For a given Pr, RBC makes transitions: laminar regime $\rightarrow$ intermediate regime $\rightarrow$ turbulent regime I $\rightarrow$ turbulent regime II. The flow is laminar for Ra $\ll 10^6$Pr^2 (see Eq. (11.35)) and it exhibits Pe $\sim$ Ra$^{0.60}$. The instabilities, patterns, and chaos discussed in Chapters 6 and 8 occur in the laminar phase. Then, in the intermediate regime, Ra $\sim 10^6$Pr^2, we obtain Pe $\sim$ Ra$^{1/2}$. With further increase of Ra to Ra $\gg 10^6$Pr^2, in the turbulent regime I, the Péclet scaling is Pe $\sim$ Ra$^{0.38}$. For very

```
┌─────────────────────────────┐          ┌─────────────────────────────┐
│      Laminar regime         │          │     Turbulent regime II     │
│  Instability, chaos, patterns│          │      Ultimate regime        │
│      (Ra ≪ 10⁶Pr²)          │          │      (Ra ≫≫ 10⁶Pr²)        │
│      Pe ~ Ra⁰·⁶⁰            │          │      Pe ~ Ra¹/²             │
└─────────────────────────────┘          └─────────────────────────────┘
              │                                         ▲
              ▼                                         │
┌─────────────────────────────┐          ┌─────────────────────────────┐
│    Intermediate regime      │          │      Turbulent regime I     │
│      (Ra ≈10⁶Pr²)          │ ───────▶ │       (Ra ≫ 10⁶Pr²)        │
│      Pe ~ Ra¹/²             │          │      Pe ~ Ra⁰·³⁸           │
└─────────────────────────────┘          └─────────────────────────────┘
```

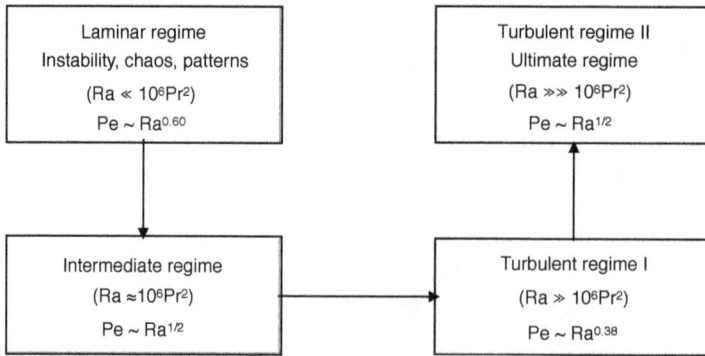

Fig. 11.5   For a given Pr, summary of different scaling of thermal convection.

large Ra (turbulent regime II), the effects of the boundary effects may disappear, and the system may behave like hydrodynamic turbulence, which is described by Kolmogorov's theory. In this regime, called the ultimate regime, c_i's may be constants, and the energy supply by buoyancy may be at large length scales. We will revisit these issues in Sec. 11.10

In the next section we will study the relative strengths of various forces in RBC.

11.3   Relative strengths of various forces in RBC

It is important to study the relative strengths of various terms of the momentum equation of RBC. Pandey *et al.* (2016a) and Pandey and Verma (2016) performed this analysis for the no-slip boundary condition in the turbulent and laminar regimes. Figure 11.6 illustrates the strengths of various forces computed using numerical data. Pandey and Verma (2016) observed that in the turbulent regime, the acceleration $\mathbf{u} \cdot \nabla \mathbf{u}$ is primarily provided by the pressure gradient $-\nabla \sigma$, while the buoyancy and viscous terms are relatively small. See Fig. 11.7(a) for an illustration.

In the turbulent regime, the Richardson number, which is the ratio of the buoyancy and nonlinear term, is

$$\mathrm{Ri} \approx \frac{c_3 \mathrm{RaPr}}{c_1 \mathrm{Pe}^2} \approx \frac{|c_1 - c_2|}{c_1} \approx 0.1, \tag{11.38}$$

which is small. Here we employed Eq. (11.32) for Pe. Note that

$$\mathrm{Ri} \approx \frac{\text{nonlinear term} - |\nabla \sigma|}{\text{nonlinear term}} \ll 1 \tag{11.39}$$

because the nonlinear term is close to $-\nabla \sigma$ in RBC. These features indicates that the turbulence in RBC and hydrodynamics are quite similar. Due to the above reasons, the energy spectrum of RBC is similar to that in hydrodynamics turbulence (Kolmogorov's spectrum), as we show in Chapter 13.

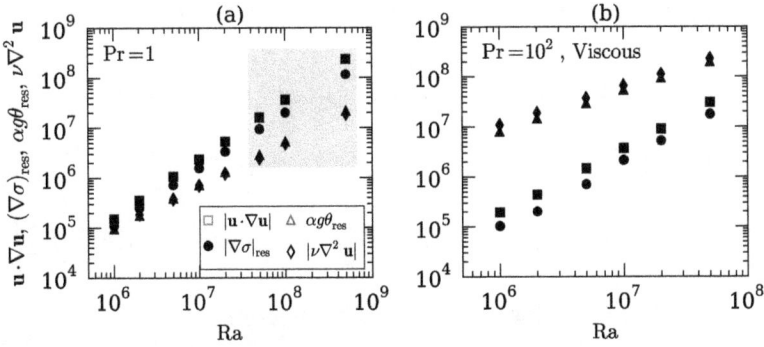

Fig. 11.6   Comparison of the rms values of $\mathbf{u} \cdot \nabla \mathbf{u}$, $(-\nabla\sigma)_{\text{res}}$, $\alpha g \theta_{\text{res}} \hat{z}$, and $\nu\nabla^2\mathbf{u}$ as function of Ra for (a) $\Pr = 1$ and (b) $\Pr = 10^2$. The shaded region of Fig. (a) corresponds to the turbulent regime, while $\Pr = 10^2$ runs are in the viscous regime. From Pandey and Verma (2016). Reprinted with permission from AIP.

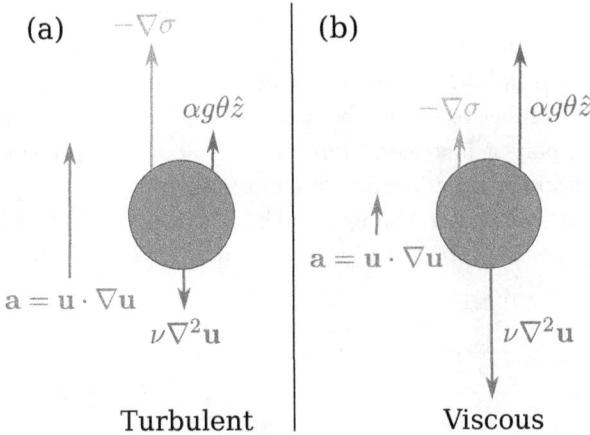

Fig. 11.7   Relative strengths of the forces acting on a fluid parcel. In the turbulent regime, the acceleration $\mathbf{u} \cdot \nabla \mathbf{u}$ is provided primarily by the pressure gradient. In the viscous regime, the buoyancy and the viscous force dominate the pressure gradient, and they balance each other. From Pandey and Verma (2016). Reprinted with permission from AIP.

In the viscous regime ($\text{Re} \lesssim 1$), the pressure gradient, $-\nabla\sigma$ is small, and the buoyancy and viscous terms balance each other. Due to this balance of forces, the acceleration $\mathbf{u} \cdot \nabla \mathbf{u}$ is quite small. See Fig. 11.7(b) for an illustration. For this case, using Eq. (11.33) we obtain

$$\text{Ri} \approx \frac{c_3 \text{Ra} \Pr}{c_1 \text{Pe}^2} \approx \frac{c_4^2 \Pr}{c_1 c_3 \text{Ra}} \approx \frac{\Pr}{\text{Ra}^{0.47}}. \tag{11.40}$$

The Reynolds number scaling can be computed using that of Péclet number:

$$\text{Laminar:} \quad \text{Re} = \frac{c_3 \text{Ra}}{c_4 \text{Pr}} \sim \frac{\text{Ra}^{0.60}}{\text{Pr}} \tag{11.41}$$

$$\text{Moderately turbulent:} \quad \text{Re} = \sqrt{\text{RaPr}}/\text{Pr} \sim \sqrt{\frac{\text{Ra}}{\text{Pr}}} \tag{11.42}$$

$$\text{Turbulent:} \quad \text{Re} = \sqrt{\frac{c_3}{|c_1 - c_2|}\text{RaPr}}/\text{Pr} \sim \text{Ra}^{-0.12}\sqrt{\frac{\text{Ra}}{\text{Pr}}} \tag{11.43}$$

$$\text{Very small Pr:} \quad \text{Re} = \frac{\text{Pe}}{\text{Pr}} \sim \text{Ra} \ (\text{Here Pe} \ll 1) \tag{11.44}$$

In the next section, we will describe the Nusselt number scaling.

11.4 Nusselt number scaling

Estimation of heat flux or Nusselt number is an important problem of RBC. Various experiments and simulations report that for $\text{Ra} \lesssim 10^{16}$, the Nusselt number exponent ranges from 0.27 to 0.38. In Fig. 11.8 we illustrate the Nu number scaling reported by Niemela and Sreenivasan (2003) and Chavanne *et al.* (2001). Cryogenic helium gas was used in both the experiments. Niemela and Sreenivasan (2003) report the Nusselt number exponent be around 0.30, while Chavanne *et al.* (2001) report that the exponent increases from 0.28 to 0.38 as Ra is increased.

The aforementioned experimental observations give a glimpse of the disagreements on the Nusselt number scaling, which is further exasperated by the elusive ultimate regime. In this section we provide some of the leading arguments on the Nusselt number scaling.

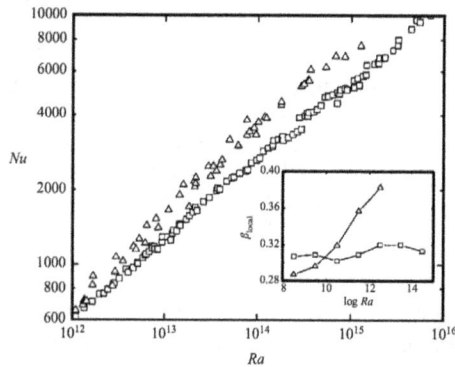

Fig. 11.8   Plot of Nusselt number Nu vs. Ra for the RBC experiment of Niemela and Sreenivasan (2003) (circles), and that of Chavanne *et al.* (2001) (triangle). The Nusselt number exponents β_{local} for the two experiments are shown in the inset. From Niemela and Sreenivasan (2003). Reprinted with permission from Cambridge University Press.

11.4.1 Arguments of Kraichnan

Using arguments similar to those presented in Sec. 11.1, Kraichnan (1962) argued that for moderate Pr, in the turbulent regime,

$$\mathrm{Nu} \sim \mathrm{Ra}^{1/2}, \tag{11.45}$$

which is called the *ultimate regime*. As described in Sec. 11.1, the arguments have certain flaws—$\langle fg \rangle \neq \langle f \rangle \langle g \rangle$, and $\Theta \neq \Delta$. Also, no experiment has observed the Nusselt number exponent closer to 1/2, though some researchers argue that the exponent shows an steady increase from Ra $\sim 10^{14}$ onwards. We will return to the discussion on the ultimate regime in Sec. 11.10.

In the next subsection, we argue how Eq. (11.15) yields Nusselt number exponent near 0.30, which is observed in RBC experiments and simulations.

11.4.2 Arguments based on correlations

The definition of Nusselt number (see Eq. (11.15)) involves $\langle u_z \theta \rangle$. Using Parseval's theorem

$$\langle u_z \theta \rangle_{xyz} = \sum_{\mathbf{k}} u_z(\mathbf{k}) \theta(\mathbf{k}) = \sum_{k_z} u_z(0,0,k_z) \theta(0,0,k_z) + \sum_{\mathbf{k}'} u_z(\mathbf{k}') \theta(\mathbf{k}'), \tag{11.46}$$

where $\mathbf{k}'$ excludes wavenumbers of the form $(0,0,k_z)$. As discussed in Sec. 9.2, $u_z(0,0,k_z) = 0$. Therefore,

$$\langle u_z \theta \rangle_{xyz} = \sum_{\mathbf{k}'} u_z(\mathbf{k}') \theta(\mathbf{k}') = \langle u_z \theta_{\mathrm{res}} \rangle_{xyz}, \tag{11.47}$$

where

$$\theta_{\mathrm{res}}(\mathbf{r}) = \theta(\mathbf{r}) - \theta_m(z) \tag{11.48}$$

with $\theta_m(z)$ representing the mean temperature profile described in Sec. 9.2. Thus, the rms value of $\theta_m(z)$ does not contribute to the Nusselt number; $\theta_{\mathrm{res}}(\mathbf{r})$, which is the fluctuation over $T_m(z)$, carries u_z that contributes to the heat convection. Hence, care is needed in extracting appropriate temperature fluctuation for the Nu computation. An application of $\theta_{\mathrm{res}}(\mathbf{r})$ leads to deviation of Nu exponent from 1/2 to ≈ 0.30.

Let us rewrite Eq. (11.15) as

$$\mathrm{Nu} \approx C_{u\theta_{\mathrm{res}}} \langle u_z'^2 \rangle_V^{1/2} \langle \theta_{\mathrm{res}}'^2 \rangle_V^{1/2}, \tag{11.49}$$

where V is a shorthand for the volume xyz, $u_z' = u_z d/\kappa$, $\theta_{\mathrm{res}}' = \theta_{\mathrm{res}}/\Delta$, and

$$C_{u\theta_{\mathrm{res}}} = \frac{\langle u_z' \theta_{\mathrm{res}}' \rangle_V}{\langle u_z'^2 \rangle_V^{1/2} \langle \theta_{\mathrm{res}}'^2 \rangle_V^{1/2}} \tag{11.50}$$

is the normalized correlation function between the vertical velocity and the residual temperature fluctuation [Verma *et al.* (2012); Pandey and Verma (2016)]. The scaling of the quantities of Eq. (11.49) are listed in Table 11.1 [Pandey and Verma

Table 11.1   Scaling of $C_{u\theta_{res}}$, $\langle\theta^2_{res}\rangle^{1/2}$, $\langle u^2_z\rangle^{1/2}$, Nu, and the global dissipation rates computed using numerical data for the no-slip boundary condition. Adopted from a table of Pandey and Verma (2016).

Quantity	Turbulent regime	Viscous regime
$\langle\theta^2_{res}\rangle^{1/2}$	$\mathrm{Ra}^{-0.13}$	$\mathrm{Ra}^{-0.18}$
$\langle u^2_z\rangle^{1/2}$	$\mathrm{Ra}^{0.51}$	$\mathrm{Ra}^{0.58}$
$C_{u\theta_{res}}$	$\mathrm{Ra}^{-0.05}$	$\mathrm{Ra}^{-0.07}$
Nu	$\mathrm{Ra}^{0.32}$	$\mathrm{Ra}^{0.33}$
ϵ_u	$(U^3/d)\mathrm{Ra}^{-0.21}$	$(\nu U^2/d^2)\mathrm{Ra}^{0.17}$
ϵ_T	$(U\Delta^2/d)\mathrm{Ra}^{-0.19}$	$(U\Delta^2/d)\mathrm{Ra}^{-0.25}$

(2016)]. In the turbulent regime, the scaling of $\langle u^2_z\rangle^{1/2} \sim \mathrm{Ra}^{1/2}$, $C_{u\theta_{res}} \sim \mathrm{Ra}^{-0.05}$ and $\langle\theta^2_{res}\rangle^{1/2} \sim \mathrm{Ra}^{-0.13}$ yield a nontrivial scaling for Nu; it takes the exponent from $1/2$ to approximately 0.32. In this section, the exponents of U, θ etc. have been derived using numerical data up to $\mathrm{Ra} \sim 10^8$. Fortunately, the derived Nu scaling, Nu $\sim \mathrm{Ra}^{0.32}$ works up to $\mathrm{Ra} \sim 10^{15}$. We need to study the exponents for larger Ra. Such analysis would help us understand the transition to the ultimate regime better.

Interestingly, Nu for the laminar regime too has a similar exponent, though $\langle u^2_z\rangle^{1/2}$, $\langle\theta^2_{res}\rangle^{1/2}$, and $C_{u\theta_{res}}$ scale differently from the turbulent regime (see Table 11.1). Thus, for both turbulent and laminar cases, careful computations of various quantities including the normalised correlation $C_{u\theta_{res}}$ yields the Nusselt number exponent to be around 0.30. In Fig. 11.9 we exhibit plots of the normalized Nusselt number, $\mathrm{NuRa}^{-0.30}$, vs. Ra for various experiments [Cioni *et al.* (1997); Xin and

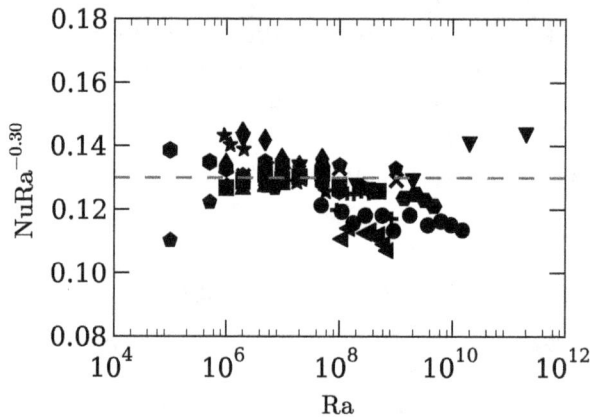

Fig. 11.9   Plots of normalized Nusselt number ($\mathrm{NuRa}^{-0.30}$) vs. Ra for experimental results [Xia *et al.* (2002); Cioni *et al.* (1997); Zhou *et al.* (2012); Xin and Xia (1997)] and Numerical results [Pandey and Verma (2016); Silano *et al.* (2010); Stevens *et al.* (2010); Scheel *et al.* (2012)]. From Pandey and Verma (2016). Reprinted with permission from AIP.

Xia (1997); Xia *et al.* (2002); Zhou *et al.* (2012)] and numerical simulations [Silano *et al.* (2010); Stevens *et al.* (2010); Scheel *et al.* (2012)], which are consistent with the aforementioned scaling.

Thus, the deviation of the Nusselt number exponent from $1/2$ to ≈ 0.30 (up to $\text{Ra} \sim 10^{15}$) is due to nontrivial scaling of $C_{u\theta_{\text{res}}}$, $\langle u_z'^2 \rangle_V^{1/2}$, and $\langle \theta_{\text{res}}'^2 \rangle_V^{1/2}$. The aforementioned scaling with Ra is primarily due to the walls and boundary layers, and they lead to $\text{Nu} \sim \text{Ra}^{0.32}$. Without these corrections, in the turbulent regime, $\text{Nu} \sim \text{Ra}^{1/2}$, as predicted by Kraichnan (1962). Lohse and Toschi (2003) simulated RBC with periodic boundary condition, thus removing the effects of the boundary layers; they observed that $\theta_{\text{res}} \sim \Delta$ and $\langle u_z \theta_{\text{res}} \rangle \sim \text{Ra}^{1/2}$, as predicted by Kraichnan. In the experimental setup of Arakeri and coworkers [Arakeri *et al.* (2000); Cholemari and Arakeri (2009)], exchange of turbulent flow takes place in a vertical pipe without boundary layers, hence the exponent of the effective Nusselt number is close to $1/2$, which is again consistent with the predictions of Kraichnan (1962).

11.4.3 *Arguments based on boundary layers*

In this subsection, we will present several models of the Nusselt number scaling that are based on boundary layer.

11.4.3.1 *Exponent of 1/3: Malkus and Spiegel*

In RBC, heat is transported as follows: Lower plate → Lower boundary layer → Bulk → Upper boundary layer → Upper plate. The boundary layer is quite thin for large Ra and $\text{Pr} \sim 1$. Based on the thinness of the thermal boundary layers for large Ra, Malkus (1954) and Spiegel (1971) argued that the heat transport would be independent of the distance d between the hot and cold plates, i.e.,

$$H = \text{Nu}(\text{Ra}, \text{Pr}) \frac{\kappa \Delta}{d} \sim \text{ independent of } d. \tag{11.51}$$

Since $\text{Ra} \sim d^3$, the above condition would hold only if

$$\text{Nu}(\text{Ra}, \text{Pr}) \sim \text{Ra}^{1/3}. \tag{11.52}$$

The above scaling is reasonably close to those observed in experiments. However, the assumption that the thermal boundary layers are thin does not hold for small Prandtl numbers. Note that for $\text{Pr} = 0$, the thermal boundary layers essentially span the whole volume. Hence, the aforementioned arguments need to be generalised, specially for small Prandtl numbers.

In the following we present two versions of the scaling arguments of Castaing *et al.* (1989), Shraiman and Siggia (1990), and Siggia (1994) who reported that $\text{Nu} \sim \text{Ra}^{2/7}\text{Pr}^{-1/7}$ for small Prandtl numbers.

11.4.3.2 *Exponent of 2/7: Castaing, Siggia, Shraiman and Siggia: Ver. 1*

Let us assume that the thermal boundary layer thickness, δ_T, and the temperature fluctuation, Θ, scale as

$$\frac{\delta_T}{d} = d\mathrm{Ra}^\beta \mathrm{Pr}^\gamma, \tag{11.53}$$

$$\frac{\Theta}{\Delta} = \Delta\mathrm{Ra}^\delta \mathrm{Pr}^\xi. \tag{11.54}$$

Using the above, the nondimensionalized vertical velocity in the turbulent bulk flow is estimated as

$$\mathrm{Pe} \approx \frac{U_z d}{\kappa} = \sqrt{\alpha g \Theta d}\frac{d}{\kappa} = \mathrm{Ra}^{(1+\delta)/2}\mathrm{Pr}^{(1+\xi)/2}, \tag{11.55}$$

where U_z is the rms value of u_z. Near the transition zone (boundary between the thermal boundary layer and the bulk), force balance between the viscous force and the buoyancy yields

$$\frac{\nu U_z}{\delta_T^2} \approx \alpha g \Delta. \tag{11.56}$$

Substitution of δ_T, Θ, U_z from Eqs. (11.53, 11.54, 11.55) into the above equation yields

$$\mathrm{Ra}^{(1-\delta)/2+2\beta} = \mathrm{Pr}^{(1+\xi)/2-2\gamma}. \tag{11.57}$$

Similarly, matching the conductive and convective heat transport near the transition zone from the thermal boundary layer to the bulk yields

$$\frac{\kappa\Delta}{\delta_T} \approx U_z\Theta. \tag{11.58}$$

On substitution of δ_T, θ, U_z as above, we obtain

$$\mathrm{Ra}^{(1+3\delta)/2+\beta} = \mathrm{Pr}^{-(1+3\xi)/2-\gamma}. \tag{11.59}$$

Since Ra and Pr are independent parameters, from Eqs. (11.57, 11.59) we deduce that

$$(1-\delta)/2 + 2\beta = 0, \tag{11.60}$$

$$(1+\xi)/2 - 2\gamma = 0, \tag{11.61}$$

$$(1+3\delta)/2 + \beta = 0, \tag{11.62}$$

$$(1+3\xi)/2 + \gamma = 0, \tag{11.63}$$

whose solution is

$$\beta = -2/7, \gamma = 1/7, \delta = -1/7, \xi = -3/7. \tag{11.64}$$

Thus we obtain the following scaling for the Nusselt and Reynolds numbers, and the temperature fluctuations:

$$\mathrm{Nu} = \frac{d}{2\delta_T} \sim \mathrm{Ra}^{2/7}\mathrm{Pr}^{-1/7} \tag{11.65}$$

$$\mathrm{Re} = \frac{\mathrm{Pe}}{\mathrm{Pr}} \sim \mathrm{Ra}^{3/7}\mathrm{Pr}^{-5/7} \tag{11.66}$$

$$\frac{\Theta}{\Delta} \sim \mathrm{Ra}^{-1/7}\mathrm{Pr}^{-3/7}. \tag{11.67}$$

11.4.3.3 *Exponent of 2/7: Castaing, Siggia, Shraiman and Siggia: Ver. 2*

We present another version of the above scaling arguments. The force balance in the bulk yields

$$\frac{U^2}{d} = \alpha g \Theta \implies \text{Pe} = \sqrt{\text{RaPr}(\Theta/\Delta)}. \tag{11.68}$$

From the definition of Nusselt number

$$\text{Nu} = \frac{\langle u'_z \theta \rangle}{\kappa \Delta/d} \approx \frac{U\Theta}{\kappa \Delta/d}, \tag{11.69}$$

we estimate that

$$\frac{\Theta}{\Delta} = \frac{\text{Nu}}{\text{Pe}}. \tag{11.70}$$

The above arguments assume perfect correlation between u_z and θ, which is a good approximation only for small Pr (see Eqs. (2.76, 11.9)). Equations (11.68, 11.70) yield

$$\text{Pe} = (\text{RaNuPr})^{1/3}. \tag{11.71}$$

A force balance between the viscous force and buoyancy near the transition zone (boundary between the thermal boundary layer and the bulk) yields

$$\frac{\nu U}{\delta_T^2} \approx \alpha g \Delta. \tag{11.72}$$

In the above we substitute $\delta_T = d/(2\text{Nu})$ that yields

$$\text{Nu}^2 \text{Pe} = \text{Ra}. \tag{11.73}$$

From Eqs. (11.71, 11.73) we deduce that

$$\text{Nu} \sim \text{Ra}^{2/7} \text{Pr}^{-1/7}, \tag{11.74}$$

$$\text{Re} \sim \text{Ra}^{3/7} \text{Pr}^{-5/7}. \tag{11.75}$$

The aforementioned relation, $\text{Re} \sim \text{Ra}^{3/7} \sim \text{Ra}^{0.43}$ is close to the scaling $\text{Pe} \sim \text{Ra}^{0.38}$ of Eq. (11.32). In addition,

$$\frac{\Theta}{\Delta} \approx \frac{\text{Nu}}{\text{Pe}} \sim \text{Ra}^{-1/7} \text{Pr}^{-3/7}. \tag{11.76}$$

The above scaling is consistent with the Ra dependence of Eq. (11.29), as well as with the experimental results of Castaing *et al.* (1989) and Niemela *et al.* (2000).

11.4.4 *Analytical bounds on* Nu *by Doering et al.*

Doering and coworkers have computed analytical bounds on Nu for various RBC configurations using *background method*. In this scheme, the background stratification $\bar{T}(z)$ of Eq. (2.15) is optimised for obtaining an upper bound on Nu. Note that $\bar{T}(z)$ is a nonlinear function in this scheme. In the following, we present an outline of Doering *et al.* (2006)'s calculation for RBC with infinite-Prandtl number.

Doering *et al.* (2006) considered infinite-Pr RBC and decomposed the temperature field into

$$T(x, y, z) = \bar{T}(z) + \theta(x, y, z) \tag{11.77}$$

The non-dimensional equations are given by

$$-\nabla\sigma + \mathrm{Ra}\theta\hat{z} + \nabla^2\mathbf{u} = 0, \tag{11.78}$$

$$\frac{\partial\theta}{\partial t} + (\mathbf{u} \cdot \nabla)\theta = -\bar{T}'u_z + \nabla^2\theta + \bar{T}'', \tag{11.79}$$

where $\bar{T}'$ and $\bar{T}''$ are the first and second derivatives of $\bar{T}$ with relative to z. The term $\bar{T}''$ appears because

$$\nabla^2 T = \nabla^2\theta + \bar{T}''. \tag{11.80}$$

In Eq. (11.78), an elimination of σ using incompressibility condition yields

$$\mathrm{Ra}\nabla_\perp^2\theta + \nabla^4 u_z = 0. \tag{11.81}$$

Shraiman and Siggia (1990)'s second exact relation in a nondimensional form is

$$\begin{aligned}
\mathrm{Nu} &= \langle(\nabla T)^2\rangle \\
&= \langle\bar{T}'^2 + (\nabla\theta)^2 + 2\bar{T}'\partial_z\theta\rangle \\
&= \langle\bar{T}'^2 + (\nabla\theta)^2 - 2\bar{T}''\theta\rangle.
\end{aligned} \tag{11.82}$$

Under a steady state, multiplication of Eq. (11.79) with 2θ and performing spatial averaging (with integration by parts) leads to

$$-2\langle\bar{T}'u_z\theta\rangle - 2\langle(\nabla\theta)^2\rangle + 2\langle\bar{T}''\theta\rangle = 0, \tag{11.83}$$

substitution of $2\langle\bar{T}''\theta\rangle$ from above in Eq. (11.82) yields

$$\mathrm{Nu} = \langle\bar{T}'^2 - [(\nabla\theta)^2 + 2\bar{T}'u_z\theta]\rangle. \tag{11.84}$$

Now, if the stratification profile $\bar{T}(z)$ is chosen in such a way that

$$Q\{\theta\} = \langle(\nabla\theta)^2 + 2\bar{T}'u_z\theta\rangle \geq 0 \tag{11.85}$$

for all $\theta(x, y, z)$ satisfying the boundary condition, then

$$\mathrm{Nu} \leq \int_0^1 \bar{T}'^2 dz, \tag{11.86}$$

thus providing an upper bound on Nu.

Doering *et al.* (2006) considered the following form of $\bar{T}(z)$:

$$\bar{T}(z) = \begin{cases} 1 - \frac{z}{\delta} & \text{for } 0 < z < \delta \\ 1/2 + \lambda(\delta)\ln(z/(1-z)) & \text{for } \delta < z < 1 - \delta \\ \frac{1-z}{\delta} & \text{for } 1 - \delta < z < 1, \end{cases} \tag{11.87}$$

where

$$\lambda(\delta) = \frac{1}{2\ln((1-\delta)/\delta)}, \tag{11.88}$$

and δ is small. For $\bar{T}(z)$ of Eq. (11.87), after a complex sets of steps, Doering *et al.* (2006) showed that the asymptotic limit of $Q\{\theta\}$ is achieved when

$$\delta \sim \left(\frac{30}{\text{Ra}\ln\text{Ra}}\right)^{1/3}. \tag{11.89}$$

Using

$$\int_0^1 \bar{T}'^2 dz = \frac{2}{\delta}\left\{1 + O([\ln\delta]^{-2})\right\} \text{ as } \delta \to 0, \tag{11.90}$$

Doering *et al.* (2006) concluded that

$$\text{Nu} \leq 2\left(\frac{\text{Ra}\ln\text{Ra}}{30}\right)^{1/3} \approx 0.643[\text{Ra}\ln\text{Ra}]^{1/3} \tag{11.91}$$

This is how Doering *et al.* (2006) obtained an upper bound on Nu for infinite-Pr RBC. Using a similar procedure, Whitehead and Doering (2011) showed that for 2D RBC with free-slip isothermal boundaries,

$$\text{Nu} \leq 0.289\text{Ra}^{5/12}. \tag{11.92}$$

It is important to note that a realistic flow may not reach the aforementioned upper bound on Nu. Also, Doering *et al.* (2006)'s computation emphasizes the role of $C_{u,\theta}$, boundary layer, and the form of $\bar{T}$ on the Nussult number scaling.

Let us contrast the models of Castaing, Siggia, Shraiman and Sigga, Doering *et al.*, and that of Pandey and Verma. In all these models, the boundary layer plays an important role. For example, Pandey and Verma (2016) showed that $C_{u,\theta} \sim \text{Ra}^{-0.05}$ and $\Theta_{\text{res}} \sim \text{Ra}^{-0.13}$, which bring down the Nusselt number exponent from $1/2$ to ≈ 0.30. The Nu exponent corrections in Doering *et al.* (2006)'s model too stems from the correlations induced by the boundary layers and walls.

11.4.5 *Arguments of Grossmann and Lohse*

Grossmann and Lohse (2000) derived the Nu scaling using the scaling of the viscous dissipation rate, ϵ_u, and the entropy dissipation rate, ϵ_θ, as well as Shraiman and Siggia (1990)'s exact relations. Since the scaling of ϵ_u and ϵ_θ will be discussed in the subsequent section, we defer discussion on Grossmann and Lohse's model to a later section.

11.5 Scaling of viscous dissipation rate

As discussed in Chapter 10, in hydrodynamic turbulence, the viscous dissipation rate is

$$\epsilon_u = \frac{U^3}{d}, \tag{11.93}$$

where U and d are the large-scale velocity and length scale respectively. Note that ϵ_u equals Π_u, as well as the energy supplied at the large scales. Let us explore whether the above law is applicable to RBC or not.

As discussed in Sec. 9.3, Shraiman and Siggia (1990) derived the following exact relation for the viscous dissipation rate, ϵ_u:

$$\epsilon_u = \nu \frac{1}{\text{Vol}} \int d\mathbf{r} |\nabla \times \mathbf{u}|^2 = \frac{\nu^3}{d^4} \frac{(\text{Nu}-1)\text{Ra}}{\text{Pr}^2} = \frac{U^3}{d} \frac{(\text{Nu}-1)\text{RaPr}}{\text{Pe}^3}, \qquad (11.94)$$

where Vol is the volume of the box. In the ultimate regime, or in the absence of boundary layer, as in a periodic box or in turbulent exchange flow of Arakeri *et al.* (2000), Nu $\sim$ Ra$^{1/2}$ and Pe $\sim$ Ra$^{1/2}$. Substitution of these relations in Eq. (11.94) yields

$$\epsilon_u \sim \frac{U^3}{d}, \qquad (11.95)$$

as in hydrodynamic turbulence. The situation however changes in the presence of walls.

In RBC with walls, in the turbulent regime, for Ra $\lesssim 10^6 \text{Pr}^2$, Pe $\sim$ Ra$^{1/2}$ and Nu $\sim$ Ra$^{0.32}$ that leads to $\epsilon_u \neq U^3/d$, rather

$$\epsilon_u \sim \frac{U^3}{d} \text{Ra}^{-0.18}. \qquad (11.96)$$

As we argued earlier, it is due to the relative suppression of the nonlinearity in RBC compared to free or unbound turbulence (see Eq. (11.31)).

As discussed in Sec. 11.2, for Ra $\gg 10^6 \text{Pr}^2$, Pe $\sim$ Ra$^{0.38}$. If we assume that Nu $\sim$ Ra$^{0.32}$ continues to hold, then the exact relation of Eq. (11.94) yields

$$\epsilon_u \sim \frac{U^3}{d} \text{Ra}^{1.32-0.38\times3} \sim \frac{U^3}{d} \text{Ra}^{0.18}. \qquad (11.97)$$

The viscous dissipation rate being larger than U^3/d appears odd, and it needs further investigation. We need to analyse the experimental and numerical data for very large Ra.

In the viscous regime, hydrodynamic flow has

$$\epsilon_u \approx \frac{\nu U^2}{d^2}, \qquad (11.98)$$

hence Shraiman and Siggia (1990)'s exact relation can be rewritten as

$$\epsilon_u = \frac{\nu^3}{d^4} \frac{(\text{Nu}-1)\text{Ra}}{\text{Pr}^2} = \frac{\nu U^2}{d^2} \frac{(\text{Nu}-1)\text{Ra}}{\text{Pe}^2}. \qquad (11.99)$$

In the laminar regime, Nu $\sim$ Ra$^{0.33}$ and Pe $\sim$ Ra$^{0.58}$ (see Table 11.1), substitution of which in Eq. (11.99) yields

$$\epsilon_u \approx \frac{\nu U^2}{d^2} \text{Ra}^{0.17}. \qquad (11.100)$$

Thus, in the laminar regime, the boundary layers in RBC enhance ϵ_u compared to unbounded flows. It will be interesting to verify if laminar RBC with periodic walls also satisfies Eq. (11.98).

We compute the following normalized dissipation rates using numerical simulations:

$$C_{\epsilon_u,1} = \frac{\epsilon_u}{U^3/d} = \frac{(\text{Nu}-1)\text{RaPr}}{\text{Pe}^3} \sim \text{Ra}^{-0.21}\text{Pr, (turbulent)} \qquad (11.101)$$

$$C_{\epsilon_u,2} = \frac{\epsilon_u}{\nu U^2/d^2} = \frac{(\text{Nu}-1)\text{Ra}}{\text{Pe}^2} \sim \text{Ra}^{0.17}, \quad \text{(viscous)} \qquad (11.102)$$

which are plotted in Fig. 11.10(a,b). We observe that for Pr $= 1$ and 6.8, $C_{\epsilon_u,1}/\text{Pr} \sim \text{Ra}^{-0.22\pm0.02}$ and $\text{Ra}^{-0.25\pm0.03}$ respectively, which are in good agreement with Eq. (11.101). The exponents of $C_{\epsilon_u,2}$ for Pr $= 6.8$ and 10^2 are 0.22 ± 0.01 and 0.19 ± 0.02 respectively, which are in reasonable accordance with Eq. (11.102). Table 11.1 lists the Ra-dependence of the above dissipation rates in the turbulent and viscous regimes. We remark that the aforementioned scaling is applicable in the intermediate regime (Ra $< 10^{15}$).

Ni *et al.* (2011) estimated $\epsilon_u \sim U^3/d \sim \text{Ra}^{1.5}$ by substituting experimentally-measured U in the above formula. This, unfortunately, is an incorrect estimate of ϵ_u of RBC because it assumes that $\epsilon_u \sim U^3/d$.

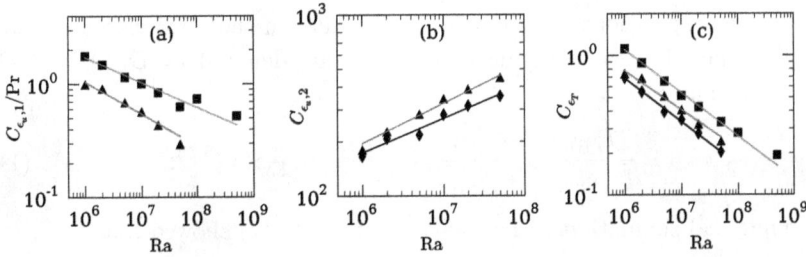

Fig. 11.10 The normalized dissipation rates: (a) $C_{\epsilon_u,1}/\text{Pr}$, (b) $C_{\epsilon_u,2}$, and (c) C_{ϵ_T} as functions of Ra for Pr $= 1$ (red squares), Pr $= 6.8$ (blue triangles), and Pr $= 10^2$ (black diamonds). See Eqs. (11.101, 11.102, 11.117). From Pandey and Verma (2016). Reprinted with permission from AIP.

In the next section, we will compare ϵ_u in the bulk and in the boundary layer.

11.6 Comparison between the dissipation rates in the bulk and boundary layer

Strong viscous dissipation takes place in the regions with large vorticity or large velocity gradients, which occurs in the boundary layer. Note however that in the turbulent regime, the volume of the boundary layer is much smaller than the bulk region. In this section we compare the dissipation rates in the bulk and in the boundary layers. Here, we limit our discussion to Pr ~ 1.

The scaling of the viscous dissipation in the bulk is similar to the total viscous dissipation rate:

$$\epsilon_{u,\text{bulk}} \sim \frac{U^3}{d}\text{Ra}^{-0.18}. \qquad (11.103)$$

Since the fluid flow in the boundary layers is laminar, we expect

$$\epsilon_{u,\mathrm{BL}} \sim \frac{\nu U^2}{\delta_u^2},\tag{11.104}$$

where δ_u is the thickness of the viscous boundary layer. Hence, the ratio of the two dissipation rates is

$$\frac{\epsilon_{u,\mathrm{BL}}}{\epsilon_{u,\mathrm{bulk}}} \sim \mathrm{Ra}^{0.18} \left(\frac{\nu U^2}{\delta_u^2}\right) / \left(\frac{U^3}{d}\right)\tag{11.105}$$

$$\sim \frac{1}{\mathrm{Re}}\left(\frac{d}{\delta_u}\right)^2 \mathrm{Ra}^{0.18} \sim \left(\frac{d}{\delta_u}\right)^2 \mathrm{Ra}^{-0.32}.\tag{11.106}$$

Here we employed $\mathrm{Re} \sim \mathrm{Ra}^{1/2}$. Note however that the boundary layers occupy much less volume than the bulk. If A is the horizontal cross-section area, and δ_u is the width of the boundary layer, then the ratio of the volumes of the boundary layer and the bulk is

$$\frac{V_{\mathrm{BL}}}{V_{\mathrm{bulk}}} \approx \frac{\delta_u A}{d A} \approx \frac{\delta_u}{d}.\tag{11.107}$$

Using the above relations, we can deduce the ratio of the total viscous dissipation in the boundary layer and in the bulk, which are denoted by $\tilde{D}_{u,\mathrm{BL}}$ and $\tilde{D}_{u,\mathrm{bulk}}$ respectively. The ratio is

$$\frac{\tilde{D}_{u,\mathrm{BL}}}{\tilde{D}_{u,\mathrm{bulk}}} \sim \frac{\epsilon_{u,\mathrm{BL}}}{\epsilon_{u,\mathrm{bulk}}} \times \frac{V_{\mathrm{BL}}}{V_{\mathrm{bulk}}} \sim \frac{d}{\delta_u}\mathrm{Ra}^{-0.32}.\tag{11.108}$$

Using numerical simulations, Bhattacharya *et al.* (2017) showed that

$$\frac{\delta_u}{d} \sim \mathrm{Re}^{-0.44} \sim (\mathrm{Ra}^{1/2})^{-0.44} \sim \mathrm{Ra}^{-0.22}.\tag{11.109}$$

This scaling differs slightly from that predicted by Prandtl-Blasius theory, according to which

$$\frac{\delta_u}{d} \sim \mathrm{Re}^{-1/2} \sim \mathrm{Ra}^{-1/4}.\tag{11.110}$$

Using Eqs. (11.108, 11.109), we deduce that

$$\frac{\tilde{D}_{u,\mathrm{BL}}}{\tilde{D}_{u,\mathrm{bulk}}} \sim \mathrm{Ra}^{-0.10}.\tag{11.111}$$

Thus, in RBC, the viscous dissipation in the boundary layer and in the bulk are comparable. For very large Ra, the bulk dissipation will dominate the dissipation in the boundary layer. This is counter to general belief that the viscous dissipation occurs primarily in the plumes of the boundary layers. Note that the aforementioned scaling may need to be modified in the ultimate regime where ϵ_u and δ_u/d are expected to scale very differently.

Bhattacharya *et al.* (2017) also showed that $\epsilon_{u,\mathrm{bulk}}$ has a log-normal distribution, while $\epsilon_{u,\mathrm{BL}}$ has stretched-exponential distribution. This is consistent with the experimental observations that in the bulk, the local dissipation is weak but spread out in larger volume. The local dissipation in the boundary layer, however, is more intense but it is localised in smaller volume. Several caveats, for very large-Pr RBC, the boundary layer extends to the whole region ($2\delta_u \approx d$), and hence $\tilde{D}_{u,\text{BL}}$ dominates $\tilde{D}_{u,\text{bulk}}$. The scaling of the dissipation rates for small-Pr RBC need closer examination.

In the next section, we will discuss the scaling of entropy dissipation rate.

11.7 Scaling of entropy dissipation rate

As discussed in Sec. 9.3, Shraiman and Siggia (1990) derived the following exact relation for the entropy dissipation rate:

$$\epsilon_\theta = \int d\mathbf{r}\kappa|\nabla\theta|^2 = \kappa\frac{\Delta^2}{d^2}(\text{Nu}-1) \approx \frac{U\Delta^2}{d}\frac{\text{Nu}}{\text{Pe}}. \tag{11.112}$$

Substitution of the Nu and Pe scaling of Table 11.1 in the above expression yields the following relations. In the turbulent regime,

$$\epsilon_\theta \approx \frac{U\Delta^2}{d}\frac{\text{Ra}^{0.32}}{\text{Ra}^{0.51}} \sim \frac{U\Delta^2}{d}\text{Ra}^{-0.19}, \tag{11.113}$$

and in the viscous regime

$$\epsilon_\theta \sim \frac{U\Delta^2}{d}\frac{\text{Ra}^{0.33}}{\text{Ra}^{0.58}} \sim \frac{U\Delta^2}{d}\text{Ra}^{-0.25}. \tag{11.114}$$

The above Ra-dependent corrections are due to the boundary layers. In the turbulent regime, for $\text{Pr}=1$, the ratio of the nonlinear term and the thermal diffusion term of the temperature equation is

$$\frac{\mathbf{u}\cdot\nabla\theta}{\kappa\nabla^2\theta} = \frac{c_5}{c_6}\frac{Ud}{\kappa} \sim \text{Ra}^{-0.30}\text{Pe}, \tag{11.115}$$

since

$$c_5 = \frac{|\mathbf{u}\cdot\nabla\theta|}{U\theta/d} \sim \text{Ra}^{0.09},$$

$$c_6 = \frac{|\nabla^2\theta|}{\theta/d^2} \sim \text{Ra}^{0.39}. \tag{11.116}$$

Thus, the nonlinearity in the temperature equation of RBC is weaker than that in unbounded flow, such as passive scalar in a periodic box. Consequently the entropy flux is weaker than that for unbounded flows, which is the reason for the weaker entropy dissipation (see Eqs. (11.113, 11.114)). This is similar to the suppression of ϵ_u and nonlinearity in the momentum equation. See Sec. 11.5.

We numerically compute the following normalized dissipation rates:

$$C_{\epsilon_T} = \frac{\epsilon_T}{U\Delta^2/d} = \frac{\text{Nu}}{\text{Pe}} \sim \begin{cases} \text{Ra}^{-0.19} & \text{for turbulent regime} \\ \text{Ra}^{-0.25} & \text{for laminar regime} \end{cases} \tag{11.117}$$

which are plotted in Fig. 11.10(c). Table 11.1 lists the Ra-dependence of the temperature dissipation rates in the turbulent and viscous regimes.

Using numerical simulations, Scheel *et al.* (2013) computed the entropy dissipation rate in the full volume and also in the bulk. They reported that for the full volume, $\epsilon_\theta \sim \text{Ra}^{0.20}$, which is closer to prediction of Eq. (11.113). In the reduced volume ($8V/27$), they reported that $\epsilon_\theta \sim \text{Ra}^{0.42}$, which is closer to the expected $U\Delta^2/d$ with $U \sim \text{Ra}^{1/2}$. Thus, the numerical results of Scheel *et al.* (2013) are in good agreement with the scaling presented here. Emran and Schumacher (2008) and Scheel *et al.* (2013) report exponential tail in the probability distribution of the entropy dissipation rate. Note that the aforementioned corrections are expected to be valid till $\text{Ra} < 10^{15}$ or so. We need to extend these arguments to the ultimate regime.

11.8 Grossmann-Lohse model for the scaling of Nusselt and Péclet numbers

Grossmann and Lohse (2000), Grossmann and Lohse (2001), Ahlers *et al.* (2009), and Stevens *et al.* (2013) derived formulas for $\text{Nu}(\text{Ra}, \text{Pr})$ and $\text{Re}(\text{Ra}, \text{Pr})$ using the properties of viscous and entropy dissipation rates, and the exact relations of Shraiman and Siggia (1990). Here we briefly describe the main assumptions and results of Grossmann-Lohse model, which is abbreviated as GL model.

11.8.1 *Grossmann-Lohse model*

Here, we briefly describe the derivation of GL model. For details refer to Grossmann and Lohse (2000) and Grossmann and Lohse (2001).

(1) The global viscous dissipation rate, $\bar{D}_u$, and thermal dissipation rate, $\bar{D}_\theta$, are sums of the respective dissipation rates in the bulk and boundary layers[2], i.e.,

$$\bar{D}_u = \bar{D}_{u,\text{BL}} + \bar{D}_{u,\text{bulk}}, \qquad (11.118)$$

$$\bar{D}_\theta = \bar{D}_{T,\text{BL}} + \bar{D}_{T,\text{bulk}}, \qquad (11.119)$$

where BL and bulk denote the boundary layer and the bulk respectively.

(2) Based on theory of hydrodynamic turbulence and properties of viscous and

---

[2]Grossmann and Lohse (2000) and Grossmann and Lohse (2001) denote the total dissipation rates by the symbols ϵ_u and ϵ_T. In our book, we use ϵ_u and ϵ_T for the average dissipation rates, and employ $\bar{D}_u$ and $\bar{D}_\theta$ for the total dissipation rates. Hence, we have adopted our notation in this section.

thermal boundary layers, Grossmann and Lohse deduced that

$$\bar{D}_{u,\text{bulk}} \sim \frac{U^3}{d} \sim \frac{\nu^3}{d^4} \text{Re}^3, \tag{11.120}$$

$$\bar{D}_{u,\text{BL}} \sim \nu \frac{U^2}{\delta_u^2} \frac{\delta_u}{d} \sim \frac{\nu^3}{d^4} \text{Re}^{5/2}, \tag{11.121}$$

$$\bar{D}_{\theta,\text{bulk}} \sim \frac{U\Delta^2}{d} \sim \kappa \frac{\Delta^2}{d^2} \text{PrRe}, \tag{11.122}$$

$$\bar{D}_{\theta,\text{BL}} \sim \kappa \frac{\Delta^2}{d^2} (\text{PrRe})^{1/2}. \tag{11.123}$$

(3) The widths of the viscous boundary layer, δ_u, and of the thermal boundary layer, δ_T, are

$$\delta_u = \frac{ad}{\sqrt{\text{Re}}}, \tag{11.124}$$

$$\delta_T = \frac{d}{2\text{Nu}}. \tag{11.125}$$

The dissipation rates of item (2) need to be appropriately scaled for which Grossmann and Lohse proposed that

$$\bar{D}_{u,\text{bulk}} = c_2 \frac{\nu^3}{d^4} \text{Re}^3, \tag{11.126}$$

$$\bar{D}_{u,\text{BL}} = c_1 \frac{\nu^3}{d^4} \frac{\text{Re}^2}{g(\sqrt{\text{Re}_c/\text{Re}})}, \tag{11.127}$$

$$\bar{D}_{\theta,\text{bulk}} = c_4 \kappa \frac{\Delta^2}{d^2} \text{PrRe} f \left[\frac{2a\text{Nu}}{\sqrt{\text{Re}_c}} g \left(\sqrt{\frac{\text{Re}_c}{\text{Re}}} \right) \right], \tag{11.128}$$

$$\bar{D}_{\theta,\text{BL}} = c_3 \kappa \frac{\Delta^2}{d^2} \sqrt{\text{RePr}} \left\{ f \left[\frac{2a\text{Nu}}{\sqrt{\text{Re}_c}} g \left(\sqrt{\frac{\text{Re}_c}{\text{Re}}} \right) \right] \right\}^{1/2}, \tag{11.129}$$

where

$$f(x) = (1 + x^n)^{-1/n}, \tag{11.130}$$

$$g(x) = x(1 + x^n)^{-1/n}, \tag{11.131}$$

and $n = 4$, $c_1 = 8.05$, $c_2 = 1.38$, $c_3 = 0.487$, $c_4 = 0.922$, $a = 0.922$, and $\text{Re}_c = (2a)^2$. These parameters have been determined using various experimental and numerical data. For details, refer to Stevens *et al.* (2013).

(4) Now using Eqs. (11.118–11.129) and the exact relations of Shraiman and Siggia (1990), Eqs. (9.31, 9.36), Grossmann and Lohse derived the following coupled equations for Re and Nu:

$$(\text{Nu} - 1)\text{RaPr}^{-2} = c_1 \frac{\text{Re}^2}{g(\sqrt{\text{Re}_c/\text{Re}})} + c_2 \text{Re}^3, \tag{11.132}$$

$$\text{Nu} - 1 = c_3 \sqrt{\text{RePr}} \left\{ f \left[\frac{2a\text{Nu}}{\sqrt{\text{Re}_c}} g \left(\sqrt{\frac{\text{Re}_c}{\text{Re}}} \right) \right] \right\}^{1/2}$$

$$+ c_4 \text{RePr} f \left[\frac{2a\text{Nu}}{\sqrt{\text{Re}_c}} g \left(\sqrt{\frac{\text{Re}_c}{\text{Re}}} \right) \right]. \tag{11.133}$$

(5) Further, Grossmann and Lohse (2000) derived the scaling relations for the limiting cases, which are listed in Fig. 11.11. Let us take the bulk dominated flows with $\delta_u < \delta_T$ as an example. For this case, $Nu \sim (RaPr)^{1/2}$ and $Re \sim Ra^{1/2}Pr^{-1/2}$, as indicated by item IV_l of Fig. 11.11.

Regime	Dominance of	BLs	Nu	Re
I_l	$\epsilon_{u,BL}, \epsilon_{\theta,BL}$	$\lambda_u < \lambda_\theta$	$Ra^{1/4}Pr^{1/8}$	$Ra^{1/2}Pr^{-3/4}$
I_u		$\lambda_u > \lambda_\theta$	$Ra^{1/4}Pr^{-1/12}$	$Ra^{1/2}Pr^{-5/6}$
I_∞		$\lambda_u = L/4 > \lambda_\theta$	$Ra^{1/5}$	$Ra^{3/5}Pr^{-1}$
II_l	$\epsilon_{u,bulk}, \epsilon_{\theta,BL}$	$\lambda_u < \lambda_\theta$	$Ra^{1/5}Pr^{1/5}$	$Ra^{2/5}Pr^{-3/5}$
II_u		$\lambda_u > \lambda_\theta$	$Ra^{1/5}$	$Ra^{2/5}Pr^{-2/3}$
III_u	$\epsilon_{u,BL}, \epsilon_{\theta,bulk}$	$\lambda_u > \lambda_\theta$	$Ra^{3/7}Pr^{-1/7}$	$Ra^{4/7}Pr^{-6/7}$
III_∞		$\lambda_u = L/4 > \lambda_\theta$	$Ra^{1/3}$	$Ra^{2/3}Pr^{-1}$
IV_l	$\epsilon_{u,bulk}, \epsilon_{\theta,bulk}$	$\lambda_u < \lambda_\theta$	$Ra^{1/2}Pr^{1/2}$	$Ra^{1/2}Pr^{-1/2}$
IV_u		$\lambda_u > \lambda_\theta$	$Ra^{1/3}$	$Ra^{4/9}Pr^{-2/3}$

Fig. 11.11   Grossmann-Lohse model: A table depicting the Re and Nu scaling for the limiting cases. From Ahlers *et al.* (2009). Reprinted with permission from APS.

The GL model is one of the most popular models for computing Re and Nu. Using the GL model, Stevens *et al.* (2013) computed Nu for the parameters used in various experiments and simulations, and observed excellent agreement between the model predictions and the experimental and numerical results. See Fig. 11.12 for an illustration. Refer to Grossmann and Lohse (2000); Ahlers *et al.* (2009); Stevens *et al.* (2013) for details.

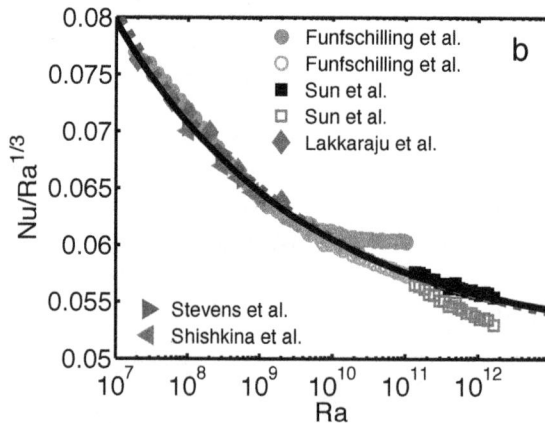

Fig. 11.12   The Nu predictions by the GL model (solid black line) match with the experimental and numerical results very well. From Stevens *et al.* (2013). Reprinted with permission from Cambridge University Press.

The GL model also predicts Pe or Re for various regimes, as shown in Fig. 11.11. According to their model, in the viscous regime the dominant $\epsilon_{u,\mathrm{BL}}$ yields Re $\sim$ Ra$^\xi$ with $\xi = 3/5$, $4/7$, and $2/3$, which are close to Re $\sim$ Ra$^{0.60}$ predicted in Sec. 11.2. On the other hand, for the turbulent regime, the GL model predicts that Re $\sim$ Ra$^{1/2}$ and Ra$^{2/5}$, which are the intermediate and turbulent regimes shown in Fig. 11.5. Thus, the GL model and that of Pandey and Verma are consistent with each other.

In the next subsection, we contrast the two models, GL and that of Pandey and Verma, in more detail.

11.8.2 *Comparison between GL model and that of Pandey and Verma*

As described above, the predictions of Re scaling by the models of GL, and Pandey and Verma are in qualitative agreement with each other. In this subsection we report a quantitative comparison between these two models (for the predictions of Re).

In Fig. 11.13, we plot several numerical and experimental results for Pr = 0.021, 0.7, 1, 6.8, 1000, and 2547.9. The figure also depicts the predictions of GL model (dashed curves) and that of Pandey and Verma (solid curves). As shown in Fig. 11.13(a), for Pr $\leq$ 1, the predictions of both the models are reasonably close to the respective numerical and experimental data. However, for Pr > 1, shown in Fig. 11.13(b), the model predictions of Pandey and Verma are somewhat better than those of GL; the predictions of GL are typically lower than the simulation and experimental results. Earlier, Ahlers *et al.* (2009) had shown that the GL model under-predicts the Reynolds number.

As described in Sec. 11.2, the model of Pandey and Verma predicts that Pe $\sim$ Ra$^{0.60}$ in the viscous regime, then Pe $\sim$ Ra$^{1/2}$ in the intermediate regime, after

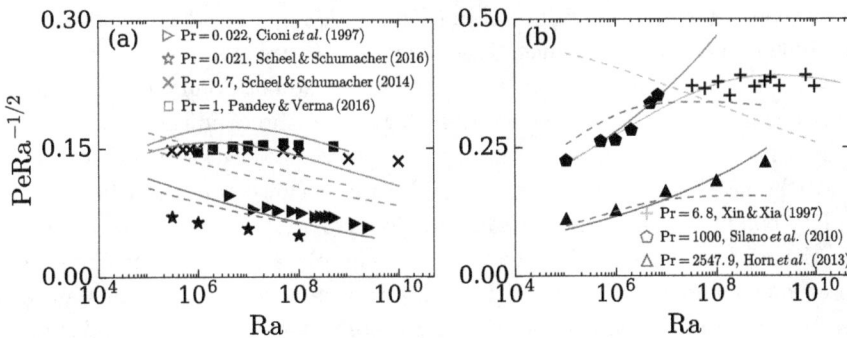

Fig. 11.13 Comparison between the model predictions of GL (dashed line) and Pandey and Verma (2016) (solid line) with the existing experimental and simulation results. The plots show that both the models work quite well for Pr $\lesssim$ 1, but GL model under-predicts the Péclet number for Pr $\geq$ 1.

which it transitions to the turbulent regime with scaling of Pe $\sim$ Ra$^{0.38}$. We observe the first two phases in Fig. 11.13(b), and expect the turbulent phase to show up at very large Ra $\gg 10^{12}$, as predicted by Eq. (11.35). For Pr $\approx 10^2$, the turbulent regime may be beyond Ra $\sim 10^{12}$. Unfortunately, the GL model does not capture the viscous regime as accurately as the model of Pandey and Verma.

Before we end our discussion on GL model, it is prudent to examine the assumptions made in the GL model. As described in Sec. 11.8.1, GL model assumes that $\epsilon_{u,\text{bulk}} \sim U^3/d$, which is not strictly valid for RBC. We show in Sec. 11.5 that $\epsilon_{u,\text{bulk}} \sim (U^3/d)\text{Ra}^{-0.18}$. In RBC, the kinetic energy flux and the viscous dissipation are lower than that for free turbulence. As shown in Sec. 11.7, the entropy dissipation rate too is suppressed compared to free turbulence. Therefore, we believe that the characterisation of $\epsilon_{u,\text{bulk}}$ and $\epsilon_{\theta,\text{bulk}}$ in GL model as in item (2) of the Sec. 11.8.1 needs further investigation.

In addition, the computation of Nu and Re in GL model using Eqs. (11.132, 11.133) is more computational intensive than the quadratic equation solver required for the model by Pandey and Verma. Note that both the models contain several coefficients that are determined using numerical and/or experiment data.

In the next section we describe the large-scale quantities of RBC with free-slip boundary condition.

11.9 Scaling of large-scale quantities for free-slip RBC

Pandey and Verma (2016) studied the scaling of Péclet and Nusselt numbers for the free-slip boundary condition and contrasted the results with those for no-slip boundaries. They performed DNS of RBC for Pr = 0.02, 1, 4.38, 10^2, 10^3, and ∞, and Ra ranging from 10^5 to 2×10^8. For the velocity field, they employed free-slip boundaries at all the walls, and for the temperature field, they employed conducting boundary at the top and bottom plates, and adiabatic boundaries at the side walls.

They computed the rms values of $|\mathbf{u} \cdot \nabla \mathbf{u}|$, $|(-\nabla \sigma)_{\text{res}}|$, $|\alpha g \theta_{\text{res}} \hat{\mathbf{z}}|$, and $|\nu \nabla^2 \mathbf{u}|$ for the turbulent and viscous regimes. These magnitudes are exhibited schematically in Fig. 11.14. In the turbulent regime, the pressure gradient dominates the buoyancy and viscous drag, similar to that observed for the no-slip condition. Also, the flows with free-slip boundaries have significant horizontal acceleration, which was not the case for the no-slip boundary condition. This is because compared to the no-slip boundary condition, the free-slip walls provide less viscous drag horizontally.

In the viscous regime, the pressure gradient plays an important role, and it opposes buoyancy (see Fig. 11.14(b)). Interestingly, the relative magnitudes of forces for the viscous regime is very similar to those observed near thermal instability for the free-slip boundary condition. See Sec. 6.1 and Fig. 6.4 for comparison. This similarity is because the flow near the instability is viscous. Note that the pressure gradient for the no-slip boundaries is negligible (see Fig. 11.7).

Pandey and Verma (2016) also computed c_i's of Eqs. (11.22) for the free-slip

(a)

(b)

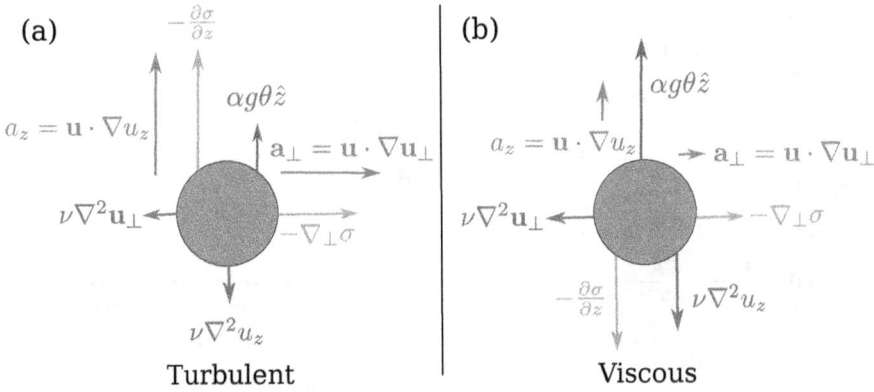

Turbulent

Viscous

Fig. 11.14 For RBC with free-slip walls, the relative strengths of the forces acting on a fluid parcel. (a) In the turbulent regime, the acceleration $\mathbf{u} \cdot \nabla \mathbf{u}$ is provided primarily by the pressure gradient. (b) In the viscous regime, the buoyancy is balanced by the pressure gradient and the viscous force along $\hat{z}$. From Pandey and Verma (2016). Reprinted with permission from AIP.

Table 11.2 Functional dependence of the coefficients c_i's on Ra and Pr for the free-slip boundary condition. Adopted from a table of Pandey and Verma (2016).

	$\mathrm{Pr} \leq 9$	$\mathrm{Pr} > 9$
c_1	$0.2\mathrm{Ra}^{0.20}$	5
c_2	$0.05(4 + 0.04\mathrm{Pr})\mathrm{Ra}^{0.15}$	$22(6 + 0.28\mathrm{Pr})\mathrm{Ra}^{-0.20}$
c_3	$1.35\mathrm{Ra}^{-0.10}\mathrm{Pr}^{-0.05}$	0.30
c_4	$2 \times 10^{-4}(1300/\mathrm{Pr} + 150)\mathrm{Ra}^{0.50}$	$0.01(1300/\mathrm{Pr} + 150)\mathrm{Ra}^{0.28}$

boundary condition. These results, listed in Table 11.2, differs from those for no-slip walls. Most notably, for large Pr, $c_2 \propto \mathrm{Pr}$, consistent with the earlier observation that the pressure gradient plays an important role for viscous RBC. We refer the reader to Pandey and Verma (2016) for a detailed discussion.

We can compute the Péclet number Pe using Eq. (11.24) and c_i's discussed above. In Fig. 11.15(a) we present the model results along with numerical results. The model captures the numerical data quite well, partly because the same data were used to compute c_i's of Table 11.2.

Here we present the scaling of the laminar regime. For the no-slip boundary condition, $\mathrm{Pe} \approx (c_3/c_4)\mathrm{Ra}$ because the pressure gradient was negligible. On the contrary, for the free-slip boundary condition, c_2 is quite important and it needs to be incorporated in the scaling analysis. Given that $c_2 = -c_2'\mathrm{Pr}$, where c_2' is a positive constant, Eq. (11.23) yields the following equation for the Péclet number:

$$c_2'\mathrm{Pe}^2 + c_4\mathrm{Pe} - c_3\mathrm{Ra} = 0. \tag{11.134}$$

In this equation, buoyancy balances the pressure gradient and viscous force, as shown in Fig. 11.14(b). Note that the above Pe is independent of Pr as observed

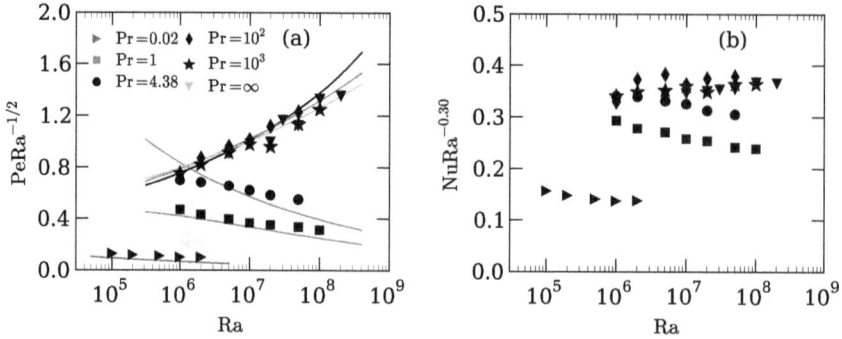

Fig. 11.15   For the free-slip boundary condition: (a) Plots of $\text{PeRa}^{-1/2}$ vs. Ra. The continuous curves represent model predictions that match well with the numerical results. (b) Plots of $\text{NuRa}^{-0.30}$ vs. Ra. From Pandey and Verma (2016). Reprinted with permission from AIP.

Table 11.3   Summary of the scaling for the free-slip boundary condition. Quantities are same as those in Table 11.1. Adopted from a table of Pandey and Verma (2016).

	Turbulent regime	Viscous regime
$C_{u\theta_{res}}$	$\text{Ra}^{-0.06}$	$\text{Ra}^{-0.17}$
$\langle \theta_{res}^2 \rangle^{1/2}$	$\text{Ra}^{-0.10}$	$\text{Ra}^{-0.12}$
$\langle u_z^2 \rangle^{1/2}$	$\text{Ra}^{0.43}$	$\text{Ra}^{0.61}$
Nu	$\text{Ra}^{0.27}$	$\text{Ra}^{0.32}$
ϵ_u	U^3/d	$(\nu U^2/d^2)\text{Ra}^{0.10}$
ϵ_T	$(U\Delta^2/d)\text{Ra}^{-0.15}$	$(U\Delta^2/d)\text{Ra}^{-0.29}$

in numerical simulations [Pandey *et al.* (2014)]. The relation that $c_2 \propto \text{Pr}$ for large Pr plays a major role for this phenomena.

Nusselt number scaling exhibited in Fig. 11.15 shows that $\text{Nu} \sim \text{Ra}^{0.30}$, which is similar to that observed for the no-slip boundary condition. For small Pr, Nu appears to drop significantly. The scaling relation for the free-slip boundary condition are summarised in Table 11.3.

11.10   Onset of ultimate regime?

Among the researchers, there are disagreements on the nature and existence of ultimate regime in RBC. Here we list only some of the issues, and make some conjectures. Note that $\text{Nu} \sim \text{Ra}^{1/2}$ in the ultimate regime. Present experiments have reached Ra up to $\approx 10^{16}$ [Niemela *et al.* (2000); He *et al.* (2012)], which may or may not be the starting point for the ultimate regime. For the RBC experiment with air, He *et al.* (2012) argued that the boundary layer becomes turbulent at

$Ra \sim 10^{15}$, which, according to Eq. (11.32), corresponds to

$$Re \sim \sqrt{7.5 Pr} Ra^{0.38} \sim 10^6. \tag{11.135}$$

Interestingly, the shear boundary layer on a flat plate becomes turbulent at $Re \sim 10^6$ [Grossmann and Lohse (2000); Landau and Lifshitz (1987)]. It is possible that in RBC, the shear induced by the large-scale circulation may yield similar transition near $Ra \sim 10^{15}$, hence a transition to the ultimate regime may occur near $Ra \sim 10^{15}$.

The aforementioned hypothesis appears to be borne out in some experiments. He *et al.* (2012) observed an increase in the Nusselt number exponent beyond $Ra \sim 10^{15}$. Chavanne *et al.* (1997) performed RBC experiment with cryogenic He gas and observed an increase in the Nusselt number exponent at $Ra \gtrsim 10^{11}$. There are several RBC experiments with rough thermal plates in which the Nu starts to increase near $Ra \approx 10^{12}$; this increase in the Nusselt number exponent could be attributed to a transition to turbulent boundary layer due to the wall roughness. We also remark that several groups disagree on the existence of the ultimate regime. For example, Niemela *et al.* (2000) report that $Nu \sim Ra^{0.30}$ till $Ra \sim 10^{16}$; Urban *et al.* (2011) argue for the nonexistence of the ultimate regime, at least till $Ra \approx 10^{16}$.

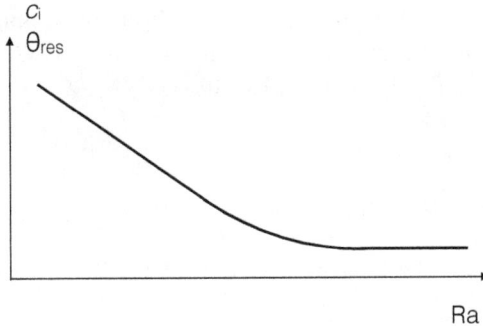

Fig. 11.16 A schematic variations of c_i's and Θ_{res} vs. Ra. These quantities become independent of Ra in the ultimate regime.

For the ultimate regime, we conjecture that the forcing via thermal plumes is active only at small wavenumbers or large scales, similar to that in hydrodynamic turbulence. The turbulence at the intermediate and small scales may be homogeneous and isotropic, and we expect c_i's and $C_{u\theta_{res}}$ to be constants, and $\Theta_{res} \sim \Delta$, leading to $U \sim \sqrt{Ra Pr}$ and $Nu \sim \sqrt{Ra Pr}$. In Fig. 11.16 we illustrate a possible variations of c_i's, $C_{u\theta_{res}}$, and Θ_{res} with Ra. These quantities decrease with Ra in the intermediate range, as shown in Fig. 11.16, but saturate to constant values in the ultimate regime. Hence, in the ultimate regime, the behaviour of thermal turbulence may become very similar to hydrodynamic turbulence, as envisaged by Kraichnan (1962). Note that in the ultimate regime, the preliminary scaling presented in Sec. 11.1 may work.

With this we end our discussion on the scaling of large-scale quantities. In the next chapter, we will discuss the spectrum and flux of stably stratified turbulence.

Further reading

There is a vast literature on the scaling of Nusselt and Reynolds number of RBC. The Péclet scaling discussed in this chapter is taken from Pandey and Verma (2016). Grossmann-Lohse model is derived in Grossmann and Lohse (2000), Grossmann and Lohse (2001), and Stevens *et al.* (2013). The review articles by Siggia (1994) and Ahlers *et al.* (2009) describe earlier results. I also recommend the papers by Kraichnan (1962), Castaing *et al.* (1989), Shraiman and Siggia (1990), and Niemela *et al.* (2000) for discussion on the Nusselt number scaling.

Exercises

(1) Derive the Reynolds number scaling of RBC from the dynamical equation.
(2) What are the key similarities and dissimilarities between the hydrodynamic and RBC turbulence?
(3) What are the physical reasons behind $\mathrm{Nu} \sim \mathrm{Ra}^{0.30}$, rather than $\mathrm{Nu} \sim \mathrm{Ra}^{1/2}$?
(4) How does the scaling relation $\mathrm{Nu} \sim \mathrm{Ra}^{0.30}$ affect the scaling of viscous dissipation rate? Note that in hydrodynamic turbulence, $\epsilon_u \sim U^3/d$.
(5) What is the role of θ_{res} in thermal turbulence?

Chapter 12

Phenomenology of Stably Stratified Turbulence

As discussed in Chapter 10, the kinetic energy (KE) spectrum of three-dimensional (3D) hydrodynamic turbulence, when forced at large scales, is

$$E_u(k) = K_{Ko}\Pi_u^{2/3}k^{-5/3}. \tag{12.1}$$

In the inertial range of hydrodynamic turbulence, the energy flux Π_u is constant due to the absence of KE feed by forcing and due to very weak dissipation. When a passive scalar is added to the flow, as discussed in Sec. 10.4, the KE spectrum is same as Eq. (12.1), and the passive scalar too exhibits $k^{-5/3}$ spectrum (see Sec. 10.4).

In buoyancy-driven flows, the mass density ρ is advected by the velocity field; and buoyancy, $\rho g\hat{z}$, acts on the velocity field. Thus, ρ acts as an active scalar. Thus the phenomenology of buoyancy-driven turbulence could differ from passive scalar turbulence. This is the topic of the present and the next two chapters. Though, the properties of buoyancy-driven turbulence are quite complex, still, many ideas of passive scalar turbulence, such as energy and entropy fluxes, provide useful tools for the construction of turbulence phenomenologies for the buoyancy-driven turbulence.

In the present chapter and the next chapter, we will study the properties of stably stratified turbulence (SST) and turbulent thermal convection respectively. In the following section we start our discussion on SST.

12.1 Various regimes of SST

The two important parameters of SST are the Reynolds number, Re, and the Richardson number, Ri. We classify the SST based on these parameters:

(1) Re $\gg$ 1 and Ri $\approx$ 1 (turbulent SST with moderate buoyancy): Comparable strength of gravity and nonlinearity yields nearly isotropic turbulence. We will show in Sec. 12.2.1 that Bolgiano-Obukhov scaling holds in this regime. Since Ri $\approx 1/\text{Fr}^2$ (see Eq. (2.45)), Fr $\approx$ 1 in this regime.
(2) Re $\gg$ 1 and Ri $\ll$ 1 (turbulent SST with weak buoyancy): In this regime, stronger nonlinearity yields behaviour similar to hydrodynamic turbulence. Here Fr $\gg$ 1.

(3) Re $\gg 1$ and Ri $\gg 1$ (turbulent SST with strong buoyancy): Strong gravity makes the flow quasi-2D with strong $u_\perp$ and weak $u_\parallel$. Such a flow is strongly anisotropic and complex, and it has certain similarities with 2D hydrodynamic turbulence. Here Fr $\ll 1$.

Also note that when the nonlinearity is weak (Re ≈ 0), strong buoyancy yields internal gravity waves, as described in Chapter 5. Flows with Re $\lesssim 1$ and Ri ≈ 0 are laminar.

In the following discussion we will describe the properties of SST for the aforementioned three regimes.

12.2   SST with moderate buoyancy

In this section, we will first describe the turbulence phenomenology by Bolgiano (1959) and Obukhov (1959), after which we will describe the numerical results of SST in this regime.

12.2.1   *Classical Bolgiano-Obukhov scaling for SST*

We start with the equation for the one-dimensional KE spectrum derived in Sec. 4.3:

$$\frac{d}{dt}E_u(k) = -\frac{d}{dk}\Pi_u(k) + \mathcal{F}_B(k) - D_u(k) + \mathcal{F}(k), \tag{12.2}$$

where $\Pi_u(k)$ is the kinetic energy flux, $\mathcal{F}_B(k)$ and $\mathcal{F}(k)$ are the energy feed by buoyancy and an external force respectively, and $D_u(k)$ is the dissipation rate. Here, the low wavenumber modes are forced by an external force in order to maintain a steady state turbulence. Bolgiano (1959) and Obukhov (1959) argued that

$$\mathcal{F}_B(k) = \sum_{k-1<k'\leq k} \langle u_z^*(\mathbf{k})\rho(\mathbf{k})\rangle < 0. \tag{12.3}$$

Also, in the inertial range, $D_u(k)$ is negligible, and $\mathcal{F}(k) = 0$. Hence, under a steady state,

$$\frac{d}{dk}\Pi_u(k) = \mathcal{F}_B(k) < 0. \tag{12.4}$$

Therefore, $\Pi_u(k)$ decreases with k, which is illustrated schematically in Fig. 12.1. Compare this figure with that for hydrodynamic turbulence, Fig. 10.1, where $\Pi_u(k)$ is constant. Bolgiano (1959) and Obukhov (1959) also proposed that $\mathcal{F}_B(k)$ becomes negligible for $k \gg k_B$, where k_B is called *Bolgiano wavenumber*. Therefore, for $k_B \ll k \ll k_d$, where k_d is the dissipation wavenumber,

$$\Pi_u(k) = \text{const.} \tag{12.5}$$

We also remark that for an internal gravity wave, $u_z^*(\mathbf{k})\rho(\mathbf{k})$ oscillates around its mean value (zero), and $D_u(k) = D_b(k) = 0$. Hence, for an internal gravity wave, the temporal average of $\mathcal{F}_B(k) = \langle u_z^*(\mathbf{k})\rho(\mathbf{k})\rangle = 0$, and hence there is no loss of kinetic energy on an average. The kinetic energy lost by a shell k at a given

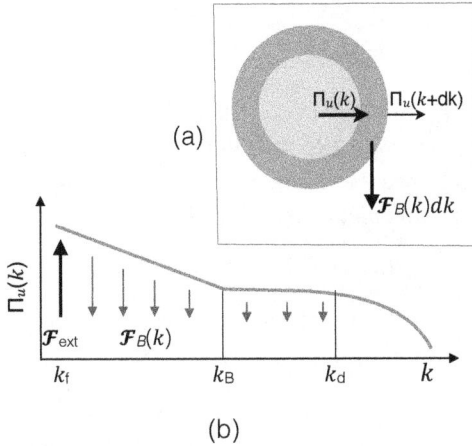

Fig. 12.1 Schematic diagrams of the kinetic energy flux $\Pi_u(k)$ of SST. According to the Bolgiano and Obukhov phenomenology, $\mathcal{F}_B(k) < 0$ for $k_f \ll k \ll k_B$, hence $\Pi_u(k)$ decreases in this band. In the band $k_B \ll k \ll k_d$, $\mathcal{F}_B(k)$ is negligible, hence $\Pi_u(k) = \text{const}$.

time returns to it at a later time, as in a harmonic oscillator. However, in SST, $\mathcal{F}_B(k) < 0$ that leads to a net loss of kinetic energy *in the inertial range*, analogous to a damped oscillator. Also note that finite ν and κ play important dissipative role in the dissipation range of SST.

Here we reproduce the arguments of Bolgiano (1959) and Obukhov (1959) phenomenology, referred to as *BO phenomenology* in short. According to the BO phenomenology, in the $k_f \ll k \ll k_B$ wavenumber regime, buoyancy is important and it is balanced by the nonlinear term. Hence,

$$\rho_k g \approx k u_k^2. \tag{12.6}$$

From the above force balance, we deduce that the Richardson number, Ri ≈ 1. In addition, the BO phenomenology assumes that the potential energy (PE) flux is an approximate constant in the band $k_f \ll k \ll k_B$, i.e.,

$$\Pi_\rho \approx k u_k \rho_k^2 \approx \epsilon_\rho, \tag{12.7}$$

where ϵ_ρ is the entropy dissipation rate. Hence, using Eqs. (12.6, 12.7), BO deduced that

$$u_k \approx \epsilon_\rho^{1/5} g^{2/5} k^{-3/5}, \tag{12.8}$$

$$\rho_k \approx \epsilon_\rho^{2/5} g^{-1/5} k^{-1/5}. \tag{12.9}$$

Therefore, the KE spectrum $E_u(k) \approx u_k^2/k$, the PE spectrum $E_\rho(k) \approx \rho_k^2/k$, and $\Pi_u(k) \approx u_k^3 k$ are

$$E_u(k) = c_1 \epsilon_\rho^{2/5} g^{4/5} k^{-11/5}, \tag{12.10}$$

$$E_\rho(k) = c_2 \epsilon_\rho^{4/5} g^{-2/5} k^{-7/5}, \tag{12.11}$$

$$\Pi_u(k) = c_3 \epsilon_\rho^{3/5} g^{6/5} k^{-4/5}, \tag{12.12}$$

$$\Pi_\rho(k) = \epsilon_\rho, \tag{12.13}$$

where c_i's are constants (different from those of Sec. 11.2).

For $k_B \ll k \ll k_d$, where k_d is the dissipation wavenumber, BO argued that buoyancy is negligible. Hence, the kinetic and potential energies have spectra similar to that in passive scalar turbulence (see Sec. 10.4), or

$$E_u(k) = K_{Ko} \epsilon_u^{2/3} k^{-5/3}, \tag{12.14}$$

$$E_\rho(k) = K_{OC} \epsilon_u^{-1/3} \epsilon_\rho k^{-5/3}, \tag{12.15}$$

$$\Pi_u(k) = \epsilon_u, \tag{12.16}$$

$$\Pi_\rho(k) = \epsilon_\rho, \tag{12.17}$$

where K_{OC} is the Obukhov-Corrsin constant. The KE spectrum, $E_u(k)$, transitions from $k^{-11/5}$ to $k^{-5/3}$ near the Bolgiano wavenumber k_B. By equating $\Pi_u(k)$ of Eq. (12.12) and Eq. (12.16), we can estimate k_B as

$$k_B \approx g^{3/2} \epsilon_u^{-5/4} \epsilon_\rho^{3/4}. \tag{12.18}$$

The BO phenomenology assumes that Re $\gg 1$ and Ri ≈ 1 (moderate buoyancy). Hence, the nonlinear term $\mathbf{u} \cdot \nabla \mathbf{u}$ and the buoyancy should be comparable to each other, and they should be much larger the viscous dissipation. We will discuss the weak and strong buoyancy cases in later sections of this chapter.

12.2.2 *BO phenomenology revisited*

In this subsection, we rewrite the spectra and fluxes in terms of u_k and b_k, which has the dimension of velocity (Eq. (2.37)). Using Eq. (2.38), the force balance equation is

$$k u_k^2 = N b_k, \tag{12.19}$$

where N is the Brunt-Väisälä frequency. In Sec. 4.3, using conservation laws, we showed that in the inertial range,

$$\Pi_u(k) + \Pi_b(k) = \epsilon = \text{const}, \tag{12.20}$$

rather than $\Pi_b(k)$ being a constant, as assumed in BO phenomenology. Here ϵ is the flux of the total energy. Using dimensional analysis, we approximate Eq. (12.20) to

$$k u_k^3 + k b_k^2 u_k = \epsilon. \tag{12.21}$$

Using Eq. (12.19) we eliminate b_k in the above equation that yields the following fifth-order polynomial in u_k:

$$k u_k^3 \left[1 + \frac{k^2 u_k^2}{N^2} \right] = \epsilon. \tag{12.22}$$

By solving the above equation, we can derive u_k and b_k as a function of k. However such solution would be quite complex. Therefore, we show consistency of the BO scaling with Eq. (12.22).

Equation (12.21) is rewritten as

$$kb_k^2 u_k \left[1 + \frac{u_k^2}{b_k^2}\right] = \epsilon. \tag{12.23}$$

Using Eqs. (12.8, 12.9), for $k_f \ll k \ll k_B$, we can deduce that $u_k^2/b_k^2 \sim (kd)^{-4/5}$, substitution of which in Eq. (12.23) yields

$$kb_k^2 u_k \left[1 + (kd)^{-4/5}\right] = \epsilon. \tag{12.24}$$

Thus, $\Pi_u(k)$ is smaller than $\Pi_b(k)$ by a factor of $(kd)^{-4/5}$, and hence $\Pi_u(k) \ll \Pi_b(k)$ for large kd. Therefore, in the wavenumber regime $k_f < k < k_B$,

$$\Pi_b(k) \approx \epsilon_b \approx \epsilon, \tag{12.25}$$

which is consistent with the BO phenomenology.

Thus, in stably stratified turbulence, the conservation law of Eq. (12.20) is consistent with the BO phenomenology. In the following we will rewrite the equations of BO phenomenology in terms of u and b. For $k_f \ll k \ll k_B$,

$$E_u(k) = c_1 \epsilon_b^{2/5} N^{4/5} k^{-11/5}, \tag{12.26}$$

$$E_b(k) = c_2 \epsilon_b^{4/5} N^{-2/5} k^{-7/5}, \tag{12.27}$$

$$\Pi_u(k) = c_3 \epsilon_b^{3/5} N^{6/5} k^{-4/5}, \tag{12.28}$$

$$\Pi_b(k) = \epsilon_b, \tag{12.29}$$

and for $k_B \ll k \ll k_d$,

$$E_u(k) = K_{Ko} \epsilon_u^{2/3} k^{-5/3}, \tag{12.30}$$

$$E_b(k) = K_{OC} \epsilon_u^{-1/3} \epsilon_b k^{-5/3}, \tag{12.31}$$

$$\Pi_u(k) = \epsilon_u, \tag{12.32}$$

$$\Pi_b(k) = \epsilon_b. \tag{12.33}$$

In the next section, we verify BO scaling using numerical simulations.

12.2.3 *Numerical verification of BO scaling*

Kumar *et al.* (2014a) simulated SST in a cubical periodic box. They took $Pr = 1$ and forced the small wavenumber modes to achieve a steady turbulence. The Richardson number of the flow was $Ri \approx 0.01$, which is $O(1)$. Note that $Fr \approx 1/\sqrt{Ri} \approx 10$.

Figure 12.2(a) demonstrates that the normalized KE spectrum $E_u(k)k^{11/5}$ is flatter than the normalised KE spectrum $E_u(k)k^{5/3}$. Thus, we conclude that $E_u(k) \sim k^{-11/5}$, consistent with Eq. (12.10). Similarly, using Fig. 12.2(b) we conclude that the PE spectrum $E_b(k) \sim k^{-7/5}$, thus confirming BO scaling for $Ri \approx 1$. Further, Kumar *et al.* (2014a) computed the KE flux and PE flux using the numerical data. As shown in Fig. 12.3,

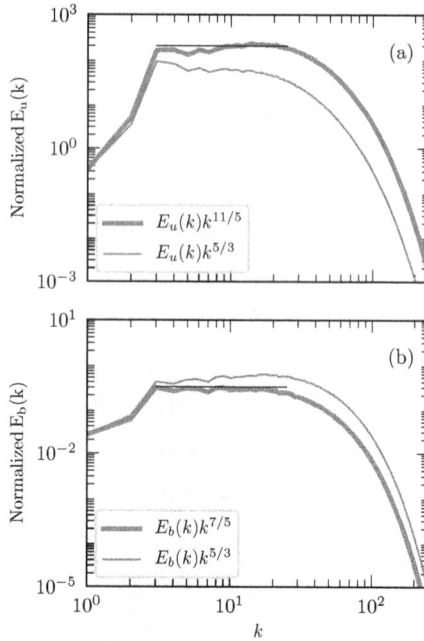

Fig. 12.2   For SST with Pr = 1, Ra = 5×10³, and Ri = 0.01, plots of (a) normalized KE spectrum and (b) normalized PE spectrum for Bolgiano-Obukhov (BO) and Kolmogorov scaling. BO scaling fits better with the data than Kolmogorov scaling. From Kumar *et al.* (2014a). Reprinted with permission from APS.

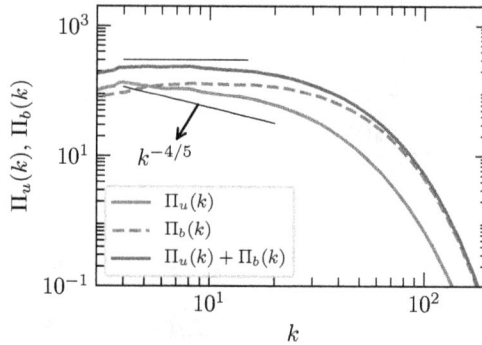

Fig. 12.3   For SST with Pr = 1, Ra = 5 × 10³, and Ri = 0.01, plots of KE flux $\Pi_u(k)$, PE flux $\Pi_b(k)$, and $\Pi_u(k) + \Pi_b(k)$. The energy fluxes are consistent with the BO phenomenology.

(1) The sum $\Pi_u + \Pi_b$ is an approximate constant in the inertial range[1].
(2) The PE flux is an approximate constant, and it dominates $\Pi_u(k)$, consistent with Eq. (12.24).

___

[1]The density fluctuation in Kumar *et al.* (2014a) is ρ, not b. However, the total energy flux is $\Pi_u + \Pi_\rho$ due to nondimensionalization.

(3) $\Pi_u(k)$ decreases with k (Eq. (12.12)], but not quite as $\Pi_u(k) \sim k^{-4/5}$, which is possibly due to the lower resolution of the simulation.

As shown in Eq. (2.39), the density field is driven by u_z, which is reasonably strong. Hence, Π_b may increase with wavenumber, therefore $E_b(k)$ may be shallower than $k^{-5/3}$ spectrum, contrary to the passive scalar turbulence. This issue needs further investigation.

The numerical data of Kumar *et al.* (2014a) yields Bolgiano wavenumber $k_B \approx 6$. This value of k_B is not distant enough from k_d to be able to exhibit $k^{-5/3}$ spectrum for $k_B \ll k \ll k_d$, which is predicted in BO phenomenology. The viscous dissipation becomes quite dominant for $k > k_B$ because of which there is no definitive crossover from $k^{-11/5}$ to $k^{-5/3}$ [Kumar *et al.* (2014a)] . One needs higher-resolution simulations to observe such a crossover.

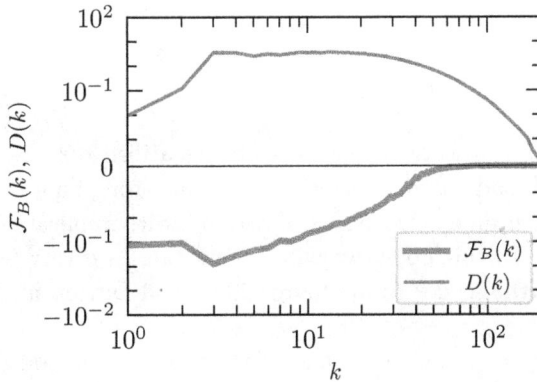

Fig. 12.4   For SST with Pr = 1, Ra = 5×10^3, and Ri = 0.01, plots of the energy supply rate by buoyancy, $\mathcal{F}_B(k)$, and the dissipation spectrum, $D_u(k)$. Note that $\mathcal{F}_B(k) < 0$.

The aforementioned numerical observations verify the BO scaling for SST with moderate stratification. Before closing this subsection we remark that SST with Ri ≈ 1, but with large or small Prandtl number may exhibit different behaviour, and they need to be investigated. The turbulence phenomenology for passive scalar discussed in Sec. 10.4 may help us formulate turbulence models for these cases.

In the next section we will discuss SST with weak buoyancy.

12.3   SST with weak buoyancy

For a turbulent flow with weak density stratification (Ri $\ll 1$), buoyancy is much weaker than the nonlinear term. In this regime we expect $E_u(k)$ to follow Kolmogorov's spectrum as in 3D hydrodynamic turbulence. Kumar *et al.* (2014a) simulated a 3D SST with Ri = 4×10^{-7} and reported Kolmogorov's spectrum for the kinetic energy, as expected. See Fig. 12.5 for an illustration.

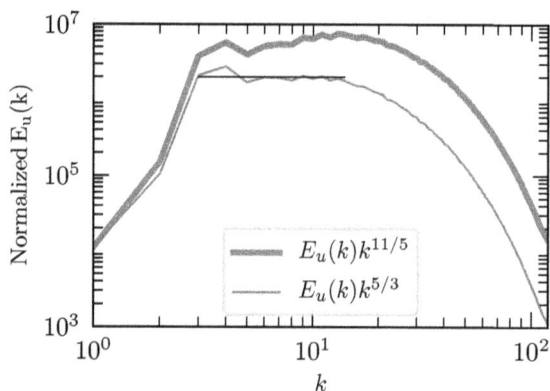

Fig. 12.5   For SST with Pr = 1, Ra = 5 × 10³, and Ri = 4 × 10⁻⁷, plot of the normalised KE spectrum. The plot clearly shows that $E_u(k) \sim k^{-5/3}$. From Kumar *et al.* (2014a). Reprinted with permission from APS.

12.4   SST with strong buoyancy

So far, we assumed near isotropy for the stably stratified turbulent, which is applicable for weak and moderate buoyancy, but not for strong buoyancy, or for Ri ≫ 1. This regime of SST is an extensive area of research with applications in atmospheric and oceanic flows. In this book, we will discuss this case very briefly due to lack of space. We refer the reader to Lindborg (2006) and Davidson (2013) for detailed discussions.

For strong buoyancy, the rms values of the velocity components perpendicular and parallel to **g** are very different; we denote these components by $u_\perp$ and $u_\parallel$ respectively. In addition, the integral length scales in these directions, defined as [Kumar *et al.* (2014b)]:

$$L_\perp = \frac{2\pi}{E_u} \int k_\perp^{-1} E_u(k) dk, \qquad (12.34)$$

$$L_\parallel = \frac{2\pi}{E_u} \int k_\parallel^{-1} E_u(k) dk, \qquad (12.35)$$

are also unequal. Following Davidson (2013), we introduce two Froude numbers:

$$\mathrm{Fr}_\perp = \frac{u_\perp}{N L_\perp}, \qquad (12.36)$$

$$\mathrm{Fr}_\parallel = \frac{u_\perp}{N L_\parallel}, \qquad (12.37)$$

where N is the Brünt Väisälä frequency. The incompressibility condition of the flow yields

$$\frac{u_\perp}{L_\perp} \approx \frac{u_\parallel}{L_\parallel} \implies \frac{u_\parallel}{u_\perp} \approx \frac{L_\parallel}{L_\perp} = \frac{\mathrm{Fr}_\perp}{\mathrm{Fr}_\parallel}. \qquad (12.38)$$

It is customary to define *gradient Richardson number* [Davidson (2013)] as

$$\text{Ri}_g = \frac{N^2}{\left[\left(\frac{\partial u_x}{\partial z}\right)^2 + \left(\frac{\partial u_y}{\partial z}\right)^2\right]}. \tag{12.39}$$

When the denominator of Eq. (12.39) is approximated as $u_\perp^2/L_\parallel^2$,

$$\text{Ri}_g = \frac{N^2 L_\parallel^2}{u_\perp^2} = \frac{1}{\text{Fr}_\parallel^2}. \tag{12.40}$$

However, when the denominator of Eq. (12.39) is dominated by the viscous dissipation,

$$\text{Ri}_g = \frac{N^2}{\epsilon_u/\nu}. \tag{12.41}$$

Substitution of the bulk dissipation rate $\epsilon_u = u_\perp^3/L_\perp$ in the above expression yields

$$\text{Ri}_g = \frac{\nu}{u_\perp L_\perp}\frac{N^2 L_\perp^2}{u_\perp^2} = (\text{Re}_\perp \text{Fr}_\perp^2)^{-1} = \text{Re}_{\text{buoy}}^{-1}. \tag{12.42}$$

Note that $\text{Re}_\perp \text{Fr}_\perp^2$ is defined as the Reynolds number based on buoyancy, Re_{buoy}.

Lindborg (2006) (also see Davidson (2013)) defined *local Froude numbers* as

$$\text{Fr}_\perp(l_\perp) = \frac{u_\perp(l_\perp)}{N l_\perp}, \tag{12.43}$$

$$\text{Fr}_\parallel(l_\perp) = \frac{u_\perp(l_\perp)}{N l_\parallel}, \tag{12.44}$$

where $l_\perp$, $l_\parallel$ are the local length scales perpendicular and parallel to $\mathbf{g}$. A question is at what $l_\perp$,

$$\text{Fr}_\perp(l_\perp) \approx 1? \tag{12.45}$$

This particular length scale, called *Ozmidov scale*, is denoted by L_O. Using Kolmogorov's theory of turbulence, we estimate $u_\perp^2 = \epsilon_u^{2/3} L_\perp^{2/3}$, substitution of which in Eq. (12.43) yields

$$\text{Fr}_\perp(L_O) \approx \frac{\epsilon_u^{1/3}}{N L_O^{2/3}} = 1 \tag{12.46}$$

or

$$L_O = \sqrt{\frac{\epsilon_u}{N^3}}. \tag{12.47}$$

Note that $\text{Fr}_\perp(l_\perp) < 1$ for $l_\perp > L_O$, and vice versa.

Now, we perform scaling analysis of the velocity fluctuations with $l_\perp > L_O$. Following Davidson (2013), and matching the dominant terms of Eqs. (2.38, 2.39) yields

$$\frac{u_\perp^2}{l_\parallel} = Nb, \tag{12.48}$$

$$\frac{u_\perp b}{l_\perp} = Nu_\parallel. \tag{12.49}$$

In Eq. (2.38), buoyancy stretches the vorticity along **g**. This is the reason for using $l_\parallel$ in Eq. (12.48). Elimination of b from the above equations, and an application of Eq. (12.38) yields

$$[\text{Fr}_\parallel(l_\perp)]^2 = \frac{u_\perp^2}{N^2 l_\parallel^2} = \frac{u_\parallel l_\perp}{u_\perp l_\parallel} \approx 1. \tag{12.50}$$

Hence, for $l_\perp > L_O$, $\text{Fr}_\parallel(l_\perp) \approx 1$. Therefore,

$$\frac{l_\parallel}{l_\perp} \approx \frac{\text{Fr}_\perp(l_\perp)}{\text{Fr}_\parallel(l_\perp)} \approx \text{Fr}_\perp < 1 \tag{12.51}$$

Thus, the flow structures in strongly-stratified turbulence are suppressed along the buoyancy direction. Also note that

$$\frac{\text{Nonlinear term}}{\text{Viscous term}} = \frac{u_\perp^2/L_\perp}{\nu u_\perp/L_\parallel^2} = \text{Re}_\perp \text{Fr}_\perp^2 = \text{Re}_{\text{buoy}}. \tag{12.52}$$

Lindborg (2006) argued that

$$E_\perp(k_\perp) \sim u_\perp^2 l_\perp \sim \epsilon_u^{2/3} k_\perp^{-5/3}, \tag{12.53}$$

and using $u_\perp = N l_\parallel \text{Fr}_\parallel \approx N l_\parallel$, Lindborg (2006) derived that

$$E_\perp(k_\parallel) \sim u_\perp^2 l_\parallel \approx N^2 l_\parallel^3 \approx N^2 k_\parallel^{-3}. \tag{12.54}$$

He also argued that E_b follows similar scaling. Vallgren *et al.* (2011) included rotation in the spectral formalism described above and obtained similar kinetic and potential energy spectra. They claimed that the dual energy spectra, k^{-3} and $k^{-5/3}$, observed in the terrestrial atmosphere [Gage and Nastrom (1986)] could be due to above dynamics. In this book, we are not in position to discuss the combined effects of rotation and stratification, except small introductory note on their wave solutions (see Chapter 19).

The scaling arguments discussed above appear quite sound, yet a closer inspection of them is in order. In Chapter 15, we will develop tools and methodology for the description of anisotropy in turbulent thermal convection. We believe that these tools could also be employed to SST.

In the next section, we discuss 2D SST.

12.5   SST in 2D

In this section we discuss 2D SST restricted in a vertical plane xz with gravity along $-\hat{z}$. Large-scale random forcing is employed for obtaining a steady state. Kumar *et al.* (2017) performed numerical simulation of 2D SST for Fr ranging from 0.16 (strong stratification) to 1.1 (moderately weak stratification). In Fig. 12.7 we present the flow profiles for the above cases [Kumar *et al.* (2017)]. Based on the these results, Kumar *et al.* (2017) reported the following properties of 2D SST:

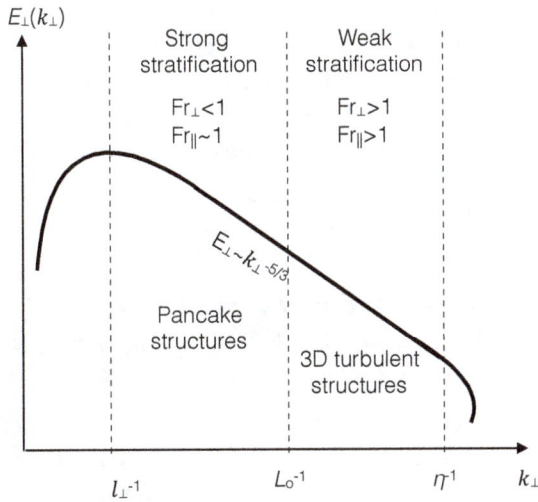

Fig. 12.6 Plot of the energy spectrum $E_\perp(k_\perp)$ vs. $k_\perp$ for strongly-stratified turbulence. Adopted from Fig. 14.3 of Davidson (2013).

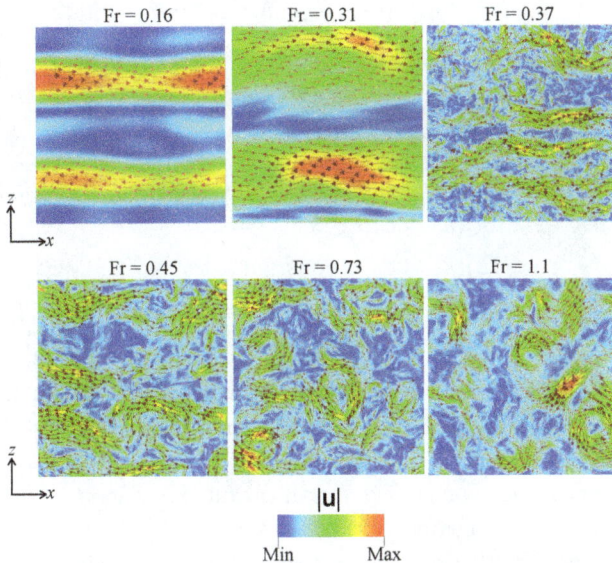

Fig. 12.7 For Fr = 0.16, 0.31, 0.37, 0.45, 0.73, and 1.1, the vector plot of $\mathbf{u}(\mathbf{r})$ superposed on the density plot of $|\mathbf{u}|$. Fluctuations are suppressed for small Fr, but they grow with the increase of Fr. Here Fr =0.16, 0.31 (strong stratification), 0.37, 0.45 (moderate stratification), and 1.1 (weak stratification). Near Fr = 0.73, transition occurs from moderate to weak stratification. From Kumar *et al.* (2017). Reprinted with permission from Taylor and Francis.

(1) *Weak stratification:* This regime has Fr > 1. In Fig. 12.7 with Fr = 1.1, we illustrate the real space flow profile for such flows. In this regime, the flow is similar to 2D scalar turbulence with some critical effects of gravity. Due to the forcing at small wavenumbers, we expect an approximate constant enstrophy flux with $E_u(k) \sim k^{-3}$. Buoyancy induces an increase in $\Pi_\omega(k)$ with k and yields the following scaling:

$$\Pi_\omega(k) \sim k^{3/4}, \tag{12.55}$$

$$E_u(k) \sim \Pi_\omega^{2/3} k^{-3} \sim k^{3/4 \times 2/3 - 3} \sim k^{-2.5}. \tag{12.56}$$

Hence, $u_k^2 \sim E_u(k)k \sim k^{-1.5}$. Kumar *et al.* (2017) observed that

$$\Pi_b(k) \approx k u_k b_k^2 \sim k^{-0.13} \implies b_k^2 \sim k^{-0.38} \tag{12.57}$$

$$E_b(k) \approx b_k{}^2/k \sim k^{-1.38}. \tag{12.58}$$

The KE and PE spectra computed numerically by Kumar *et al.* (2017) are in good agreement with the above scaling. See Table 12.1 for a summary.

(2) *Moderate stratification:* In this regime, Fr ≈ 1. Refer to Fr = 0.37 of Fig. 12.7 for illustration of the real space flow profile. Here, we expect Bolgiano-Obukhov scaling to work with appropriate modification due to the 2D nature of the flow. Following the arguments of Sec. 12.2.2, for $k_f < k < k_B$ we expect the following scaling:

$$E_u(k) = c_1 \epsilon_b^{2/5} N^{4/5} k^{-11/5}, \tag{12.59}$$

$$E_b(k) = c_2 \epsilon_b^{4/5} N^{-2/5} k^{-7/5}, \tag{12.60}$$

$$\Pi_u(k) = c_3 \epsilon_b^{3/5} N^{6/5} k^{-4/5}, \tag{12.61}$$

$$\Pi_b(k) = \epsilon_b. \tag{12.62}$$

However, due to two-dimensionality of the flow, for $k_B < k < k_d$,

$$E_u(k) = c_5 \Pi_\omega^{2/3} k^{-3} \implies u_k \sim k^{-1}, \tag{12.63}$$

$$\Pi_\omega(k) = \epsilon_\omega = \text{constant}, \tag{12.64}$$

$$\Pi_b(k) = k u_k b_k^2 = \epsilon_b \implies b_k \sim \text{const}, \tag{12.65}$$

$$E_b(k) = b_k{}^2/k \sim k^{-1}. \tag{12.66}$$

Kumar *et al.* (2017) observed the above scaling with some shifts in the exponents due to the observed $\Pi_u(k) \sim k^{-0.3}$, rather than $\Pi_u(k) \sim k^{-4/5}$. See Table 12.1 for the numerical results.

(3) *Strong stratification:* For strong stratification, Kumar *et al.* (2017) observed large-scale vertically sheared horizontal flow (VSHF) along with small scale turbulence. See Fig. 12.7 with Fr = 0.16 and 0.31 for an illustration of the flow patterns. A VSHF is a combination of internal gravity waves, and turbulent fluctuations whose properties are given by:

$$E_u(k) = c_5 \Pi_\omega^{2/3} k^{-3} \implies u_k \sim k^{-1}, \tag{12.67}$$

$$\Pi_\omega(k) = \epsilon_\omega = \text{constant}, \tag{12.68}$$

$$\Pi_b(k) = k u_k b_k^2 \sim k^{-2} \implies b_k \sim k^{-1}, \tag{12.69}$$

$$E_b(k) = b_k{}^2/k \sim k^{-3}. \tag{12.70}$$

Note the difference in the scaling $\Pi_b(k)$ in Eqs. (12.65, 12.69). For moderate stratification, $\Pi_b(k) \sim$ const, but for the strong stratification, $\Pi_b(k) \sim k^{-2}$. These differences in the PE flux behaviour lead to different PE spectra.

We summarise the above results in Table 12.1. The numerical results of Kumar *et al.* (2017) are in general agreement with the aforementioned scaling. In the next section we describe the experimental and observational results of SST.

Table 12.1 For 2D stably stratified turbulence, scaling of $E_u(k)$, $E_b(k)$, $\Pi_K(k)$, $\Pi_b(k)$, and $\Pi_\omega(k)$ for different levels of stratification. Adopted from a table of Kumar *et al.* (2017).

Strength of stratification	Spectrum	Flux
Weak	$E_u(k) \sim k^{-2.5}$ $E_b(k) \sim k^{-1.6}$	$\Pi_u(k) \sim$ const. (Negative) $\Pi_\omega(k) \sim k^{3/4}$ $\Pi_b(k) \sim k^{-0.13}$
Moderate	For $5 \le k \le 90$: $E_u(k) \sim k^{-2.2}$ $E_b(k) \sim k^{-1.64}$	$\Pi_u(k) \sim k^{-0.98}$ (Negative) $\Pi_\omega(k) \sim$ const. $\Pi_b \sim k^{-0.3}$
	For $90 \le k \le 400$: $E_u(k) \sim k^{-3}$ $E_b(k) \sim k^{-1.64}$	$\Pi_u(k)$: weak (positive) $\Pi_\omega(k) \sim$ const. $\Pi_b \sim k^{-0.3}$
Strong	Large scale VSHF Small scale turbulence: $E_u(k) \sim k^{-3}$ $E_b(k) \sim k^{-3}$	$\Pi_u(k) \sim 0$ $\Pi_\omega(k) \sim$ const. $\Pi_b(k) \sim k^{-2}$

12.6   Experimental and observational results

There are only a small number of laboratory experiments on stably stratified flows, leading ones being on soap films. See Zhang *et al.* (2005) and Thorpe (1973) for a detailed discussion on laboratory experiments. The planetary and stellar atmospheres are good examples of stably stratified flows. Earth's oceans too exhibit some phenomena of stably stratified flows. These systems are good platforms for studying waves and turbulence in stably stratified flows [Vallis (2006)]. Due to lack of space, we do not detail these results, except a brief discussion on the KE spectrum of the terrestrial atmosphere.

Many geophysicists and atmospheric physicists attempt to model some of the atmospheric and oceanic phenomena using physics of stably stratified flows. In particular, scientists have measured the KE spectrum of the Earth's atmosphere and related it to the theoretical predictions of strongly-stratified turbulence. Gage and Nastrom (1986) reported that the kinetic energy of the upper terrestrial atmosphere has dual spectrum with branches of k^{-3} and $k^{-5/3}$. Lindborg (2006), Vallgren *et al.*

(2011), and Davidson (2013) derived Eqs. (12.53, 12.54) and related them to the observations of Gage and Nastrom (1986).

With this we close our discussion on SST. In the next chapter we will describe the phenomenology of turbulent thermal convection.

Further reading

The Bolgiano-Obukhov scaling for Fr $\sim$ 1 first appeared in Bolgiano (1959) and Obukhov (1959). Refer to Kumar *et al.* (2014a) for the numerical verification of the above scaling. Lindborg (2006) and Davidson (2013) cover the scaling analysis of stably stratified flows under strong gravity or Fr $\ll$ 1; refer to Vallgren *et al.* (2011) and Davidson (2013) for numerical simulations of such flows. Kumar *et al.* (2017) describe two-dimensional stably stratified turbulence.

Exercises

(1) Estimate Froude, Richardson, and Reynolds numbers for the atmospheric flow on the Earth.
(2) What is the nature of kinetic and potential energy fluxes in moderately-stratified stably stratified turbulence?
(3) Consider the stably stratified turbulence with moderate Fr and with small Fr. Contrast the two kinds of turbulence.
(4) What are the differences between 2D and 3D stably stratified turbulence with moderate stratification?
(5) In stably stratified turbulence, what is the impact of conservation laws on the kinetic and potential energy fluxes?

Chapter 13

Phenomenology for Turbulent Thermal Convection

The equations of RBC and stably stratified flows are identical, as described in Chapter 2. The only difference between the two systems is the sign of density stratification $\bar{\rho}(z)$. In stably stratified turbulence (SST), $d\bar{\rho}/dz < 0$, but in turbulent thermal convection, $d\bar{\rho}/dz > 0$. Since the set of equations for the two systems are identical, one may expect Bolgiano-Obukhov or BO scaling to be applicable to both SST and RBC. Procaccia and Zeitak (1989), L'vov and Falkovich (1992), and Rubinstein (1994) employed field-theoretic techniques to thermal convection and concluded that BO scaling should hold for turbulent thermal convection.

However, the energetics of turbulent convection, to be discussed in this chapter, differs from that of SST. In Rayleigh-Bénard convection (RBC), buoyancy feeds energy to the velocity field because the hot plumes near the bottom plate accelerate the flow, so do the cold plumes near the top plate. This is unlike SST in which buoyancy drains the kinetic energy leading to flow stability. This difference in the energetics of SST and turbulent convection have very different effects on the energy spectra and fluxes of these two systems. In this chapter, we will describe the turbulence phenomenology of turbulent thermal convection, as well as its associated numerical and experimental results.

We start with a discussion on the phenomenology of turbulent convection.

13.1 Energy flux based phenomenological arguments for turbulent convection

We start with the energy equation

$$\frac{d}{dt}E_u(k) = -\frac{d}{dk}\Pi_u(k) + \mathcal{F}_B(k) - D_u(k), \tag{13.1}$$

where $\Pi_u(k)$ is the KE flux, $\mathcal{F}_B(k)$ is the KE feed by buoyancy, and $D_u(k)$ is the viscous dissipation rate. For RBC,

$$\mathcal{F}_B(k) = \sum_{k-1 < k' \le k} \langle u_z^*(\mathbf{k})\theta(\mathbf{k}) \rangle > 0. \tag{13.2}$$

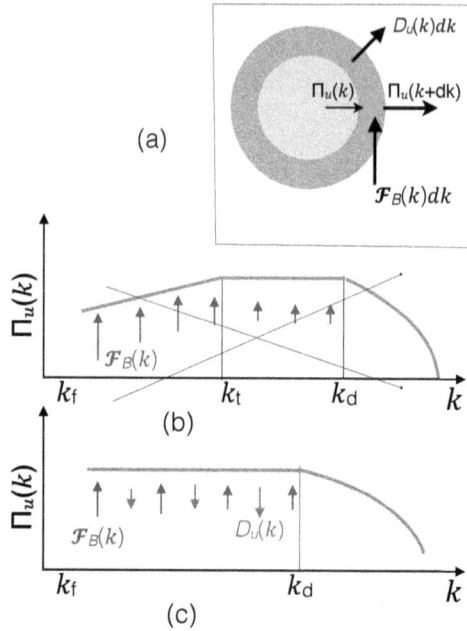

Fig. 13.1   For turbulent thermal convection, schematic diagrams of the KE flux $\Pi_u(k)$. Here $\mathcal{F}_B(k) > 0$ and $D_u(k) < 0$, but they are relatively small. Hence, $\Pi_u(k) \approx$ constant.

This is because hot plumes $(\theta > 0)$ ascends $(u_z > 0)$, while cold plumes $(\theta < 0)$ descend $(u_z < 0)$, thus u_z and θ are positively correlated[1]. Under a steady state, $dE_u(k)/dt \approx 0$. Hence,

$$\frac{d}{dk}\Pi_u(k) = \mathcal{F}_B(k) - D_u(k). \tag{13.3}$$

See Fig. 13.1 for an illustration. In the inertial range, we may expect that $D_u(k) \to 0$. Therefore, positive $\mathcal{F}_B(k)$ yields

$$\frac{d}{dk}\Pi_u(k) > 0. \tag{13.4}$$

Hence we expect $\Pi_u(k)$ to increase with k, but recent numerical simulations of Kumar *et al.* (2014a) and Verma *et al.* (2017b) show that $\Pi_u(k)$ of RBC is nearly constant, which is because

(1) As discussed in Chapter 11, in the inertial range of RBC, buoyancy and viscous dissipation are much weaker than the pressure gradient. Also, buoyancy and viscous drag have opposite signs, and hence they tend to cancel each other. Thus, in turbulent thermal convection, the acceleration $\mathbf{u} \cdot \nabla \mathbf{u}$ is primarily provided by the pressure gradient, as in 3D hydrodynamic turbulence. This

---

[1]Note however that $\langle u_z(\mathbf{r})\theta(\mathbf{r})\rangle > 0$ does not imply that $\Re[u_z^*(\mathbf{k})\rho(\mathbf{k})] > 0$, however this is the case for most $\mathbf{k}$'s (see Sec. 4.1 and Eq. (4.18)).

is consistent with the fact that the Richardson number of turbulent thermal convection is much less than unity (see Sec. 11.3).

(2) The kinetic energy feed by buoyancy is stronger for low wavenumbers, similar to that in 3D hydrodynamic turbulence (to be discussed in the next subsection).

Therefore, the turbulence phenomenology of RBC is very similar to 3D hydrodynamic turbulence. See Fig. 13.1 for an illustration of $\Pi_u(k)$ for turbulent convection.

In the next section we will describe the numerical results that demonstrate similarities turbulent convection and 3D hydrodynamic turbulence.

13.2 Numerical demonstration of Kolmogorov-like scaling for RBC

Verma *et al.* (2017b) performed a very high-resolution (4096^3 grid) RBC simulations that demonstrates Kolmogorov-like scaling for RBC. In the following discussion we will report the KE spectrum and flux, and the shell-to-shell KE transfers obtained from simulations.

Verma *et al.* (2017b) simulated RBC with $\Pr = 1$ and $\mathrm{Ra} = 1.1 \times 10^{11}$. For the velocity field, they employed free-slip boundary condition at the top and bottom plates, and periodic boundary condition at the side walls. The temperature field satisfies conducting boundary condition at the top and bottom plates, and periodic boundary condition at the side walls. Verma *et al.* (2017b) computed the spectra and fluxes of the KE and entropy ($\theta^2/2$).

In Fig. 13.2(a) we exhibit the KE spectrum normalized with $k^{11/5}$ and $k^{5/3}$. Near constancy of $E_u(k)k^{5/3}$ in the inertial range demonstrates Kolmogorov-like scaling for turbulent thermal convection. In Fig. 13.2(b) we plot the KE and entropy fluxes which are approximate constants. In Sec. 4.3, using conservation laws, we showed that in the inertial range,

$$\Pi_u(k) - \frac{\alpha g}{|d\bar{T}/dz|}\Pi_\theta(k) = \text{const.} \tag{13.5}$$

In simulations of Verma *et al.* (2017b), $(\alpha g)/|d\bar{T}/dz| = 1$, hence the above equation becomes

$$\Pi_u(k) - \Pi_\theta(k) = \text{const.} \tag{13.6}$$

The flux difference plotted in Fig. 13.2(b) is nearly flat in the inertial range, consistent with the above formula. We compute the Kolmogorov's constant using

$$K_{\mathrm{Ko}} = \frac{E(k)}{\Pi^{2/3}\Pi_u^{2/3}}, \tag{13.7}$$

and find its value to be approximately 1.5. Based on these observations, we deduce that the turbulence in RBC and 3D hydrodynamic are similar. We also remark that the formulas of Pao (1965), Eqs. (10.24, 10.25), describe the KE spectrum and flux quite well.

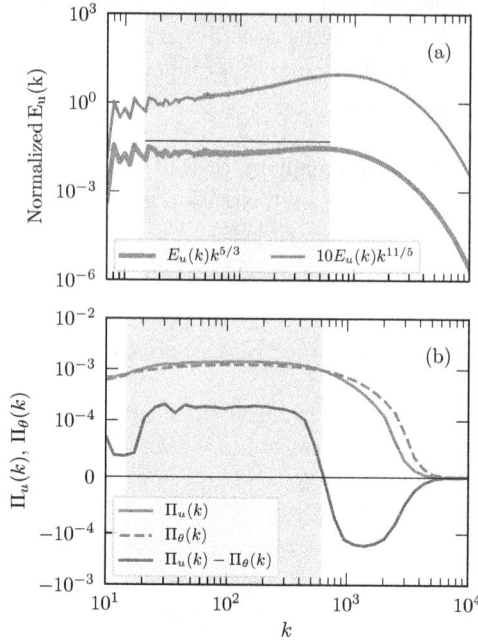

Fig. 13.2 For the RBC simulation with Pr = 1 and Ra = 1.1×10^{11} on 4096^3 grid: (a) plots of normalized KE spectra reveal that $E(k) \sim k^{-5/3}$. (b) KE flux $\Pi_u(k)$ and entropy flux $\Pi_\theta(k)$ are constants in the inertial range, which is the shaded region in the plot. The flux difference, $\Pi_u(k) - \Pi_\theta(k)$, is also quite flat in the inertial range. Adopted from a figure of Verma *et al.* (2017b).

As shown in Fig. 13.3(a), $\mathcal{F}_B(k) > 0$ consistent with the discussion of Sec. 13.1. Also, $\mathcal{F}_B(k) \sim k^{-5/3}$, thus energy feed by buoyancy decreases rather sharply with wavenumbers [Verma *et al.* (2017b)], thus the energy feed is mostly at small wavenumbers. Also, in the inertial range, $\mathcal{F}_B(k)$ and $D_u(k)$ are of the same order and they approximately cancel other. Hence, $d\Pi_u(k)/dk \approx 0 \implies \Pi_u(k) \approx$ constant (see Fig. 13.3).

Strong similarities between thermal convection and 3D hydrodynamic turbulence prompt us to explore whether the shell-to-shell energy transfers in RBC are local and forward. The shell-to-shell energy transfers [Eq. (4.55)] computed using the steady-state data for 40 concentric wavenumber shells are shown in Fig. 13.4. In the plot, the indices of the x, y axes represent the receiver and giver shells respectively. The plot demonstrates that shell m gives energy to shell $(m+1)$, and it receives energy from shell $(m-1)$. Thus the energy transfer in RBC is local and forward, analogous to the 3D hydrodynamic turbulence. We also remark that the KE distribution in the Fourier space is nearly isotropic, as in 3D hydrodynamic turbulence (to be discussed in Chapter 15).

The aforementioned results are for the free-slip RBC that does not have viscous boundary layer. However, the KE spectrum and flux for the no-slip RBC too

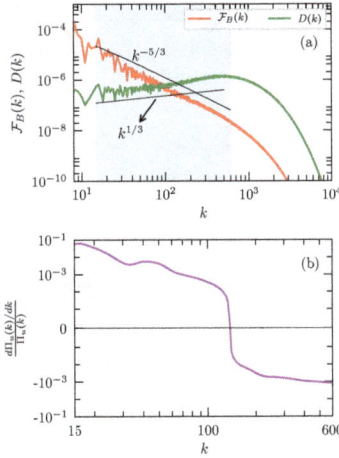

Fig. 13.3   For the RBC simulation with Pr $= 1$ and Ra $= 1.1 \times 10^{11}$: (a) Plots of $\mathcal{F}_B(k)$ and $D_u(k)$. (b) Plot of $[d\Pi_u(k)/dk]/\Pi_u(k)$ in the inertial range $15 < k < 600$ that indicates $\Pi_u(k)$ variation is insignificant in the inertial range. From Verma *et al.* (2017b). Reprinted with permission from Institute of Physics.

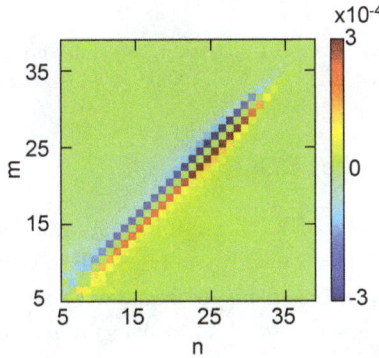

Fig. 13.4   For the RBC simulation for Pr $= 1$ and Ra $= 1.1 \times 10^{11}$, plot of the shell-to-shell energy transfers $T_{u,n}^{u,m}$ of Eq. (4.55), where m, n represent the giver and receiver shell indices respectively, shows that the shell-to-shell energy transfer in turbulent convection is local and forward. From Verma *et al.* (2017b). Reprinted with permission from Institute of Physics.

follow similar scaling, as demonstrated by Kumar and Verma (2017) using no-slip RBC simulations for Rayleigh number Ra $= 10^8$ and Pr $= 1$. See Fig. 13.5 for an illustration of $E_u(k)$ and $\Pi_u(k)$. We remark that the flow structures in the boundary layer are small in size, and they contribute to $E_u(k)$ in the viscous regime. Thus, the inertial range spectrum for the no-slip and free-slip boxes are expected to be similar.

In the next section we will describe the spectrum and flux of entropy.

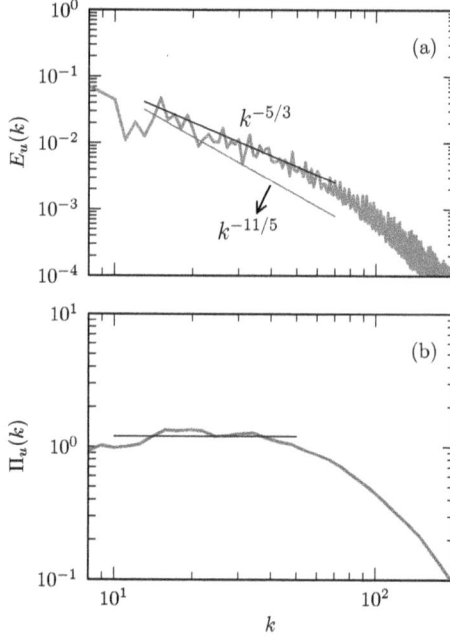

Fig. 13.5   For RBC simulation for Pr $= 1$ and Ra $= 10^8$ under no-slip boundary condition, plots of $E_u(k)$ and $\Pi_u(k)$ reveal that $E_u(k) \sim k^{-5/3}$ and $\Pi_u(k) \approx$ const. Adopted from a figure of Kumar and Verma (2017).

13.3   Entropy spectrum of RBC

The temperature fluctuations of RBC exhibit a unique spectrum. In Fig. 13.6 we plot $E_\theta(k)$ for RBC with Pr $= 1$ and Ra $= 1.1 \times 10^{11}$ [Verma *et al.* (2017b)] . The entropy spectrum $E_\theta(k)$ exhibits a bispectrum, with the upper branch exhibiting k^{-2} spectrum, while the lower branch exhibiting a scatter that fits neither with $k^{-5/3}$ (Kolmogorov's scaling) nor with $k^{-7/5}$ (BO scaling). When we analyse the spectrum carefully, we observe that the upper branch arises due to the mean temperature profile shown in Fig. 9.3 that yields $\theta(0,0,k) \approx -1/(k\pi)$. These modes lead to $E_\theta(k) = |\theta(0,0,k)|^2 \sim k^{-2}$. Let us relate the above spectrum to the $T_m(z)$ profile of Sec. 9.2.

From the discussion of Sec. 9.2, we have the following set of equations:

$$T(\mathbf{r}) = T_m(z) + \theta_{\text{res}}(\mathbf{r}), \tag{13.8}$$

$$T_m(z) = T_c(z) + \theta_m(z) = 1 - z + \theta_m(z), \tag{13.9}$$

$$\theta(\mathbf{r}) = \theta_m(z) + \theta_{\text{res}}(\mathbf{r}), \tag{13.10}$$

where $T(\mathbf{r})$ is the total temperature, and $T_m(z)$ and $\theta_m(z)$ are the planar averages of $T(\mathbf{r})$ and $\theta(\mathbf{r})$ respectively. Note that $\theta_m(z)$ creates $T_m(z)$, which is nearly $1/2$ in the bulk and that has steep variations in the boundary layer; $\theta_{\text{res}}(\mathbf{r})$ is fluctuation over $T_m(z)$, and it couples with the velocity field. The Fourier transform of $\theta_m(z)$ yields

$$\theta_m(0,0,k_z) \approx \begin{cases} -\frac{1}{\pi k_z} & \text{for even } k_z \\ 0 & \text{otherwise,} \end{cases} \tag{13.11}$$

whose spectrum forms the upper branch of $E_\theta(k)$. The lower branch of $E_\theta(k)$ is formed by θ_{res} fluctuations. It is important to note that $\theta_m(z)$ and θ_{res} interact nonlinearly and collectively generate the entropy flux $\Pi_\theta(k)$. Recall the nonlinear interactions among $\theta(0,0,2)$, $\theta(1,0,1)$, and $\mathbf{u}(1,0,1)$ in the seven-mode model of Chapter 8.

We remark that the an experimental probe typically measures $T(\mathbf{r})$ in the bulk [Niemela *et al.* (2000)]. Hence, from the above equations, the temperature fluctuations, $T(\mathbf{r}) - T_m(z) = \theta_{\text{res}}(\mathbf{r})$. Therefore, the power spectrum of this experimental data would correspond to the lower branch of $E_\theta(k)$ of Fig. 13.6. Hence, care is required in $E_\theta(k)$ computation from the experimental data.

Also note that the bispectrum arises for the temperature fluctuations but not for the velocity fluctuations. This is because the velocity field does not exhibit a nonzero mean profile ($\bar{\mathbf{u}}(z) = \langle \mathbf{u}(z) \rangle = 0$), unlike the temperature field that has a mean profile $\bar{T}(z)$, as shown Fig. 9.1. We conjecture that the Couette and Taylor-Couette flows could exhibit bispectrum for the velocity field because $\bar{\mathbf{u}}(z)$ in such flows have a mean profiles (vertical or radial).

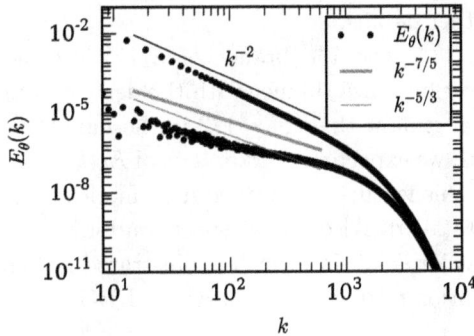

Fig. 13.6 For RBC simulation with Pr = 1 and Ra = 1.1×10^{11}, the entropy exhibits a bispectrum. The upper branch varies as k^{-2}, while the lower part matches neither with $k^{-7/5}$ nor with $k^{-5/3}$. From Verma *et al.* (2017b). Reprinted with permission from Institute of Physics.

The KE and entropy spectra and fluxes discussed so far are applicable to Pr ∼ 1. Unfortunately, these arguments do not carry over to RBC with extreme Prandtl numbers (Pr ≫ 1 or Pr ≪ 1). In the next two sections we report turbulence properties for small-Pr and large-Pr RBC.

13.4   Turbulence in zero- and small-Pr RBC

In this section we report the energy and entropy spectra for RBC when $\mathrm{Pr} \to 0$ and $\mathrm{Pr} \ll 1$ [Mishra and Verma (2010)].

RBC with zero or very-small Prandtl numbers have the thermal diffusivity $\kappa \to \infty$. To analyse this case, we start with the nondimensionalized Eq. (2.76)[2]:

$$u_z + \nabla^2 \theta = 0 \implies u_z(\mathbf{k}) = \theta(\mathbf{k})k^2. \tag{13.12}$$

Therefore, the energy feed by buoyancy,

$$\mathcal{F}_B(\mathbf{k}) \sim \mathrm{Ra}\langle \theta(\mathbf{k})u_z^*(\mathbf{k})\rangle \sim \mathrm{Ra}\langle |u_z(\mathbf{k})|^2\rangle/k^2 \tag{13.13}$$

is very steep. Hence, for $\mathrm{Pr} \to 0$, buoyancy is active at small wavenumbers only, as assumed in the Kolmogorov's phenomenology in which the forcing acts at small k's. Therefore, we expect Kolmogorov's phenomenology for hydrodynamic turbulence to be applicable to zero or small-Pr RBC, i.e.,

$$E_u(k) = K_{\mathrm{Ko}}\Pi_u^{2/3}k^{-5/3}. \tag{13.14}$$

Since $\mathrm{Pr} \to 0$, the thermal boundary layer spans the whole volume with $\bar{T}(z) = 1-z$ and $\theta_m(z) \to 0$ (see Fig. 9.3). Hence $E_\theta(k)$ has only one branch with

$$E_\theta(k) = \frac{E_z(k)}{k^4} \approx \Pi_u^{2/3}k^{-17/3}. \tag{13.15}$$

The mean profile $\theta_m(z)$ starts to appear for finite but small Pr. Here we expect that $E_\theta(k)$ with bispectrum, similar to Fig. 13.6, would start to appear. Since the nonlinear term of the θ equation is small, we expect $E_\theta(k)$ and $\Pi_\theta(k)$ to have the form $\exp(-k/\bar{k}_c)$, as in Eqs. (10.46, 10.47). The exponential branch, which is the lower branch, of $E_\theta(k)$ is illustrated in the subfigure of Fig. 13.7. See Mishra and Verma (2010) for further details.

Note that in Fig. 13.7, the k^{-2} branch of $E_\theta(k)$ is rather insignificant. The thermal boundary layer becomes thinner with the decrease of κ (or increase of Pr), leading to significant regime with $\bar{T}(z) = 1/2$ in the bulk that strengthens the k^{-2} branch of $E_\theta(k)$. Thus, we expect the k^{-2} branch of $E_\theta(k)$ to become significant and dominant for $\mathrm{Pr} \geq 1$; this feature is observed in numerical simulations, for example in Mishra and Verma (2010). Also, $E_\theta(k)$ spectrum for $\mathrm{Pr} = 0$ [Eq. (13.15)] differs significantly from that for small Pr, which is illustrated in Fig. 13.7.

Now we turn attention to infinite and large-Pr RBC.

13.5   Turbulence for infinite and large-Pr RBC

In the limit of infinite or very-large Prandtl number ($\nu \to \infty$), the momentum equation is linear [Pandey *et al.* (2014)]. See Sec. 2.7 for details. By equating the viscous and nonlinear terms, we obtain

$$\alpha g \theta_k \approx \nu k^2 u_k. \tag{13.16}$$

---

[2]Note that $\mathrm{Pr} = 0$ corresponds to $\mathrm{Pe} = 0$. Hence, as described in Chapter 2, Oberbeck-Boussinesq approximation is not applicable in this case. Still, $\mathrm{Pr} = 0$ RBC provides valuable inputs for low-Pr thermal convection.

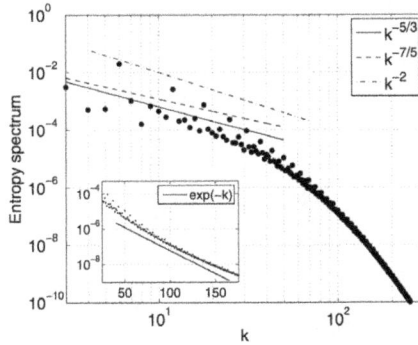

Fig. 13.7 For RBC simulation with Pr = 0.02, plot of $E_\theta(k)$ exhibits a bispectrum with an exponential lower branch ($E_\theta(k) \sim \exp(-k/k_c)$) and a weak k^{-2} upper branch. From Mishra and Verma (2010). Reprinted with permission from APS.

When the Péclet number is large, the temperature equation is nonlinear and it yields a constant entropy flux leading to

$$\Pi_\theta \approx k\theta_k^2 u_k. \tag{13.17}$$

The aforementioned equations yield the following formulas for the energy and entropy spectra[3]:

$$E_u(k) = \left(\frac{\alpha g}{\nu}\right)^{4/3} \epsilon_\theta^{2/3} k^{-13/3}, \tag{13.18}$$

$$E_\theta(k) = \left(\frac{\alpha g}{\nu}\right)^{-2/3} \epsilon_\theta^{2/3} k^{-1/3}. \tag{13.19}$$

In Fig. 13.8 we plot $E_u(k)$ and $E_\theta(k)$ for Pr = ∞ and 100. From Fig. 13.8(a) we deduce that $E_u(k) \sim k^{-13/3}$. The entropy spectrum however is more complex. The upper branch k^{-2} is due to the $\theta_m(z)$, as discussed in Sec. 13.3. The lower branch is somewhat flat. Note that the k^{-2} branch becomes strong for large-Pr RBC due to the thinning down of the thermal boundary layer and prominence of $\bar{T}(z) \approx 1/2$ in the bulk.

It is important to contrast the above scaling with that for the passive scalar turbulence with large Schmidt number for which $E_u(k) \sim k^{-1} \exp(-k/\bar{k}_c)$ (see Eq. (10.46)). In scalar turbulence, the velocity field is not affected by the scalar. But in RBC, buoyancy drives the velocity field leading to Eq. (13.16). Thus, in RBC, temperature acts as an active scalar. Also, note that the entropy spectrum in RBC is very different from that of passive scalar turbulence. This is due to the presence of walls.

In the next section we will discuss the structure function for RBC.

---

[3]Interestingly, elastic turbulence exhibits $E_u(k) \sim k^{-4}$ [Majumdar and Sood (2011)]. Such flows are laminar with weak nonlinearity for the velocity field (Re $\lesssim$ 1). However, the nonlinearity in the polymer equation is reasonably strong. On comparison, the elastic turbulence appears to be quite similar to RBC turbulence for very large Pr. We conjecture that the flux of the polymer tensor is constant, just like $\Pi_\theta \approx$ const. This would trivially yield $E_u(k) \sim k^{-13/3} \approx k^{-4}$, as in Eq. (13.18), thus explaining the kinetic energy spectrum for the elastic turbulence.

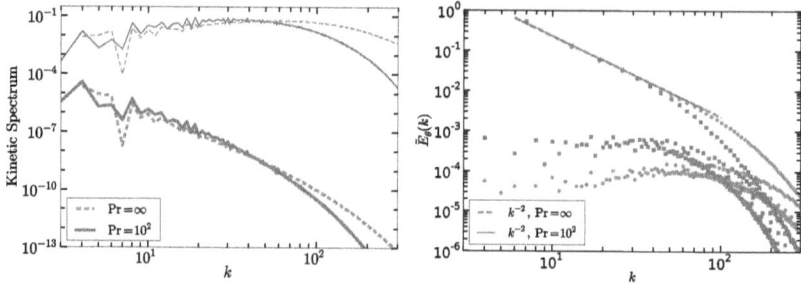

Fig. 13.8   For RBC simulation with Pr $= \infty$ and 100: (Left panel) Plots of $E_u(k)$ (thick lines) and normalised $E_u(k)k^{13/3}$ (thin lines) indicating that $E_u(k) \sim k^{-13/3}$. (Right panel) Plots of $E_\theta(k)$ indicating bispectrum. From Pandey *et al.* (2014). Reprinted with permission from APS.

13.6   Structure functions for RBC

Structure function is another popular diagnostics that is often used to describe turbulence. Here we make a brief presentation of the structure function of turbulent thermal convection. For a detailed discussion, refer to Frisch (1995) and Ching (2013).

The structure functions for the velocity and temperature fluctuations are defined as

$$S_q^u(r) = \langle [\delta u_\parallel(r)]^q \rangle, \tag{13.20}$$

$$S_q^\theta(r) = \langle [\delta\theta(r)]^q \rangle, \tag{13.21}$$

where

$$\delta u_\parallel(r) = [\mathbf{u}(\mathbf{x}+\mathbf{r}) - \mathbf{u}(\mathbf{r})] \cdot \frac{\mathbf{r}}{r}, \tag{13.22}$$

$$\delta\theta(r) = \theta(\mathbf{x}+\mathbf{r}) - \theta(\mathbf{r}), \tag{13.23}$$

and $\langle . \rangle$ represents an ensemble average. Therefore,

$$S_2^u(r) = \langle [u_\parallel(\mathbf{x}+\mathbf{r})]^2 \rangle + \langle [u_\parallel(\mathbf{r})]^2 \rangle - 2\langle u_\parallel(\mathbf{x}+\mathbf{r})u_\parallel(\mathbf{r}) \rangle$$
$$= \text{const} - 2C_\parallel(r), \tag{13.24}$$

where $C_\parallel(r)$ is the correlation of the parallel components of the velocity fields at the two points $(\mathbf{x}+\mathbf{r})$ and $\mathbf{x}$. The correlation

$$C(r) = \langle \mathbf{u}(\mathbf{x}+\mathbf{r}) \cdot \mathbf{u}(\mathbf{r}) \rangle \tag{13.25}$$

has a similar scaling as $C_\parallel(r)$, and the Fourier transform of $C(r)$ yields the corresponding power spectrum. For 3D turbulent flows,

$$E(k) = \frac{1}{4\pi k^2} \int C(r) \exp(i\mathbf{k} \cdot \mathbf{r}) d\mathbf{r}. \tag{13.26}$$

From the above relation, we deduce that

$$\text{if } C(r) \sim r^{2\xi} \implies E(k) \sim k^{-1-2\xi}. \tag{13.27}$$

Therefore, for Kolmogorov-like passive scalar turbulence,

$$S_2^u(r) = \langle \epsilon_u \rangle^{2/3} r^{2/3}, \tag{13.28}$$

$$S_2^\theta(r) = \langle \epsilon_u \rangle^{-1/3} \langle \epsilon_\theta \rangle r^{2/3}. \tag{13.29}$$

However, if Bolgiano-Obukhov model is valid for RBC, then for $L_B \ll r \ll L_f$, where L_B is the Bolgiano length, and L_f is forcing length, the structure functions are

$$S_2^u(r) = \langle \epsilon_\theta \rangle^{2/5} N^{4/5} r^{6/5}, \tag{13.30}$$

$$S_2^\theta(r) = \langle \epsilon_\theta \rangle^{4/5} N^{-2/5} r^{2/5}, \tag{13.31}$$

where N is the Brunt-Väisälä frequency. For $\eta < r < L_B$, the second-order structure function is expected to be given by Eqs. (13.28, 13.29). These relations are consistent with Eqs. (12.26–12.29) for the BO scaling.

Assuming linear scaling for the exponents, it is extrapolated that

$$S_q^u(r) = \langle \epsilon_u \rangle^{q/3} r^{q/3}, \tag{13.32}$$

$$S_q^\theta(r) = \langle \epsilon_u \rangle^{-q/6} \langle \epsilon_\rho \rangle^{q/2} r^{q/3}, \tag{13.33}$$

and the corresponding structure functions for the BO scaling are

$$S_2^u(r) = \langle \epsilon_\theta \rangle^{q/5} N^{2q/5} r^{3q/5}, \tag{13.34}$$

$$S_2^\theta(r) = \langle \epsilon_\theta \rangle^{2q/5} N^{-q/5} r^{q/5}. \tag{13.35}$$

Any deviation of the exponents from the above predictions is attributed to the *intermittency effects*. Ching and Cheng (2008) computed the anomalous scaling for the turbulent RBC.

From the discussion of earlier sections that favours Kolmogorov-like phenomenology for RBC, we expect $S_2^u(r) \propto l^{2/3}$. The temperature fluctuation θ exhibits bispectrum (see Sec. 13.3) that may have effects on $S_2^\theta(r)$. This issue needs further investigation.

Ching (2007) and Ching (2013) studied the structure functions for the velocity and temperature fluctuations of turbulent convection, and claimed consistency with the BO scaling. Ching et al. (2004) also studied the structure function associated with the cross correlation between the velocity and temperature fluctuations, and reported that the cross-scaling exponents are different from the predictions of the BO scaling. As described in the earlier section, we believe that the BO is not applicable to RBC, and the aforementioned analysis of the structure function may need further examination.

In the next two sections, we will describe experimental results on $E_u(k)$ and $E_\theta(k)$ of turbulent convection.

13.7 Taylor's frozen-in hypothesis for RBC?

In fluid experiments, high-resolution measurements of three-dimensional velocity field $\mathbf{u(r)}$ is very difficult. However, it is much easier to measure the velocity field at fixed locations in a flow using hot-wire probes, laser Doppler velocimetry (LDV), and similar techniques, and then compute the frequency spectrum $E_u(f)$ from the signal. The frequency spectrum $E_u(f)$ is related to $E_u(k)$:

$$E_u(k) = E_u(f)\frac{df}{dk} = \frac{U_0}{2\pi}E_u(f), \tag{13.36}$$

where $\mathbf{U}_0$ is mean velocity of the flow, which is advecting the turbulent fluctuations. This is called *Taylor's frozen-in hypothesis*. Wind tunnel experiments invoke this principle to compute $E_u(k)$. Unfortunately, RBC does not have such a mean flow. But before going to RBC, we review Taylor's frozen-in hypothesis.

Recently, Verma and Kumar (2014) derived $E_u(f)$ of Eq. (13.36) for $U_0 \gg u_{\mathrm{rms}}$, and $E_u(f) \sim f^{-2}$ for $U_0 = 0$. For this derivation, they employed the following Green's function:

$$G(\mathbf{k}, \omega) = \frac{1}{-i\omega + \nu(k)k^2 + i\mathbf{U}_0 \cdot \mathbf{k}}, \tag{13.37}$$

where

$$\nu(k) = (K_{\mathrm{Ko}})^{1/2}\Pi_u^{1/3}k^{-4/3}\nu_* \tag{13.38}$$

is the wavenumber-dependent viscosity (also called renormalized viscosity) [Leslie (1973); Yakhot and Orszag (1986); Verma (2004)]. Here K_{Ko} is the Kolmogorov's constant, $\omega = 2\pi f$ is the angular frequency, $\nu_* \approx 0.38$ is a constant, and Π_u is the KE flux. From the above definition of Green's function, we obtain the dominant

$$\omega = \mathbf{U}_0 \cdot \mathbf{k} - i\nu(k)k^2. \tag{13.39}$$

When $\mathbf{U}_0 \cdot \mathbf{k} \gg \nu(k)k^2$, we obtain

$$\omega = U_0 k_z. \tag{13.40}$$

Therefore, using $E_u(k) = K_{\mathrm{Ko}}\Pi^{2/3}k^{-5/3}$, we obtain

$$E_u(\omega) = E_u(k)\frac{dk}{d\omega} \sim K_{\mathrm{Ko}}\Pi^{2/3}(\omega/U_0)^{-5/3}(1/U_0)$$
$$\sim K_{\mathrm{Ko}}(U_0\Pi)^{2/3}\omega^{-5/3}, \tag{13.41}$$

consistent with the principle of Taylor's frozen-in turbulence hypothesis. For simplification, in Eq. (13.40) we have replaced k_z with k.

On the contrary, when $\mathbf{U}_0 \cdot \mathbf{k} \ll \nu(k)k^2$ (for zero or small U_0), we obtain

$$\omega \approx \nu(k)k^2 = \nu_*\sqrt{K_{\mathrm{Ko}}}\Pi^{1/3}k^{2/3}. \tag{13.42}$$

Therefore,

$$E_u(\omega) = E_u(k)\frac{dk}{d\omega}$$
$$= \frac{K_{\mathrm{Ko}}\Pi^{2/3}k^{-5/3}}{\nu_*\sqrt{K_{\mathrm{Ko}}}\Pi^{1/3}(2/3)k^{-1/3}}$$
$$= \frac{3}{2}\nu_*(K_{\mathrm{Ko}})^{3/2}\Pi\omega^{-2}. \tag{13.43}$$

Thus, the aforementioned definition of Green's function helps us deduce both $\omega^{-5/3}$ and ω^{-2} frequency spectra depending on the strength of $\mathbf{U}_0$. Verma and Kumar (2014) performed numerical simulations of hydrodynamic turbulence for $\mathbf{U}_0 = 0$, $0.4\hat{z}$, and $10\hat{z}$ (compared to fluctuations, which has rms value of the order of unity). They observed that $E_u(f)$'s are consistent with the aforementioned predictions (see Fig. 13.9). Equation (13.43) has been derived earlier by Landau and Lifshitz (1987) using dimensional analysis.

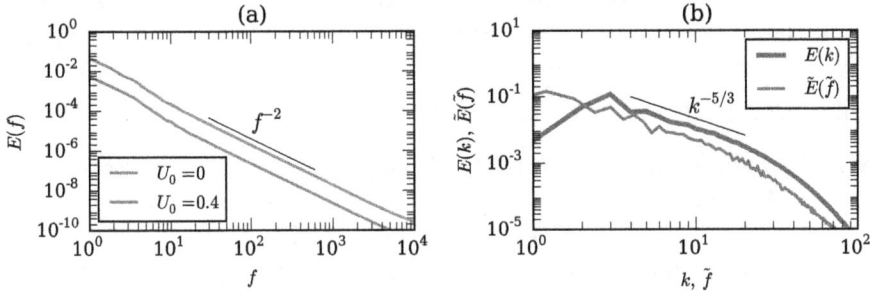

Fig. 13.9 (a) For $U_0 = 0$ and 0.4, $E_u(f) \sim f^{-2}$ for the velocity time series measured by a real space probe. (b) For $U_0 = 10$, $E_u(f) \sim f^{-5/3}$. Adopted from Verma and Kumar (2014).

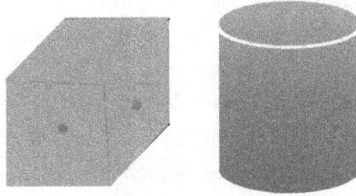

Fig. 13.10 In RBC experiments in a cube or in a cylinder, velocity or thermal probes, shown by red dots, are placed near the lateral walls.

In a typical RBC experiment, we may like to employ velocity probes inside the experimental container. Probes are typically placed near the walls, as shown in Fig. 13.10, to avoid their back-reactions on the flow. The probe locations are shown as red dots. Note however that RBC has no mean flow with constant $\mathbf{U}_0$, but typically RBC exhibits a large scale circulation (LSC). Therefore, we may be tempted to treat the mean velocity of the LSC at the probe location as $\mathbf{U}_0$. This is however problematic because LSC is dynamic and it may exhibit flow reversals (change of direction).

Also, in a cylindrical geometry, the LSC rotates azimuthally, hence $\mathbf{U}_0$ at the probe changes with time. In Chapter 17, we describe these phenomena in detail. Due to these reasons, in RBC, it is difficult to deduce $E_u(k)$ from the $E_u(f)$ using Eq. (13.36). This is one of the reasons for divergent experimental results on $E_u(k)$. In Sec. 17.3 we point out that it may be somewhat safer to apply Taylor-hypothesis to non-reversing RBC flows in rectangular geometries.

In the next section, we describe the experimental results on $E_u(k)$ of turbulent thermal convection.

13.8   Experimental results on RBC spectrum

Experimental investigation of the KE and entropy spectra for RBC is quite tricky. Researchers have computed KE and entropy spectra using either 2D Particle Image Velocitmetry (PIV) or local field measurements. In the following discussion we describe the results from such experiments.

13.8.1   *Results from PIV measurements*

In a PIV measurement, the experimental fluid is seeded with tiny passive tracer particles. A region of the fluid is illuminated with laser beam, and its pictures are taken by a single camera or multiple high-speed cameras. A single camera yields only two components of the velocity field, but several cameras provide all three components of the velocity field. Since 3D PIV is more expensive, most RBC experiments investigating $E_u(k)$ employ 2D PIV. Note however the spectra computed using 2D PIV assume homogeneity and isotropy of the flow, which are not strictly valid for RBC due to the presence of LSC, flow reversals, and azimuthal reorientations of the rolls (see Chapter 17). In the following, we describe some of the results of PIV-based RBC experiments.

Kunnen *et al.* (2008) measured the velocity field in rectangular regions of 9cm × 12cm near the mid plane, and 12cm × 15cm near the top lid of a cylinder. They computed the structure function of the velocity field using the experimental data and concluded in favour of BO scaling. On the contrary, using a similar experimental configuration, Sun *et al.* (2006) obtained results consistent with 3D hydrodynamic turbulence.

Using laser beam sweeping technique, Chillà *et al.* (1993) measured local temperature gradients in a RBC experiment with water, and reconstructed temperature profiles in the xz planes. Using this data, they computed the temperature spectrum, and reported $k^{-5/3}$ power law for a wide range of wavenumbers. Using ultrasonic velocimetry, Mashiko *et al.* (2004) measured the velocity field in mercury along a line and computed the second-order structure function and $E_u(k)$; they concluded that the system exhibits BO scaling. Note that PIV technique cannot be used for mercury because it is opaque.

13.8.2   *Results from local field measurements*

A large number of RBC experiments employ local field measurements of the velocity and/or temperature fields to compute $E_u(k)$. The velocity fields at selected locations are measured using laser Doppler velocimetry (LDV), in which a Doppler shift in a laser beam is used to measure the velocity of a tracer embedded in a transparent or semi-transparent fluid. The probes are typically placed near the lateral walls, as in Fig. 13.10. The frequency spectrum $E_u(f)$ is computed using the measured signal, after which $E_u(k)$ is computed using the Taylor's frozen-in hypothesis, which is however questionable for RBC due to the presence of LSC, flow reversal, and azimuthal reorientation of LSC.

Ashkenazi and Steinberg (1999) employed LDV to measure the velocity field in a RBC experiment with SF_6. They concluded in favour of BO scaling. Shang and Xia (2001) performed similar experiment with water and found that $E_u(f) \sim f^{-1.35}$ at lower frequencies and $\sim f^{-11/5}$ at higher frequencies.

A large number RBC experiments involve temperature measurements using thermistors or temperature-sensitive dyes. As remarked in Sec. 13.3, the probes measure $T(\mathbf{r})$, whose fluctuations are $\theta_{\mathrm{res}}(\mathbf{r})$. The spectrum of $\theta_{\mathrm{res}}(\mathbf{r})$ is essentially the lower branch of $E_\theta(k)$ of Fig. 13.6. Thus, the measured $E_\theta(k)$ cannot differentiate between Bolgiano-Obukhov and Kolmogorov's spectrum. We believe this factor to be one of the reasons for discrepancies in the experimental results on $E_\theta(k)$.

Cioni *et al.* (1995) performed such experiment with mercury as a working fluid. They observed that the temperature spectrum $E_T(f) \sim f^{-5/3}$, similar to the passive scalar turbulence. Zhou and Xia (2001) performed similar experiments with water, and reported $E_T(f) \sim f^{-5/3}$ in the mixing zone where the mean flow is relatively strong, and $E_T(f) \sim f^{-7/5}$ (BO scaling) in the core region. Wu *et al.* (1990) measured the temperature field in thermal convection involving Helium gas, and reported $E_T(f) \sim f^{-7/5}$. However, Castaing (1990) argued in favour of scaling similar to passive scalar turbulence. Niemela *et al.* (2000) performed convection experiment on Helium gas and reported dual spectrum for the temperature: $E_T(f) \sim f^{-7/5}$ at small frequencies, and $E_T(f) \sim f^{-5/3}$ at large frequencies.

In summary, RBC experiments yield divergent results, some favouring BO scaling, while others favouring Kolmogorov's $k^{-5/3}$ spectrum. The divergence arise due to various reasons, which are described below:

(1) We need full 3D high-resolution velocity and temperature fields for accurate determination of the KE and entropy spectra. Unfortunately such measurements are very expensive, and, to best of our knowledge, none have been performed so far for RBC.

(2) As an alternative, researchers have performed 2D PIV experiments that yield two components of velocity field in a plane. Unfortunately, the fluid flow in RBC are inhomogeneous, anisotropic, and dynamic due to the presence of LSC and its reorientations. Hence, definitive conclusions cannot be drawn from the 2D PIV measurements.

(3) As described in Sec. 13.7, Taylor's frozen-in hypothesis is not applicable to RBC due to the absence of a mean flow. Hence, the velocity field measured by real-space probes cannot be used to compute the KE spectrum accurately. If LSC is somewhat steady, then we possibly could use the mean velocity of the LSC as $\mathbf{U}_0$ for the Taylor's hypothesis. Experiments however indicate that LSC is

highly dynamic in the turbulent regime. This is one of the prime reasons for the divergent results from RBC experiments.

(4) The LSC is absent in the central zone of a RBC setup. Therefore we expect that $U_0 = 0$ in this region, and hence $E_u(f) \sim f^{-2}$, as described in Sec. 13.7. Some researchers report $E_u(f) \sim f^{-11/5}$ in the central region; we believe the exponent to be really -2, not $-11/5$.

(5) The temperature exhibits bispectrum due to wall effects (see Fig. 13.6). As described earlier, the experimental probes provide limited inputs on the spectrum.

The energetics arguments described in Sec. 13.1 clearly indicate that the spectral properties of RBC are very similar to those of 3D hydrodynamic turbulence, with the temperature field behaving like a passive scalar. Hence, we expect $E_u(k) \sim \Pi_u^{2/3} k^{-5/3}$. The entropy spectrum however does not have a single power law, rather it follows a bispectrum as discussed in Sec. 13.3.

These arguments clearly demonstrate that the spectral results of RBC from experiments are not reliable diagnostics at present. In the next section, we describe some of the numerical results on the spectra of RBC.

13.9 Numerical results on RBC

Numerical simulations of RBC provide access to complete velocity and temperature fields, unlike experiments where these fields could be measured at most in a plane. However, present day numerical simulations cannot reach as high Reynolds or Rayleigh numbers as in experiments. For example, the maximum Rayleigh number achieved in numerical simulations is 2×10^{12} [Stevens *et al.* (2011)], in contrast to Ra $\sim 10^{16}$ achieved in laboratory experiments. Note that in solar convection, Ra $\sim 10^{24}$. In addition, numerical simulation in complex geometries is very difficult. Thus, experiments and numerical simulations have their limitations, but they complement each other.

Early spectral simulations of RBC were on much coarser grids and for periodic boundary condition (also see Sec. 13.12). Grossmann and Lohse (1991) performed simulation using twelve modes derived under Fourier-Weierstrass approximation, and reported Kolmogorov's scaling ($E_u(k) \sim k^{-5/3}$). For periodic boundary conditions, Borue and Orszag (1997) and Skandera *et al.* (2007) reported $k^{-5/3}$ spectra for both the velocity and temperature fields.

Kumar *et al.* (2014a), Verma *et al.* (2017b), and Kumar and Verma (2017) simulated turbulent RBC for free-slip and no-slip boundary conditions with Pr ≈ 1, and showed that turbulent thermal convection has behaviour very similar to 3D hydrodynamic turbulence. The details are given in Secs. 13.2 and 13.3. In Secs. 13.4 and 13.5 we discussed the numerical results for RBC with very small and very large Prandtl numbers respectively. Refer to Sec. 13.6 for the numerical results related to the structure functions of turbulent convection.

In the aforementioned discussions, we described the spectral properties of turbulent convection without differentiating the bulk and the boundary layers. In the following discussion, we will describe the properties of turbulence in the boundary layers of RBC.

13.10   Turbulence in the boundary layers of RBC

The arguments presented in the earlier sections deal with $E_u(k)$ and $E_\theta(k)$ for the whole box. In this section, we briefly discuss the turbulence phenomenology of the boundary layer, which is an important research topic of turbulent convection. Note however that the turbulent properties of boundary layers remain largely unresolved at present.

For Pr $\sim$ 1 and large Ra, the viscous and thermal boundary layers occupy a much smaller volume than the bulk. In addition, the fluctuations in the boundary layers are of small size, hence they contribute to the spectra at large k's (in the dissipative and diffusive wavenumber regimes). Hence, the inertial range $E_u(k)$ and $E_\theta(k)$ described in the earlier part of this chapter are essentially for the bulk flow.

We can extend the energetics argument discussed in Sec. 13.1 to the boundary layer where $u_z \ll u_\perp$, thus the flow is quasi-2D. Hence within a boundary layer, we expect $\Pi_u(k) < 0$ for small k due to two-dimensionalization. Following the arguments of Bolgiano (1959) and Obukhov (1959), we deduce that

$$d\Pi_u(k)/dk \approx \mathcal{F}_B(k) > 0 \qquad (13.44)$$

or

$$\Pi_u(k+dk) > \Pi_u(k), \quad \text{or} \quad |\Pi_u(k+dk)| < |\Pi_u(k)|, \qquad (13.45)$$

as shown in Fig. 13.11. Thus $|\Pi_u(k)|$ decreases as k increases, as in stably stratified turbulence. Since the buoyancy is strong at small k, we expect BO scaling for $k \ll k_B$, where k_B is the Bolgiano wavenumber. See Fig. 12.1 and Eq. (12.10). This is unlike the bulk flow that exhibits Kolmogorov's $k^{-5/3}$ spectrum for KE.

For $k_B \ll k \ll k_d$, the KE spectrum may exhibit either $k^{-5/3}$ in the regime where $\Pi_u(k) < 0$, or k^{-3} where $\Pi_\omega(k) > 0$. The above scaling depends critically on where the effective forcing band lies in relation to k_B. Thus, the KE spectrum depends on the magnitude of k_B and the regime of inverse cascade. These arguments indicate that the energy spectrum in the boundary layer needs a careful investigation. We remark that using structure function computations, Calzavarini *et al.* (2002) reported BO scaling for the boundary layer.

In the next section, we briefly describe the properties for 2D turbulent RBC.

13.11   Turbulence in two-dimensional RBC

In this section we make brief remark on 2D RBC that has not been studied in detail. We assume that the flow is confined in xz plane with z as the buoyancy

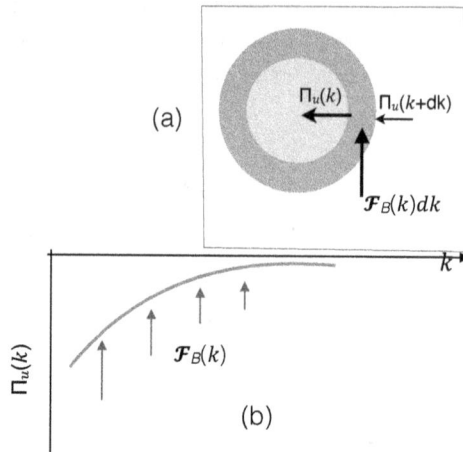

Fig. 13.11   A possible schematic diagram of the kinetic energy flux $\Pi_u(k)$ for two-dimensional RBC, and in the boundary layers of RBC.

direction. The two-dimensionality of the flow may lead to significant regime where $\Pi_u(k) < 0$. As argued in Sec. 13.10, due to strong buoyancy, we expect BO scaling in this regime. This conjecture is consistent with the numerical results of Mazzino (2017) in which they argue that 2D RBC and Rayleigh-Taylor turbulence (to be discussed in Chapter 14) exhibit BO scaling, i.e., $E_u(k) \sim k^{-11/5}$, and the structure function $S_q(r) \sim r^{3q/5}$.

The BO scaling would however breakdown in the absence of wavenumber band with inverse energy cascade. Thus, 2D RBC needs to be studied carefully, specially its energy and entropy flux. In the next section, we describe turbulence properties of turbulent convection in a periodic box.

13.12   Simulation of turbulent convection in a periodic box

Thermal plates in RBC induce boundary layers that affects the turbulence properties. For example, the bispectrum of entropy is due to the wall effects. In this section, we describe the results of RBC simulations in a periodic box with a constant gradient $d\bar{T}/dz$. In the absence of boundary layers, the velocity and temperature fields exhibit $k^{-5/3}$ spectra, as shown by Borue and Orszag (1997), Skandera *et al.* (2007), Lohse and Toschi (2003), and Calzavarini *et al.* (2005). Earlier, Grossmann and Lohse (1991) simulated RBC flow under Fourier-Weierstrass approximation and reported Kolmogorov's scaling. These results are consistent with the turbulence phenomenology of RBC presented in Sec. 13.1.

Several important points to note for an RBC flow in a periodic box are: (a) Turbulent thermal convection in a periodic box is numerically unstable; the system exhibits non-diverging solution only for a carefully chosen set of initial conditions;

(b) The temperature spectrum for the periodic box exhibits $E_\theta(k) \sim k^{-5/3}$, similar to that in passive scalar turbulence. Hence it is very different from RBC with walls that exhibits bispectrum as shown in Fig. 13.6; (c) The Nusselt number Nu $\sim$ Ra$^{1/2}$ [Lohse and Toschi (2003); Calzavarini *et al.* (2005)], as in the ultimate regime when the boundary effects are negligible.

In the next section, we briefly discuss the field-theoretic results on RBC and stably stratified turbulence.

13.13 Field theoretic treatment of RBC

In the earlier sections, we described the energetics argument of RBC, which is based on KE flux. Also, in chapter 12 we described the Bolgiano-Obukhov's (BO) scaling for stably stratified turbulence. These arguments however do not start from the basic equations, which are the Navier-Stokes and temperature equations, for example, Eqs. (2.10–2.12). One of the difficult yet popular analytic approach to study turbulence is field-theory. Some of the main attempts in this direction are direct interaction approximation (DIA) of Kraichnan (1959) (also see Leslie (1973)), and the renormalization group analysis of Yakhot and Orszag (1986), and McComb (1990). Several researchers have employed these tools to RBC. In this book, we are not focussing on such works, hence we describe them very cursorily.

Procaccia and Zeitak (1989) started with RBC equations with forcing added to the temperature equation. They employed mean field approximation and obtained Bolgiano-Obukhov scaling. Rubinstein (1994) used Yakhot and Orszag (1986)'s renormalization group procedure and reported Bolgiano-Obukhov scaling for RBC.

L'vov (1991), and L'vov and Falkovich (1992) made use of conservation laws, and KE and entropy fluxes in their framework. They deduced that $\Pi_u(k) \sim k^{-4/5}$ due to conversion of KE to PE, and that RBC follows Bolgiano-Obukhov scaling. Grossmann and Lohse (1991) employed Fourier-Weierstrass mode analysis and argued against the conversion of KE to PE, contrary to the proposal by L'vov (1991) and L'vov and Falkovich (1992). They argued in favour of $u_k \sim k^{1/3}$ scaling [Shraiman and Siggia (1990)]. Recently, Bhattacharjee (2015) used the global energy balance for the stratified fluid and argued that the BO scaling could be observed in stably stratified flows at high Richardson number.

In the next chapter, we will describe turbulence phenomenologies of several related buoyancy-driven systems—Rayleigh-Taylor turbulence, unstably stratified turbulence, Taylor-Couette turbulence.

Further reading

Kumar *et al.* (2014a) and Verma *et al.* (2017b) describe the energetics (flux-based arguments) of thermally-driven turbulence that shows close similarities between thermally-driven and hydrodynamic turbulence. They also discuss the numerical simulations that verifies the above scaling. Lohse and Xia (2010) review earlier results on energy spectrum and structure function. For experimental results, refer to the original articles referred to in Sec. 13.8. Kumar and Verma (2017) discuss whether Taylor's hypothesis is applicable to RBC or not.

Exercises

(1) How does the energy flux of turbulent thermal convection differ from that of hydrodynamic turbulence and stably stratified turbulence?
(2) Contrast the turbulent thermal convection in 2D and 3D.
(3) Contrast the turbulent thermal convection for small Pr, moderate Pr, and large Pr.
(4) What are the difficulties in computing energy spectrum of turbulent thermal convection using experimental data?
(5) A velocity probe measures the velocity field in the middle of a cube exhibiting thermal convection. What kind of frequency spectrum is expected?
(6) Consider RBC experiment in two identical cylinder containing water and mercury. The Rayleigh number for both the flows are same. Which of the two systems is more turbulent?

Chapter 14

Turbulence Phenomenology of Other Buoyancy-Driven Flows

The energetics arguments described in the last two chapters are quite general. For flows with stable stratification, in the inertial range, $\Pi_u(k)$ decreases with k, i.e., $\Pi_u(k) \sim k^{-\xi}$ with $\xi > 0$. Hence, we expect the kinetic energy spectrum as

$$E_u(k) \sim [\Pi_u(k)]^{2/3} k^{-5/3} \rightarrow k^{-5/3 - 2\xi/3}, \tag{14.1}$$

or $E_u(k)$ is steeper than $5/3$ spectrum. However, for unstable stratification, the kinetic energy gains from buoyancy. Therefore, we expect $\Pi_u(k) \sim k^{\xi}$ ($\xi > 0$) in the inertial range. Consequently, we expect

$$E_u(k) \sim [\Pi_u(k)]^{2/3} k^{-5/3} \rightarrow k^{-5/3 + 2\xi/3} \tag{14.2}$$

to be shallower than $5/3$ spectrum. In RBC, we show that buoyancy is much weaker than the pressure gradient, hence $\xi \rightarrow 0$ and $E(k) \sim k^{-5/3}$. Thus, in general, for stable stratification, we expect $E(k)$ to be steeper than $k^{-5/3}$ spectrum, while, for unstable stratification, $E(k) \sim k^{-5/3}$, or shallower than $k^{-5/3}$ spectrum. Note however that the aforementioned arguments hinge on the assumption of forward kinetic energy flux and isotropy, and on near isotropy of the flow.

In the last two chapters we discussed turbulence phenomenology of two buoyancy-driven flows: RBC and stably stratified flow. There are many more buoyancy-driven flows, some of which will be described in this chapter. In particular, we will discuss the turbulence properties of Rayleigh-Taylor, Taylor-Couette, unstably stratified, and bubbly flows, as well as in hemispherical convection.

14.1 Rayleigh-Taylor turbulence (RTT)

In Sec. 6.6, we introduced Rayleigh-Taylor instability in which a heavy fluid sits on top of a light fluid. As the instability grows, the heavy fluid descends and the light fluid ascends. The gain in the kinetic energy is at the expense of the potential energy, i.e.,

$$\frac{d}{dt} \frac{U^2}{2} = g\bar{\rho}U, \tag{14.3}$$

where $U, \bar{\rho}$ are the large-scale velocity and density respectively [Boffetta and Mazzino (2017)]. The solution of the above equation provides the velocity of the fingers of the light and heavy fluids:

$$U(t) \approx g\bar{\rho}t. \tag{14.4}$$

An integration of the above equation yields the height of the fingers as

$$h(t) \approx g\bar{\rho}t^2/2. \tag{14.5}$$

If we assume the vertical extent of the fluids to be such that $-H/2 < z < H/2$ with large H (see Fig. 6.6), then the growth of $U(t)$ and $h(t)$ would continue until $h(t) \approx H$. Therefore,

$$T_{\text{final}} \approx \sqrt{\frac{H}{g\bar{\rho}}}. \tag{14.6}$$

The width of the fingers, δ, can be estimated by setting

$$\text{Ra}_\delta = \frac{g(\rho_2 - \rho_1)\delta^3}{\nu\kappa(\rho_2 + \rho_1)} \sim 1 \tag{14.7}$$

that yields

$$\delta \sim d \left[\frac{g(\rho_2 - \rho_1)d^3}{\nu\kappa(\rho_2 + \rho_1)}\right]^{-1/3} \sim d\text{Ra}^{-1/3}. \tag{14.8}$$

The flow becomes turbulent when the Reynolds number $\text{Re} = UL/\nu \gg 1$. In this section, we focus on this regime, which is also termed as *Rayleigh-Taylor turbulence* or *RTT*. In the following discussion, we will describe properties of RTT. RTI and RTT are analogous to a system with a cold fluid sitting above a hot fluid. Hence we can define Nusselt number for RTT. Boffetta and Mazzino (2017) showed that

$$\text{Nu} \sim (\text{RaPr})^{1/2}, \tag{14.9}$$
$$\text{Re} \sim \text{Ra}^{1/2}\text{Pr}^{-1/2}, \tag{14.10}$$

which corresponds to the ultimate regime. This is expected because unlike RBC, RTT does not have any rigid boundary. Recall that in RBC, the boundary layers near walls bring down the Nusselt number exponent from $1/2$ to ≈ 0.30.

Chertkov (2003) first proposed that a fully-developed 3D RTT will exhibit Kolmogorov's spectrum due to the large-scale pumping by buoyancy. This conjecture is consistent with the energetics arguments presented in the last chapter. For unstably stratified flows, the energy flux scales as $\Pi(k) \sim k^\xi$. However, in RBC and RTT with $\text{Pr} \approx 1$, energy feed by buoyancy scales as $\mathcal{F}_B(k) \sim k^{-5/3}$, hence buoyancy feeds the kinetic energy at large scales, similar to that in Kolmogorov's model of hydrodynamic turbulence. Therefore, $\xi \approx 0$, and we expect RTT to be similar to 3D hydrodynamic turbulence. This conjecture has been validated by Banerjee *et al.* (2010) and Akula and Ranjan (2016) who observed $E_u(k) \sim k^{-5/3}$ in their RTT experiments, and by Young *et al.* (2001) who observed similar spectrum in their high-resolution simulations.

2D RTT however has a different scaling. Following similar arguments as that for 2D RBC (Sec. 13.11), we expect that the kinetic energy flux of 2D RTT is negative, and hence we expect $|\Pi_u(k)|$ to decrease with k, as shown in Fig. 13.11. Such energetics would yield Bolgiano-Obukhov scaling, i.e., $E_u(k) \sim k^{-11/5}$ and $E_\rho(k) \sim k^{-7/5}$. In a quasi-2D box $(L_y \ll L_x)$, Boffetta *et al.* (2011) observed Bolgiano-Obukhov scaling for 2D RTT, consistent with the above arguments.

In the next section, we will describe turbulence properties of unstably stratified flows.

14.2   Unstably stratified turbulence

Arakeri and coworkers [Arakeri *et al.* (2000); Cholemari and Arakeri (2009)] performed an interesting experiment involving unstable density stratification. They took a long vertical tube, and placed brine solution and fresh water at the top and bottom of the tube respectively. See Fig. 14.1 for an illustration. This system is analogous to RBC with Rayleigh number Ra_g:

$$\mathrm{Ra}_g = \frac{g}{\rho_0} \frac{(d\bar{\rho}/dz)d^4}{\nu\kappa}, \tag{14.11}$$

where g is the acceleration due to gravity, ρ_0 is the mean density, $\bar{\rho}(z)$ is the density stratification in the tube, κ is the salt diffusivity, and d is the diameter of the tube. Here, Schmidt number $\mathrm{Sc} = \nu/\kappa$ takes the role of the Prandtl number.

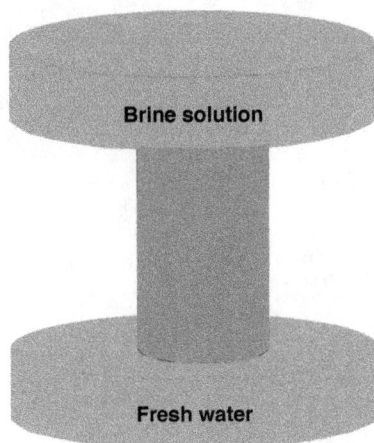

Fig. 14.1   Unstable stratification in a long thin tube with brine solution at the top and pure water at the bottom. This system is similar to RBC in a periodic box, which is described in Sec. 13.12.

This relatively simple experiment exhibits many interesting features. For Rayleigh number beyond 10^8, Cholemari and Arakeri (2009) observed that

$$\text{Re} \sim \text{Ra}^{1/2}\text{Sc}^{-1/2}; \quad \text{Nu} \sim \text{Ra}^{1/2}\text{Sc}^{1/2}, \tag{14.12}$$

an expected scaling for the ultimate regime (see Sec. 11.1, 11.10). In this system, which is similar to the RBC in a periodic box, the ultimate regime is achieved quite easily due to the absence of boundary layers at the top and bottom surfaces.

Pawar and Arakeri (2016) also computed the spectra of the kinetic energy and entropy (ρ^2) and reported that the kinetic energy exhibits $k^{-5/3}$ spectrum, while the entropy spectrum is closer to $k^{-7/5}$. We believe that the present system should yield Kolmogorov's spectrum for both kinetic energy and entropy due to its similarities with RBC. The two systems have same energetics due to unstable stratification. The discrepancy in the entropy spectrum of Pawar and Arakeri (2016) may be due to the experimental limitations in computing $E_\rho(k)$ using 2D PIV measurements (see Sec. 13.8.1).

In the next section, we will describe the properties of Taylor-Couette turbulence.

14.3 Taylor-Couette turbulence

In a Taylor-Couette (TC) system, a viscous fluid is confined between two rotating cylinders as shown in Fig. 14.2. The inner and outer cylinders, whose radii are r_i and r_o respectively, rotate with angular frequencies of ω_i and ω_o respectively, and $d = r_o - r_i$. This system exhibits very rich behaviour depending on the frequencies of rotation and the ratio r_i/r_o [Chandrasekhar (2013); Drazin (2002); Grossmann *et al.* (2016)]. Here we will provide a brief introduction to the turbulent behaviour of Taylor-Couette flow. For details refer to [Chandrasekhar (2013); Grossmann *et al.* (2016)]

The equations for the incompressible Taylor-Couette flow are [Grossmann *et al.* (2016)]

$$\frac{\partial \mathbf{u}}{\partial t} + (\mathbf{u} \cdot \nabla)\mathbf{u} = -\nabla\sigma + \text{Ro}^{-1}\hat{z} \times \mathbf{u} + \frac{f(\eta)}{\sqrt{\text{Ta}}}\nabla^2\mathbf{u}, \tag{14.13}$$

$$\nabla \cdot \mathbf{u} = 0, \tag{14.14}$$

where

$$\eta = \frac{r_i}{r_0}, \tag{14.15}$$

$$a = -\frac{\omega_o}{\omega_i}, \tag{14.16}$$

$$\text{Ta} = \frac{(1+\eta)^4}{64\eta^2}\frac{(r_o^2 - r_i^2)^2(\omega_i - \omega_o)^2}{\nu^2}, \tag{14.17}$$

$$\text{Ro} = \frac{2\omega d}{|\omega_i - \omega_o|r_i}, \tag{14.18}$$

$$\text{Re}_{i,o} = \frac{r_{i,o}\omega_{i,o}d}{\nu}. \tag{14.19}$$

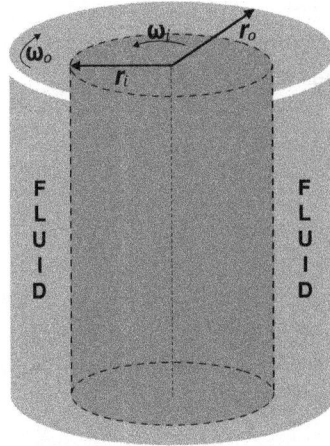

Fig. 14.2   Setup of Taylor-Couette flow: fluid is confined between two concentric cylinders that rotate with angular frequencies of ω_i and ω_o. Taylor-Couette flow shows interesting patterns, and chaotic as well as turbulent behaviour.

Here Ta, Ro, $\mathrm{Re}_{i,o}$ are the Taylor number, Rossby number, and Reynolds number based on the parameters of inner and outer cylinders respectively. TC flow and RBC have similarities because the centrifugal force in TC flow acts like gravitational force, and hence TC flow exhibits instabilities, patterns, and turbulence similar to those in RBC. See Fig. 2 of Grossmann *et al.* (2016) for an illustration of various states of TC flow.

The heat flux of RBC is related to the following angular velocity flux of TC [Grossmann *et al.* (2016)]:

$$J^\omega = r^3(\langle u_r \omega \rangle_{A,t} - \nu \partial_r \langle \omega \rangle_{A,t}), \tag{14.20}$$

where $\langle . \rangle_{A,t}$ represents averaging over time and over the cylinder of radius r. Using J^ω, we can construct a quantity related to the Nusselt number of RBC as

$$\mathrm{Nu}_\omega = \frac{J^\omega}{J^\omega_{\mathrm{lam}}}, \tag{14.21}$$

where $J^\omega_{\mathrm{lam}} = 2\nu r_i^2 r_o^2 (\omega_i - \omega_o)/(r_o^2 - r_i^2)$ is the angular velocity flux for the laminar case. Huisman *et al.* (2012) and Grossmann *et al.* (2016) concluded that turbulent TC flow exhibits scaling similar to the ultimate regime of RBC. In particular, $\mathrm{Re}_\omega \propto \mathrm{Ta}^{1/2}$ and $\mathrm{Nu}_\omega \propto \mathrm{Ta}^{0.38}$.

Considering strong similarities between the TC and RBC, we expect that for very large Re, $E_u(k)$ of TC turbulence should exhibit Kolmogorov-like spectrum, i.e.,

$$E_u(k) = K_{\mathrm{Ko}} \Pi^{2/3} k^{-5/3}, \tag{14.22}$$

as in RBC. We are not aware of any study on the energy spectrum of Taylor-Couette flow. We hope that future analysis of the experimental or simulation data of TC flows will clarify this issue. Note however that for the spectral analysis, the cylindrical shell need to be mapped to a rectangular box.

In the next section, we will describe the turbulence properties of bubbly turbulence.

14.4   Bubbly turbulence

Another important buoyancy-driven flow is bubbly turbulence. In such flows, bubbles rise due to buoyancy, and hence they push the fluid in the upward direction and setup a strong convective flow, similar to that in RBC. Thus, bubbles feed kinetic energy to the fluid. Hence, $\Pi_u(k) \sim k^\xi$ with $\xi > 0$, as in unstably stratified flows. However, the energy feed and the exponent ξ crucially depends on the bubble size. Larger bubbles feed kinetic energy at the large scales, and hence we expect $\xi \approx 0$. Therefore, we expect Kolmogorov-like spectrum ($k^{-5/3}$) for such flows. For bubbles of smaller sizes, the energy feed could be at the intermediate and dissipative scales. For such flows, there may be complex energy transfers, for example, nonlocal transfers from small scales to large scales. However, from the flux arguments given earlier, we expect the energy spectrum to be shallower than $k^{-5/3}$, but such flows need closer investigation, as envisaged by Rensen *et al.* (2005).

Prakash *et al.* (2016) performed an experiment on bubbly turbulence and reported $k^{-5/3}$ energy spectrum for $k < 1/b$, and k^{-3} for $k > 1/b$, where b is the bubble size. The $k^{-5/3}$ spectrum is consistent with the above arguments. Prakash *et al.* (2016) explained the k^{-3} energy spectrum for $k > 1/b$ by invoking equipartition between the energy dissipation and energy feed by the buoyancy. An investigation of energy flux and energy transfers in such flows would yield interesting insights. Refer to Lakkaraju *et al.* (2013) for discussion on the Nusselt number scaling in bubbly turbulence. It will be interesting to extend the $\langle u_z \theta \rangle$ correlation study of Sec. 11.4.2 to bubbly turbulence.

In the next section we discuss turbulence properties in a thermally-driven hemisphere.

14.5   Turbulence in thermally-driven hemisphere

Bruneau *et al.* (2017) performed a convection experiment and numerical simulations of buoyancy-driven flows on a hemispherical surface. As shown in Fig. 14.3, the equatorial ring, placed horizontally, is heated at a constant temperature. The fluid cools down as it rises towards to the pole. When the temperature difference between the ring and the pole crosses a critical value, patterns and turbulence emerge in the flow. Bruneau *et al.* (2017) defined Rayleigh number similar to that in RBC, with Δ as the temperature difference between the pole and the equator. This system appears to be closely analogous to RBC, hence we expect similarities between flow properties of thermally-driven hemisphere and RBC [Bruneau *et al.* (2017)].

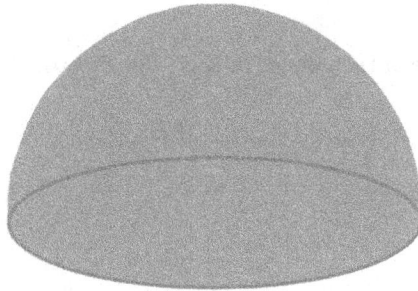

Heated ring

Fig. 14.3   A soap film attached to a hot equatorial ring. The flow in the soap film exhibits behaviour similar to that in RBC.

Bruneau *et al.* (2017) observed that Nusselt number Nu $\sim$ Ra$^{0.31}$, similar to that observed for RBC. They also reported that Re $\sim$ Ra$^{0.55}$ for Ra $< 10^8$, and Re $\sim$ Ra$^{0.43}$ for Ra $> 10^8$; these observations match closely with the predictions of Pandey and Verma (2016) (see Sec. 11.2). In addition, Bruneau *et al.* (2017) reported bi-spectrum for the entropy, which is similar to that described in Sec. 13.3 for RBC. For the kinetic energy, Seychelles *et al.* (2008) reported $k^{-11/5}$ spectrum, which is consistent with the results of 2D turbulent convection described in Sec. 13.11. The inverse cascade of kinetic energy essentially leads to a steeper kinetic energy flux that leads to Bolgiano-Obukhov spectrum in 2D RBC. This study is quite attractive due to its strong similarities with the convective flows on the Earth's atmosphere.

There are many more examples of buoyancy-driven flows, e.g., horizontal convection, free convection as in the atmosphere, non-Boussinesq flows, stellar convection, mantle convection etc. But we will not discuss them here due to lack of space. In the next chapter we will discuss the anisotropy in thermally-driven turbulence.

Further reading

Chertkov (2003) described Kolmogorov-like spectrum for Rayleigh-Taylor turbulence. Boffetta and Mazzino (2017) review Rayleigh-Taylor turbulence, both in 2D and 3D. Pawar and Arakeri (2016) and references therein describe the properties of unstably stratified turbulence in long cylindrical tube. Refer to Chandrasekhar (2013) for Taylor-Couette instability, and Grossmann *et al.* (2016) for Taylor-Couette turbulence. Refer to Prakash *et al.* (2016) and reference therein for discussion on bubbly turbulence, and Bruneau *et al.* (2017) and reference therein for turbulence in thermally-driven hemisphere.

Exercises

(1) Contrast the nature of turbulence in Rayleigh-Bénard convection and Rayleigh-Taylor turbulence.
(2) What kind of $E_u(k)$ is expected in 2D Rayleigh-Taylor turbulence?

Chapter 15

Anisotropy in Thermally-Driven Turbulence

Buoyancy-driven flows are anisotropic due to the gravitational force. In Rayleigh-Bénard convection, buoyancy drives the hot plumes preferentially along the direction of gravity, thus $u_\parallel > u_\perp$, where $u_\parallel$, $u_\perp$ are the velocity components parallel and perpendicular to the buoyancy direction. In stably stratified flows, $u_\parallel < u_\perp$ due to the transfer of the kinetic energy to the potential energy. In real space, anisotropy can be quantified using directional measures of flow structures (such as flow alignment along the gravity direction) or by angular dependence of quantities like kinetic energy (u^2).

Real-space visualisation and global quantities like energy do not capture the flow anisotropy at the intermediate and small scales. For the same, we employ diagnostics based on Fourier modes. In Chapter 4, we studied the shell spectrum and energy flux, which are averaged over the polar and azimuthal angles, hence they do not capture the flow anisotropy. New spectral quantities, ring spectrum and ring-to-ring energy transfers overcome these deficiencies, and they provide scale-dependent and angular-dependent energy distribution and energy transfers. We will describe these quantities as well as other anisotropy measures in the present chapter.

We start with anisotropy description in real space.

15.1 Anisotropy quantification in real space

In this section we describe several popular measures of anisotropy in real space.

15.1.1 *Flow visualisation*

Flow visualisation is the simplest way to describe flow anisotropy in a qualitative manner. An isotropic flow or object appears to be the same in all directions. For example, a sphere is isotropic about its centre; a vortex exhibits azimuthal symmetry. A turbulent flow however is not as isotropic as a sphere or a vortex, hence we introduce a concept of *statistical isotropy*. A statistically isotropic flow is not identical in all details along every direction, but the flow appears to be the same on an average. See for example the enstrophy $|\boldsymbol{\omega}|^2$ of a turbulent flow shown in Fig. 15.1(a). Since

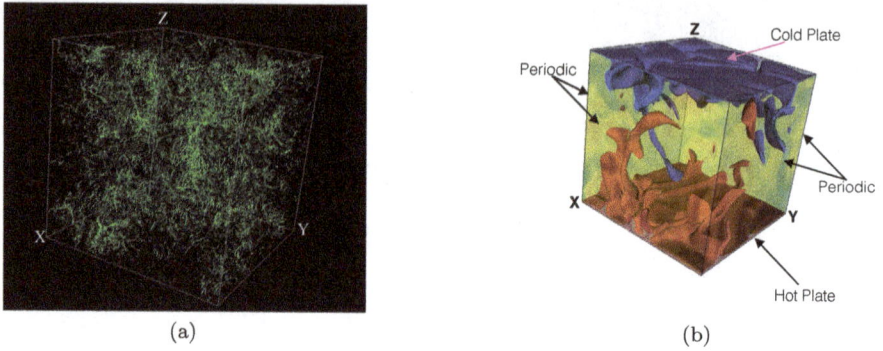

(a)   (b)

Fig. 15.1   (a) Isosurface plots of enstrophy $|\omega|^2$ of a hydrodynamic turbulent flow. The flow is statistically isotropic. (b) Isosurface plots of the temperature in a turbulent thermal convection with hot (red) and cold (blue) plumes; here the plumes are aligned along $\mathbf{g}$ thus indicating anisotropy.

the enstrophy in the figure is randomly distributed with no preferential angular dependence, we term this turbulent flow as *statistically isotropic*.

Most flows in the universe are under the influence of some external force, for example, gravity, rotation, or external magnetic field. When the external force is as strong as the nonlinear term $\rho \mathbf{u} \cdot \nabla \mathbf{u}$, it leaves its signature on the flow structures, and the flow becomes anisotropic. For example, strong gravity makes buoyancy-driven flows anisotropic. The real space anisotropy in such flows could be illustrated using isocontours of density or temperature, as in Fig. 15.1(b). The hot (red) and cold (blue) plumes are aligned along the vertical, which is the direction of gravity. These are examples of visual representations of anisotropic flows.

15.1.2   *Structure function, $E_{u,\perp}/((D-1)E_{u,\|})$, and $L_\perp/L_\|$*

Visual interpretations are qualitative and subjective. One of the quantitative measures of anisotropy in real space is the structure function, which is defined as [Lesieur (2008); Frisch (1995)]

$$S_q(\mathbf{r}) = \langle [(\mathbf{u}(\mathbf{x}+\mathbf{r}) - \mathbf{u}(\mathbf{x})) \cdot \hat{r}]^q \rangle, \tag{15.1}$$

where q is a real number. For isotropic flows, $S_q(\mathbf{r}) = f(r)$, a function only of r, but independent of θ. However, for anisotropic flows, the structure function $S_q(\mathbf{r}) = f(r, \theta)$. We refer to textbooks on turbulence for further discussion on structure function [Davidson (2004); Lesieur (2008)].

Another quantitative measure of anisotropy is the following ratio of the energies parallel and perpendicular to the anisotropy direction:

$$A_1 = \frac{E_{u,\perp}}{(D-1)E_{u,\|}} = \frac{u_\perp^2/2}{(D-1)u_\|^2/2}, \tag{15.2}$$

where D is the dimensionality of the system. For 3D flows with $\hat{z}$ as the buoyancy direction,

$$A_1 = \frac{E_{u,\perp}}{2E_{u,\|}} = \frac{v_x^2 + v_y^2}{2v_z^2}, \tag{15.3}$$

and for 2D flows,

$$A_1 = \frac{E_{u,\perp}}{E_{u,\parallel}} = \frac{v_x^2}{v_z^2}. \tag{15.4}$$

For isotropic flows, $A_1 = 1$. Thus deviation of A_1 from unity provides a measure of anisotropy of the flow.

We can also quantify anisotropy using the ratio of the integral lengths perpendicular and parallel to the direction of anisotropy:

$$A_2 = \frac{L_\perp}{L_\parallel}, \tag{15.5}$$

where

$$L_\parallel = 2\pi \frac{\int d\mathbf{k} E_u(\mathbf{k})/k_\parallel}{\int d\mathbf{k} E_u(\mathbf{k})}, \tag{15.6}$$

$$L_\perp = 2\pi \frac{\int d\mathbf{k} E_u(\mathbf{k})/k_\perp}{\int d\mathbf{k} E_u(\mathbf{k})}. \tag{15.7}$$

Turbulent flows have structures at multiple length scales, hence we need to specify anisotropy at different scales. The structure function can capture anisotropy at multiple scales, but the anisotropy measures A_1 and A_2 capture anisotropy only at the large length scale. Visualization of a single picture typically provide information about structures at large scales only. Of course, zoomed views of a picture could describe anisotropy at different scales.

In the next section we will discuss anisotropy measures in Fourier space.

15.2 Anisotropy quantification in Fourier space

Fourier transform captures the strengths of multi-scale structures. In earlier chapters we discussed the 1D energy spectrum, energy flux, and shell-to-shell energy transfers that capture the nonlinear interactions among the structures at different scales. However, due to averaging over the polar angle, these measures do not capture the angular anisotropy. In this section, we will describe quantities that describe anisotropy in Fourier space.

15.2.1 $E_{u,\perp}(k)/(2E_{u,\parallel}(k))$

The scale-dependent anisotropy can be quantified using

$$A(k) = \frac{E_{u,\perp}(k)}{2E_{u,\parallel}(k)}, \tag{15.8}$$

where

$$E_{u,\perp}(k) = \sum_{k-1<k'\leq k} \frac{1}{2}|\mathbf{u}_\perp(\mathbf{k}')|^2, \tag{15.9}$$

$$E_{u,\parallel}(k) = \sum_{k-1<k'\leq k} \frac{1}{2}|u_\parallel(\mathbf{k}')|^2. \tag{15.10}$$

We will show later that the anisotropy in RBC is scale-dependent, which is captured quite nicely by the aforementioned parameter $A(k)$.

15.2.2 *Ring spectrum for spherical rings*

To quantify scale- and angular-dependent energy distribution, we divide a wavenum-ber shell into rings as shown in Fig. 15.2. A ring is characterised by two indices—the shell index k and the sector index β [Teaca *et al.* (2009); Reddy and Verma (2014); Verma (2017)]. The polar angle is denoted by ζ. The rings containing $\zeta = 0$ and $\zeta = \pi/2$ are termed as the *polar* and the *equatorial* rings respectively. The gravita-tional field is aligned along $\zeta = 0$. The energy spectrum of a ring, called the *ring spectrum*, is defined as

$$E_u(k, \beta) = \frac{1}{C_\beta} \sum_{\substack{k-1<k'\leq k; \\ \angle\mathbf{k}'\in(\zeta_{\beta-1},\zeta_\beta]}} \frac{1}{2}|\mathbf{u}(\mathbf{k}')|^2, \tag{15.11}$$

where $\angle\mathbf{k}'$ is the angle between $\mathbf{k}'$ and the unit vector $\hat{z}$. The sector β contains the modes between the polar angles $\zeta_{\beta-1}$ to ζ_β, as shown in Fig. 15.2(b).

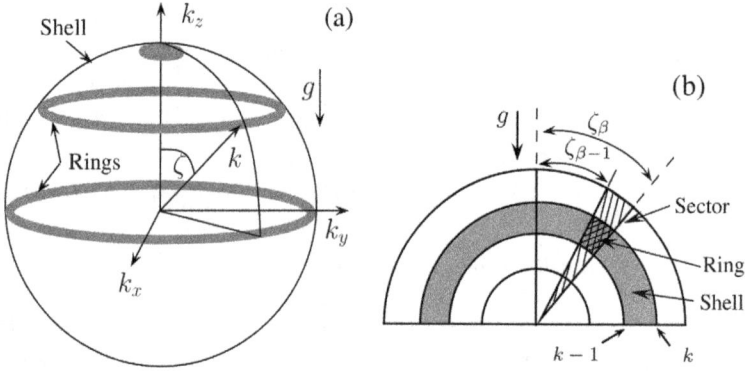

Fig. 15.2   (a) A schematic diagram depicting rings. (b) A vertical cross section of the wavenumber sphere depicting shells, sectors, and rings. From Nath *et al.* (2016). Reprinted with permission from APS.

For a uniform $\Delta\zeta = \zeta_\beta - \zeta_{\beta-1}$, the sectors near the equator contain larger number of modes than those near the poles. To compensate for the above, we divide the sum $\sum_k |\mathbf{u}(\mathbf{k}')|^2/2$ of Eq. (15.11) by $C(\beta)$:

$$C_\beta = |\cos(\zeta_{\beta-1}) - \cos(\zeta_\beta)|. \tag{15.12}$$

We obtain further details on the flow anisotropy by computing the ring spectra of the perpendicular and parallel components of the velocity:

$$E_{u,\perp}(k, \beta) = \frac{1}{C_\beta} \sum_{\substack{k-1<k'\leq k; \\ \angle(\mathbf{k}')\in(\zeta_{\beta-1},\zeta_\beta]}} \frac{1}{2}|\mathbf{u}_\perp(\mathbf{k}')|^2, \tag{15.13}$$

$$E_{u,\|}(k, \beta) = \frac{1}{C_\beta} \sum_{\substack{k-1<k'\leq k; \\ \angle(\mathbf{k}')\in(\zeta_{\beta-1},\zeta_\beta]}} \frac{1}{2}|\mathbf{u}_\|(\mathbf{k}')|^2. \tag{15.14}$$

Note that the total energy

$$E_u(k, \beta) = E_{u,\perp}(k, \beta) + E_{u,\parallel}(k, \beta). \tag{15.15}$$

We can also define energy contents of a sector β as

$$E_u(\beta) = \sum_k E_u(k, \beta); \quad E_{u,\perp,\parallel}(\beta) = \sum_k E_{u,\perp,\parallel}(k, \beta). \tag{15.16}$$

Teaca *et al.* (2009) proposed the aforementioned normalisation. However, if we normalise Eq. (15.11) by

$$C'_\beta = k|\cos(\zeta_{\beta-1}) - \cos(\zeta_\beta)|. \tag{15.17}$$

then $E(k, \beta)$ would provide an averaged energy content of a Fourier mode in a ring. The additional factor k is employed to take into account the fact that the number of Fourier modes in a ring is proportional to the radius of the ring.

In the subsequent sections we will report the above spectrum for RBC.

15.2.3 *Ring spectrum for cylindrical rings*

We can divide the Fourier space into cylindrical rings, as illustrated in Fig. 15.3. A cylindrical ring is specified by a shell index k and a height index i. The energy spectrum of a cylindrical ring is defined as

$$E_u(k, i) = \frac{1}{k} \sum_{\substack{k-1<k'\leq k \\ H_{i-1}<h'\leq H_i}} \frac{1}{2}|\mathbf{u}(\mathbf{k}')|^2, \tag{15.18}$$

We divide the sum by the normalisation factor k to compensate for the larger number of modes in a ring of larger radius. We can define $E_{u,\perp}(k, i)$ and $E_{u,\parallel}(k, i)$ in a similar manner.

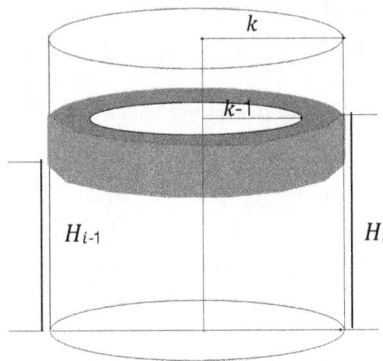

Fig. 15.3 A schematic diagram depicting a cylindrical ring. Its inner and outer radii are $k - 1$ and k respectively, while its vertical extent is from H_{i-1} to H_i.

15.3 Measures of Energy transfers

In Chapter 4, we described the energy transfers in Fourier space using the energy flux and shell-to-shell energy transfers. Unfortunately, these quantities do not capture the anisotropic effects since they are computed by averaging over the polar angle ζ. In the next two subsections we discuss two quantitative measures of the anisotropic energy transfers—ring-to-ring energy transfers, and the energy exchange between the perpendicular and parallel components of the velocity field.

15.3.1 *Ring-to-ring energy transfers*

As described in Sec. 15.2.2, we divide the Fourier space into a set of rings, and study the energy transfers among these rings. The modes in a ring (m, α), where m and α represent the shell and sector indices of the ring, interacts with all other rings. The ring-to-ring kinetic energy transfer from ring (m, α) to ring (n, β) is

$$T^{(u,m,\alpha)}_{(u,n,\beta)} = \sum_{\mathbf{k}\in(n,\beta)} \sum_{\mathbf{p}\in(m,\alpha)} S^{uu}(\mathbf{k}|\mathbf{p}|\mathbf{q}). \tag{15.19}$$

The calculations of the ring-to-ring energy transfers for all m and n's are very time-consuming since they involve a large number of rings.

The above formula can be adopted for cylindrical rings. The energy transfer from ring (m, h_1) to ring (n, h_2) is given by

$$T^{(u,m,h_1)}_{(u,n,h_2)} = \sum_{\mathbf{k}\in(n,h_2)} \sum_{\mathbf{p}\in(m,h_1)} S^{uu}(\mathbf{k}|\mathbf{p}|\mathbf{q}). \tag{15.20}$$

Here m, h_1 represent the shell and height indices respectively.

15.3.2 *Energy exchange between* $u_\parallel$ *and* $\mathbf{u}_\perp$ *in RBC*

In RBC, buoyancy drives $u_\parallel$, hence $u_\parallel \gg \mathbf{u}_\perp$. In this section, we will demonstrate how energy is exchanged between $u_\parallel$ and $\mathbf{u}_\perp$.

Using Eq. (3.8, 3.11) we derive the following equations for $\mathbf{u}_\perp(\mathbf{k})$ and $u_\parallel(\mathbf{k})$:

$$\frac{d}{dt}u_\parallel(\mathbf{k}) = i\sum_{\mathbf{q}}[\mathbf{k}\cdot\mathbf{u}(\mathbf{q})]u_\parallel(\mathbf{p}) - ik_\parallel\sigma(\mathbf{k}) + \alpha g\theta(\mathbf{k}) - \nu k^2 u_\parallel(\mathbf{k}), \tag{15.21}$$

$$\frac{d}{dt}\mathbf{u}_\perp(\mathbf{k}) = i\sum_{\mathbf{q}}[\mathbf{k}\cdot\mathbf{u}(\mathbf{q})]\mathbf{u}_\perp(\mathbf{p}) - ik_\perp\sigma(\mathbf{k}) - \nu k^2\mathbf{u}_\perp(\mathbf{k}'), \tag{15.22}$$

where $\mathbf{q} = \mathbf{k} - \mathbf{p}$, and σ is the pressure. The corresponding equations for their energy are:

$$\frac{\partial E_\parallel(\mathbf{k}')}{\partial t} = \sum_{\mathbf{p}} S^{uu}_\parallel(\mathbf{k}'|\mathbf{p}|\mathbf{q}) + \mathcal{P}_\parallel(\mathbf{k}') - 2\nu k'^2 E_\parallel(\mathbf{k}') + \alpha g\Re(u_\parallel^*(\mathbf{k}')\theta(\mathbf{k}')), \tag{15.23}$$

$$\frac{\partial E_\perp(\mathbf{k}')}{\partial t} = \sum_{\mathbf{p}} S^{uu}_\perp(\mathbf{k}'|\mathbf{p}|\mathbf{q}) + \mathcal{P}_\perp(\mathbf{k}') - 2\nu k'^2 E_\perp(\mathbf{k}'), \tag{15.24}$$

where $\mathbf{k}' + \mathbf{p} + \mathbf{q} = 0$, $\mathbf{k}' = -\mathbf{k}$, and

$$S_\perp^{uu}(\mathbf{k}'|\mathbf{p}|\mathbf{q}) = -\Im\left\{[\mathbf{k}' \cdot \mathbf{u}(\mathbf{q})]\,[\mathbf{u}_\perp(\mathbf{k}') \cdot \mathbf{u}_\perp(\mathbf{p})]\right\}, \tag{15.25}$$

$$S_\parallel^{uu}(\mathbf{k}'|\mathbf{p}|\mathbf{q}) = -\Im\left\{[\mathbf{k}' \cdot \mathbf{u}(\mathbf{q})]\,[u_\parallel(\mathbf{k}')u_\parallel(\mathbf{p})]\right\}, \tag{15.26}$$

$$\mathcal{P}_\perp(\mathbf{k}') = -\Im\left\{[\mathbf{k}'_\perp \cdot \mathbf{u}_\perp(\mathbf{k}')]\,\sigma(\mathbf{k}')\right\}, \tag{15.27}$$

$$\mathcal{P}_\parallel(\mathbf{k}') = -\Im\left\{\left[k'_\parallel u_\parallel(\mathbf{k})\right]\sigma(\mathbf{k}')\right\}. \tag{15.28}$$

Here we use the fact that $E_{\perp,\parallel}(\mathbf{k}') = E_{\perp,\parallel}(\mathbf{k}) = |\mathbf{u}_{\perp,\parallel}(\mathbf{k})|^2/2$. Note that in RBC, $\alpha g \Re(u_\parallel^*(\mathbf{k})\theta(\mathbf{k})) > 0$, hence buoyancy feeds energy to $u_\parallel(\mathbf{k})$. Hence, $E_{u,\parallel}(\mathbf{k}) > E_x(\mathbf{k})$ and $E_{u,\parallel}(\mathbf{k}) > E_y(\mathbf{k})$. Also,

$$S_\perp^{uu}(\mathbf{k}'|\mathbf{p}|\mathbf{q}) + S_\perp^{uu}(\mathbf{p}|\mathbf{k}'|\mathbf{q}) = \Im\left\{[\mathbf{k}' \cdot \mathbf{u}(\mathbf{q})]\,[\mathbf{u}_\perp(\mathbf{k}') \cdot \mathbf{u}_\perp(\mathbf{p})]\right\}$$
$$+ \Im\left\{[\mathbf{p} \cdot \mathbf{u}(\mathbf{q})]\,[\mathbf{u}_\perp(\mathbf{k}') \cdot \mathbf{u}_\perp(\mathbf{p})]\right\}$$
$$= 0 \tag{15.29}$$

because $\mathbf{k}' + \mathbf{p} = -\mathbf{q}$ and $\mathbf{q} \cdot \mathbf{u}(\mathbf{q}) = 0$. Therefore

$$\sum_{\mathbf{k}'}\sum_{\mathbf{p}} S_\perp^{uu}(\mathbf{k}'|\mathbf{p}|\mathbf{q}) = 0 \tag{15.30}$$

for the modes in a triad $(\mathbf{k}', \mathbf{p}, \mathbf{q})$, as well as for the whole system. The above proof is similar to that discussed in Sec. 4.1. We sum the terms of Eq. (15.24) over all $\mathbf{k}'$. Using Eqs. (15.30), we derive the temporal evolution of $E_{u,\perp}$ as

$$\frac{dE_{u,\perp}}{dt} = \sum_{\mathbf{k}'}\mathcal{P}_\perp(\mathbf{k}') - \sum_{\mathbf{k}'} 2\nu k'^2 E_{u,\perp}(\mathbf{k}). \tag{15.31}$$

Thus, the viscous term dissipates $E_{u,\perp}$. Hence, for a steady state, $E_{u,\perp}$ needs an energy source, which is provided by $\mathcal{P}_\perp$, without which $E_{u,\perp}$ would vanish. The term $\mathcal{P}_\perp$ is the energy supply by $u_\parallel$ to $\mathbf{u}_\perp$,

Thus, Eq. (15.24) reveals that $E_\perp(\mathbf{k})$ receives energy from $E_{u,\parallel}(\mathbf{k})$ with a rate of $\mathcal{P}_\perp(\mathbf{k})$. Thus pressure plays an important role in the energy exchange between $u_\parallel$ and $\mathbf{u}_\perp$. Interestingly, the incompressibility condition, $\mathbf{k} \cdot \mathbf{u}(\mathbf{k}) = 0$, yields

$$\mathcal{P}_\perp(\mathbf{k}) = -\mathcal{P}_\parallel(\mathbf{k}). \tag{15.32}$$

That is, the energy gained by $\mathbf{u}_\perp(\mathbf{k})$ via pressure equals the energy lost by $u_\parallel(\mathbf{k})$.

Following the arguments similar to those in Sec. 4.3, we define the energy flux for $\mathbf{u}_\perp$ using $S_\perp^{uu}(\mathbf{k}'|\mathbf{p}|\mathbf{q})$. The $\Pi_{u,\perp}(k_0)$ is defined as the transfer of $u_\perp^2/2$ from all the modes residing inside a wavenumber sphere of radius k_0 to the modes outside the sphere, and it is given by

$$\Pi_{u,\perp}(k_0) = \sum_{|\mathbf{k}|>k_0}\sum_{|\mathbf{p}|\leq k_0} S_\perp^{uu}(\mathbf{k}'|\mathbf{p}|\mathbf{q}). \tag{15.33}$$

Similarly, using the arguments used in Sec. 4.1 as well as those used to derive Eq. (15.30), we can show that

$$\sum_{\mathbf{k}'}\sum_{\mathbf{p}} S_\parallel^{uu}(\mathbf{k}'|\mathbf{p}|\mathbf{q}) = 0 \tag{15.34}$$

for the modes in a triad $(\mathbf{k'}, \mathbf{p}, \mathbf{q})$, as well as for the whole system. Using the above identity and following the arguments similar to those in Sec. 4.3, we define the energy flux for $u_{\parallel}$ as

$$\Pi_{u,\parallel}(k_0) = \sum_{|\mathbf{k}|>k_0} \sum_{|\mathbf{p}|\leq k_0} S_{\parallel}(\mathbf{k'}|\mathbf{p}|\mathbf{q}). \tag{15.35}$$

It is important to note that there is no direct energy transfer between $\mathbf{u}_{\perp}(\mathbf{k})$ and $u_{\parallel}(\mathbf{k})$. It follows from the fact the nonlinear transfer does not have a term of the type $[\mathbf{u}_{\perp}^{*}(\mathbf{k}) \cdot \mathbf{u}_{\parallel}(\mathbf{p})]$ (which would anyway vanish since they are perpendicular to each other). The total kinetic energy flux Π_u is a sum of $\Pi_{u,\perp}$ and $\Pi_{u,\parallel}$:

$$\Pi_u(k_0) = \Pi_{u,\perp}(k_0) + \Pi_{u,\parallel}(k_0). \tag{15.36}$$

The aforementioned fluxes are computed using the same method as that outlined in Sec. 4.5. In the next section we will describe the above transfers for stably stratified flows.

15.3.3 *Energy exchange between $u_{\parallel}$ and $\mathbf{u}_{\perp}$ in stably stratified flows*

The equations for the energy exchange between $u_{\parallel}$ and $\mathbf{u}_{\perp}$ in stably stratified flows are same as those for RBC. However, there are subtle differences, which are listed below:

(1) For stably stratified flows, $\mathcal{F}_B < 0$. Hence, the total energy, $E_{u,\perp} + E_{u,\parallel}$ would decay in the absence of external force. Therefore, to maintain a steady state, we need to inject energy using an external force.
(2) Since $\mathcal{F}_B < 0$ for stably stratified flows, we expect that $E_{u,\parallel}(\mathbf{k}) < E_{u,\perp}(\mathbf{k})$. Therefore, pressure would transfer energy from $E_{u,\perp}(\mathbf{k})$ to $E_{u,\parallel}(\mathbf{k})$. In RBC, the energy transfer is in the opposite direction.

In the next section we will describe the numerical results on the anisotropy in RBC.

15.4 Numerical results on the anisotropy in RBC

Nath *et al.* (2016) performed numerical simulations of RBC with free-slip boundary conditions. The size of the closed box was chosen as unity, and the grid resolution of the simulation as 512^3. The simulation parameters are listed in Table 15.1.

As listed in Table 15.1, the anisotropy parameter $A_1 = E_{u,\perp}/(2E_{u,\parallel})$ is reasonably close to unity for a wide range of Pr—from 0.02 to ∞. Based on these observations, Nath *et al.* (2016) concluded that the flow in RBC is nearly isotropic. This is because the buoyancy is much weaker than the pressure gradient. Note that Richardson number is much less than unity for turbulent RBC (see Sec. 11.3).

When we analyse A_1 is more detail, we observe that the flow is least anisotropic for $\mathrm{Pr} = 1$ with $A_1 \approx 0.73$, and the degree of anisotropy increases as Pr is decreased

Table 15.1   Anisotropy results of Nath *et al.* (2016): Pr, Ra, Reynolds number Re, anisotropic parameter $A_1 = E_\perp/(2E_\parallel)$, integral length scale L, vertical integral length scale $L_\parallel$. Adopted from a table of Nath *et al.* (2016).

Pr	Ra	Re	A_1	L	$L_\parallel$
0.02	2×10^6	7.05×10^3	0.63	0.478	0.553
1	10^8	3.11×10^3	0.73	0.468	0.547
6.8	10^8	9.08×10^2	0.59	0.484	0.591
100	10^8	1.25×10^2	0.49	0.531	0.635
∞	2×10^8	0	0.30	0.449	0.716

or increased from unity. For very small Pr (strongly diffusive), the thermal boundary layer is broader than that for Pr $= 1$. But for very large Pr, the momentum diffusion spans over a large space region. These are the reasons for higher anisotropy for small-Pr or large-Pr RBC. In Table 15.1, we also list the integral lengths L and $L_\parallel$. Clearly $L_\parallel > L$ due to the stretching of plumes by buoyancy.

Nath *et al.* (2016) also computed $A(k)$ defined in Eq. (15.8) and showed that for small and unit Prandtl number, $A(k) \sim 1$ for all k's. However, for very large and infinite Pr, $A(k) \sim 1$ for small k, but $A(k) \ll 1$ for large k, which is due to the fact that the thin plumes for large Pr have dominant u_z.

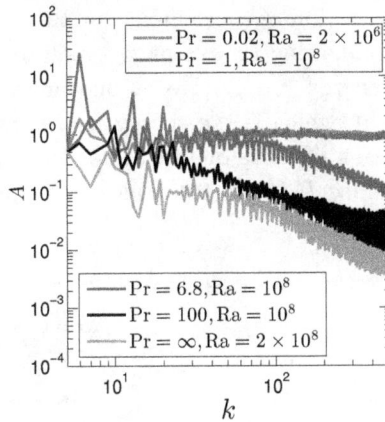

Fig. 15.4   Plot of $A(k) = E_{u,\perp}(k)/[2E_{u,\parallel}(k)]$ vs. k for Pr $= 0.02, 1, 6.8, 100$, and ∞. For small Pr, $A(k) \sim 1$ for all k's. For Pr $= 100, \infty$, $A(k) \sim 1$ for small k, but $A(k) \ll 1$ for large k. From Nath *et al.* (2016). Reprinted with permission from APS.

Nath *et al.* (2016) computed the ring spectrum $E(k, \beta)$ for Pr $= 1$. Figure 15.5 exhibits the ring spectrum of the total kinetic energy, and the kinetic energies corresponding to $\mathbf{u}_\perp$ and $u_\parallel$, which are $E_{u,\perp}(k, \beta)$, $E_{u,\parallel}(k, \beta)$ respectively. The figures show that the energy distribution is nearly isotropic (independent of polar angle). Therefore we can claim that the RBC flow is nearly isotropic at all scales.

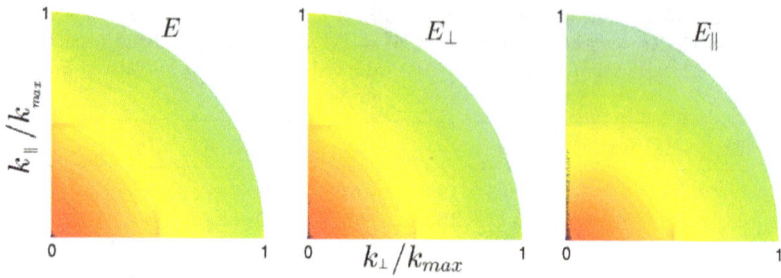

Fig. 15.5   The ring spectra $E(k, \beta)$, $E_{u,\perp}(k, \beta)$, $E_{u,\parallel}(k, \beta)$ (from left to right) for turbulent RBC with $\text{Pr} = 1$ and $\text{Ra} = 10^8$. From Nath *et al.* (2016). Reprinted with permission from APS.

This property of RBC, shared with 3D hydrodynamic turbulence, is one of the key features that is responsible for the Kolmogorov-like energy spectrum for thermally driven turbulence. We however remark that the ring spectrum for no-slip RBC may have somewhat different ring spectrum.

Regarding the energy transfers, we are not aware of any works that report ring-to-ring energy transfers in RBC. However, we expect these transfers to be local, and forward in k. It will be interesting to investigate how the energy flows along the angular direction, that is, whether it is from polar region to the equator, or vice versa.

Nath *et al.* (2016) computed the energy flux $\Pi(k)$, as well as $\Pi_\perp(k)$ and $\Pi_\parallel(k)$. See Fig. 15.6(a) for an illustration of these fluxes for $\text{Pr} = 1$. Both $\Pi_\perp(k)$ and $\Pi_\parallel$ are positive, with $\Pi_\perp(k)$ dominating $\Pi_\parallel$ except for small wavenumbers. Nath *et al.* (2016) also computed the energy supply rate by buoyancy, $\mathcal{F}_B(k)$, the dissipation rate, $D(k)$, as well as, $P_\parallel(k)$, which is the energy transferred from $\mathbf{u}_\perp$ to $\mathbf{u}_\parallel$. See Fig. 15.6(b) for an illustration of these quantities for $\text{Pr} = 1$. As expected, $\mathcal{F}_B(k)$ and $D_u(k)$ are positive. Since $P_\parallel(k) < 0$ for most k's, we conclude that energy

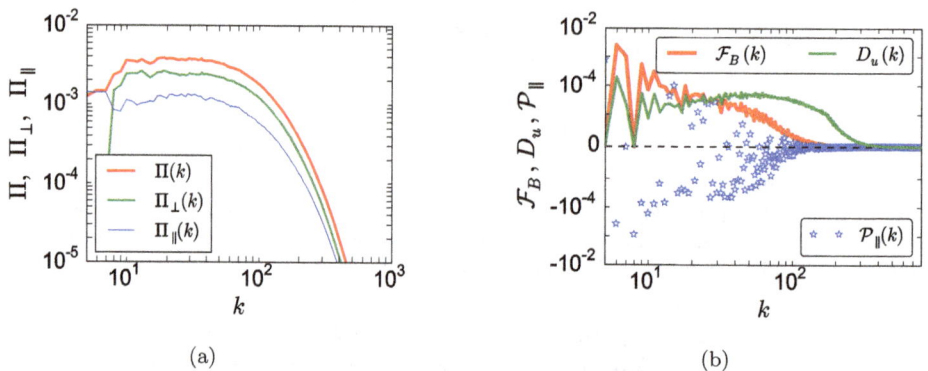

(a)   (b)

Fig. 15.6   For turbulent RBC with $\text{Pr} = 1$ and $\text{Ra} = 10^8$: (a) Plots of $\Pi_u(k)$, $\Pi_\perp(k)$, and $\Pi_\parallel(k)$; (b) Plots of buoyancy feed $\text{F}_B(k)$, viscous dissipation rate $D_u(k)$, and $\mathcal{P}_\parallel(k)$, the energy transfer occurs from $u_\perp(k)$ to $\mathbf{u}_\parallel$. Adopted from two figures of Nath *et al.* (2016).

flows from $u_{\|}(k)$ to $\mathbf{u}_{\perp}(k)$. Thus $\mathcal{P}_{\|}$ helps maintain the transverse flow. Without $\mathcal{P}_{\|}$, $E_{u,\perp}$ will decay to zero since it has no forcing.

In the next section, we briefly describe flow anisotropy in stably stratified flows.

15.5 Numerical results on anisotropy in stably stratified flows

Stably stratified flows (SSF) too are anisotropic due to the presence of buoyancy. Kumar and Verma [Kumar *et al.* (2014a)] simulated such flows for Richardson number Ri $\ll 1$ and showed that A_1 varies from 0.41 to 1.2. Hence, SSF with small and moderate Richardson numbers are nearly isotropic; this feature plays an important role in the maintaining the kinetic energy spectrum as $E_u(k) \sim k^{-11/5}$.

For Ri $\gg 1$, the flow is anisotropic with $E_\perp \gg E_\|$ or $u_\perp \gg u_\|$, which is due to the suppression of $u_\|$ by buoyancy. Under a steady state, energy transfer takes place from $\mathbf{u}_\perp$ to $u_\|$ via pressure with mechanism described in Sec. 15.3.2.

In Table 15.2 we summarise the differences between the anisotropy properties of RBC and stably stratified flow.

Table 15.2 Differences between the anisotropy properties of RBC and stably stratified flow (SSF).

Measure	RBC	SSF
$A_1 = E_\perp/(2E_\|)$	$A_1 \approx 1$	$A_1 \approx 1$ for Ri $\lesssim 1$ $A_1 \gg 1$ large Ri
Role of buoyancy	Feeds $u_\|$	Suppresses $u_\|$
Energy transfer by pressure	From $u_\|$ to $u_\perp$	From $u_\perp$ to $u_\|$

In the next chapter we will describe the shell models for buoyancy-driven flows.

Further reading

The anisotropic measures—ring spectrum, ring-to-ring transfers, energy exchange between $u_\|$ and $u_\perp$—were proposed earlier by Teaca *et al.* (2009), Reddy and Verma (2014), Reddy *et al.* (2014), and Verma (2017) in the context of magnetohydrodynamic (MHD) turbulence and quasi-static MHD turbulence. Nath *et al.* (2016) studied anisotropy in RBC in the above framework. Zimin and Frick (1989) discuss some aspects of turbulent convections.

Exercises

(1) In this chapter we demonstrate that turbulent RBC is nearly isotropic. What are the physical reasons for this behaviour?

(2) For very strong gravity, stably stratified turbulence is strongly anisotropic, but RBC is not. Give reasons for the same.

(3) In RBC, the velocity component along the gravity, $u_{\parallel}$, is forced. How does the energy flow from $u_{\parallel}$ to $u_{\perp}$?

Chapter 16

Shell Models for Buoyancy-Driven Turbulence

Direct numerical simulation of a turbulent flow is very expensive computationally due to the large degrees of freedom (e.g., number of $\mathbf{u}(\mathbf{k})$ modes) present in the system. Therefore, scientists often employ shell models that have far fewer variables. In this chapter we present shell models for buoyancy-driven flows. These shell models help us compute the spectra and fluxes of the velocity and temperature fields. The results of the shell model are consistent with the simulation results described in the earlier chapters. In this chapter we also compare various shell models proposed for buoyancy-driven flows.

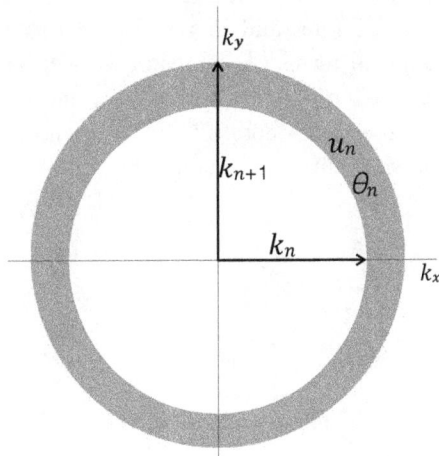

Fig. 16.1   Schematic diagram of a shell of a shell model. The velocity and temperature fields of the shell is denoted by complex number u_n and θ_n respectively.

In shell model studies, a wavenumber sphere is divided into shells as shown in Fig. 16.1. In shell models for turbulent convection, variables u_n and θ_n represent the velocity and temperature modes of shell n collectively. For stably stratified turbulence (SST), θ_n is replaced by the density variable ρ_n. This process reduces the number of variables significantly. Note however that the structure of the two models presented in Secs. 16.1 and 16.2 are identical. We detail the shell models and their results in subsequent sections. We start with the shell model for turbulent convection.

16.1   Shell model for turbulent convection

In this section we present the shell model for turbulent convection proposed by Kumar and Verma (2015). In this model, the equations for u_n and θ_n are [Kumar and Verma (2015)]

$$\frac{du_n}{dt} = M_n[u, u] + \alpha g \theta_n - \nu k_n^2 u_n, \tag{16.1}$$

$$\frac{d\theta_n}{dt} = N_n[u, \theta] + \left|\frac{d\bar{T}}{dz}\right| u_n - \kappa k_n^2 \theta_n, \tag{16.2}$$

where $M_n[u, u]$ and $N_n[u, \theta]$ are the nonlinear terms, $k_n = k_0 \lambda^n$ is the wavenumber of the n-th shell, ν and κ are the kinematic viscosity and thermal diffusivity respectively. We choose $\lambda = (\sqrt{5} + 1)/2$, the golden mean [Ditlevsen (2010)]. Note that the thickness of the shells increases with k, which is because of the power-law physics of turbulent flow. The term $\alpha g \theta_n$ represents buoyancy, while $|d\bar{T}/dz|$ is the adverse temperature gradient. Also, the functional forms of the nonlinear terms $M_n[u, u]$ and $N_n[u, \theta]$ are different, and they will be described below.

We construct the nonlinear terms $M_n[u, u]$ and $N_n[u, \theta]$ so that the kinetic energy $E_u = \int d\mathbf{r}(u^2/2)$, kinetic helicity $H_K = \int d\mathbf{r}(\mathbf{u} \cdot \boldsymbol{\omega})$, and entropy $E_\theta = \int d\mathbf{r}(\theta^2/2)$ are conserved when buoyancy is absent, $|d\bar{T}/dz| = 0$, and $\nu = \kappa = 0$. For the shell model, the aforementioned qualities are defined as [Ditlevsen (2010)]:

$$E_u = \sum_n |u_n|^2/2, \tag{16.3}$$

$$H_K = \sum_n (-1)^n k_n |u_n|^2, \tag{16.4}$$

$$E_\theta = \sum_n |\theta_n|^2/2. \tag{16.5}$$

The above constraints yield the following nonlinear term $M_n[u, u]$ [Ditlevsen (2010)]

$$M_n[u, u] = -i(a_1 k_n u_{n+1}^* u_{n+2} + a_2 k_{n-1} u_{n-1}^* u_{n+1} - a_3 k_{n-2} u_{n-1} u_{n-2}) \tag{16.6}$$

with constraints

$$a_1 + a_2 + a_3 = 0, \tag{16.7}$$

$$a_1 - \lambda a_2 + \lambda^2 a_3 = 0. \tag{16.8}$$

Equations (16.8, 16.8) provide 2 equations for 3 unknowns: a_1, a_2, a_3. Hence, there is no unique solution for a_i's. Therefore, for our computation, we choose a solution: $a_1 = 1$, $a_2 = \lambda - 2$, and $a_3 = 1 - \lambda$ [Kumar and Verma (2015)].

For the construction of the nonlinear term $N_n[u, \theta]$, we invoke the constraints that

(1) the nonlinear term of the temperature equation is a bilinear product of the temperature fluctuation θ_n and the velocity fluctuation u_n.
(2) The entropy is conserved when $|d\bar{T}/dz| = 0$ and $\kappa = 0$ that yields a condition

$$\text{Re}\left(\sum_n \theta_n^* N_n[u, \theta]\right) = 0. \tag{16.9}$$

A combination of the above constraints yields the following form for $N_n[u, \theta]$:

$$\begin{aligned}
N_n[u, \theta] = -i[k_n(d_1 u_{n+1}^* \theta_{n+2} + d_3 \theta_{n+1}^* u_{n+2}) \\
+ k_{n-1}(d_2 u_{n-1}^* \theta_{n+1} - d_3 \theta_{n-1}^* u_{n+1}) \\
- k_{n-2}(-d_1 u_{n-1} \theta_{n-2} - d_2 \theta_{n-1} u_{n-2})]
\end{aligned} \tag{16.10}$$

with arbitrary d_1, d_2, and d_3. For our shell model, we choose $d_1 = 1$, $d_2 = \lambda - 2$, and $d_3 = 1 - \lambda$.

A nondimensionalized version of the above shell model with free-fall speed as the velocity scale and $d|d\bar{T}/dz|$ (where d is the distance between the plates) as the temperature scale is

$$\frac{du_n}{dt} = M_n[u, u] + \theta_n - \sqrt{\frac{\text{Pr}}{\text{Ra}}} k_n^2 u_n, \tag{16.11}$$

$$\frac{d\theta_n}{dt} = N_n[u, \theta] + u_n - \frac{1}{\sqrt{\text{RaPr}}} k_n^2 \theta_n, \tag{16.12}$$

where Pr is the Prandtl number, and Ra is the Rayleigh number. See Sec. 2.7 for further details.

Following the usual convention followed for the shell model, we choose the boundary conditions:

$$u_{-1} = u_0 = \theta_{-1} = \theta_0 = 0, \tag{16.13}$$

$$u_{N+1} = u_{N+2} = \theta_{N+1} = \theta_{N+2} = 0, \tag{16.14}$$

where N is the total number of shells. Also note that the aforementioned shell model is based on the Sabra model [L'vov *et al.* (1998)] that exhibits weaker fluctuations for the energy and entropy spectrum than the GOY model [Gledzer (1973); L'vov *et al.* (1998); Biferale (2003)].

For the shell model, the kinetic energy spectrum ($E_u(k)$) and the entropy spectrum ($E_\theta(k)$) are defined as

$$E_u(k) = \frac{|u_k|^2}{k}, \tag{16.15}$$

$$E_\theta(k) = \frac{|\theta_k|^2}{k}. \tag{16.16}$$

We can also define the energy and entropy fluxes for a sphere of radius k. For the same we need to compute the cumulative energy transfers from all the shell with

$k' \leq k$ to the shells with $k' > k$. Kumar and Verma (2015) showed that the energy and entropy fluxes for a sphere of radius k are (see Ditlevsen (2010)):

$$\Pi_u(k) = \sum_{n>k} \sum_{m \leq k} \sum_p -k_p \mathrm{Im}(u_p u_m u_n^*), \tag{16.17}$$

$$\Pi_\theta(k) = \sum_{n>k} \sum_{m \leq k} \sum_p -k_p \mathrm{Im}(u_p \theta_m \theta_n^*). \tag{16.18}$$

In the next section, we will describe the shell model for stably stratified turbulence.

16.2   Shell model for stably stratified turbulence

Based on Eqs. (2.38, 2.39) the equations of the shell model for stably stratified turbulence (SST) are [Kumar and Verma (2015)]:

$$\frac{du_n}{dt} = M_n[u, u] - N b_n - \nu k_n^2 b_n + f_n, \tag{16.19}$$

$$\frac{db_n}{dt} = N_n[u, b] + N u_n - \kappa k_n^2 b_n. \tag{16.20}$$

The conservation laws for RBC and SST are identical, hence, the nonlinear terms $M_n[u, u]$ for RBC and SST are same, and

$$N_n[u, \theta] = N_n[u, b]. \tag{16.21}$$

In SST, the kinetic energy is transferred to the potential energy. Hence, we need to force the flow to get a steady state. For the same, we force a set of small wavenumber shells randomly that provides a constant energy supply ε to the flow. We assume that each forcing shell receives an equal amount of energy. For n_f forcing shells, the forcing applied to shell n is [Stepanov and Plunian (2006)]

$$f_n = \sqrt{\frac{\varepsilon}{n_f \Delta t}} e^{i\phi_n}, \tag{16.22}$$

where ϕ_n is the random phase chosen from a uniform distribution in $[0, 2\pi]$.

Kumar and Verma (2015) simulated the aforementioned shell models. In the following sections we summarise their results.

16.3   Shell-model results for turbulent convection

Kumar and Verma (2015) simulated the shell model for RBC of Sec. 16.1 for $\mathrm{Pr} = 1$ and $\mathrm{Ra} = 10^{12}$. In Fig. 16.2 we exhibit the energy and entropy spectra, as well as their fluxes. The figures clearly demonstrate that both the energy and entropy spectra vary as $k^{-5/3}$, and the energy and entropy fluxes are constants. These results are consistent with the theoretical arguments presented in Chapter 13.

In RBC, buoyancy feeds kinetic energy, hence $\mathcal{F}_B(k) > 0$, as exhibited in Fig. 16.3(a). Another important point in convective turbulence is that though

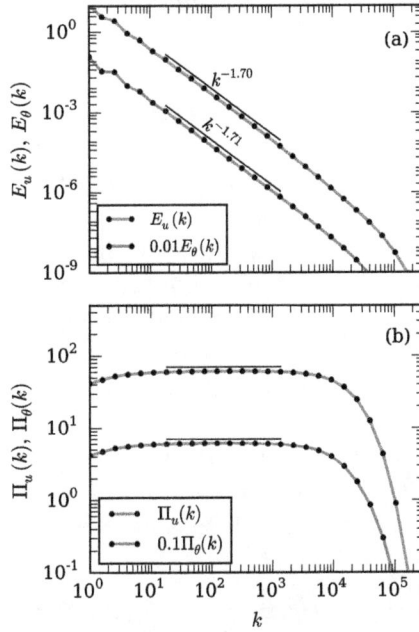

Fig. 16.2   Shell model simulation for turbulent convection with $\mathrm{Pr} = 1$ and $\mathrm{Ra} = 10^{12}$: (a) kinetic energy and entropy spectra $\sim k^{-5/3}$; (b) Plots of kinetic energy flux $\Pi_u(k)$ and entropy flux $\Pi_\theta(k)$ exhibiting constant fluxes. From Kumar and Verma (2015). Reprinted with permission from APS.

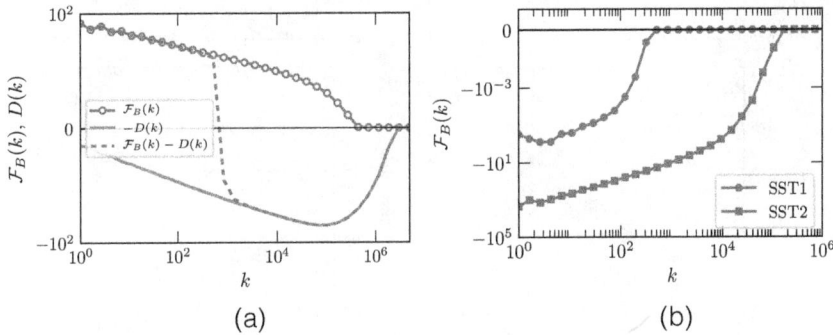

Fig. 16.3   Plots of $\mathcal{F}_B(k)$ and $D_u(k)$ computed using the shell model simulation: (a) for turbulent convection, (b) for stably stratified turbulence (SST). Adopted from figures of Kumar and Verma (2015).

$\mathcal{F}_B(k) > 0$, it is much weaker than the nonlinear term and pressure gradient. Due to these reasons, convective turbulence has behaviour similar to passive scalar turbulence with temperature acting like a passive scalar. Refer to Sec. 13.1 for a detailed discussion on the turbulence phenomenology of thermal convection.

In the next section we describe the numerical results of the shell model for stably stratified turbulence.

16.4 Shell model results for stably stratified turbulence

Kumar and Verma (2015) simulated the shell model for SST for moderate and small Richardson numbers, i.e., Ri = 0.10 and 1.6×10^{-7}. The Prandtl number was chosen as unity. For Ri = 0.10, as expected, they obtained Bolgiano-Obukhov scaling because buoyancy and nonlinear terms are comparable for such flows. As shown in Fig. 16.4(A.a), the kinetic and potential energy spectra $E_u(k)$ and $E_b(k)$ are approximately $k^{-2.17}$ and $k^{-1.47}$ respectively, which are reasonably close to the theoretical predictions of Bolgiano-Obukhov scaling ($-11/5$ and $-7/5$ respectively).

The energy and potential fluxes, exhibited in Fig. 16.4(A.b), show that potential energy flux $\Pi_b(k)$ is constant. However, $\Pi_u(k)$ decreases with k, but not quite as $k^{-4/5}$, which is a deviation from the predictions of Bolgiano-Obukhov phenomenology. Another important point to note is that Kumar and Verma (2015) could not

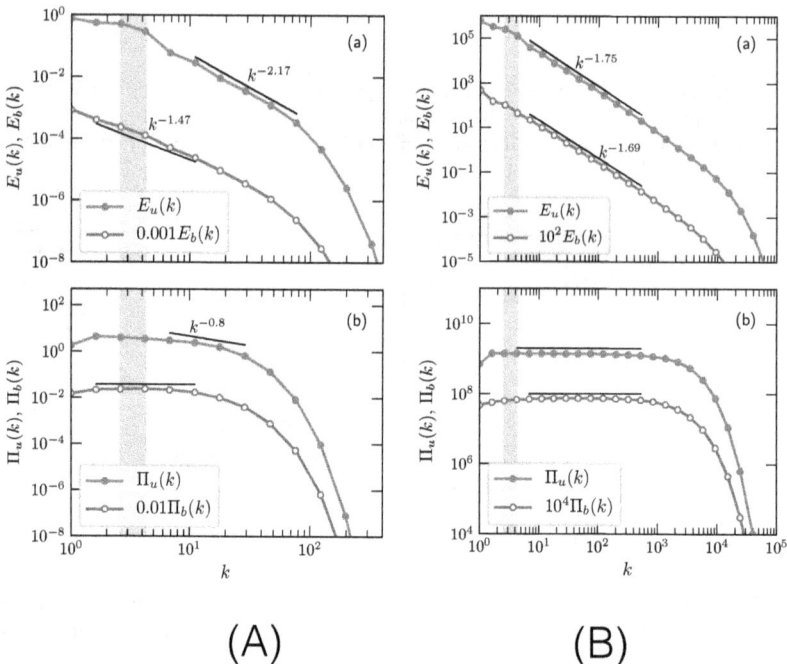

(A)   (B)

Fig. 16.4 For the shell model for SST of Sec. 16.2 with Pr = 1, Ri = 0.10, and forcing at the green band of k's: (A.a) plots of kinetic energy and entropy spectra; (A.b) plots of kinetic energy flux $\Pi_u(k)$ and entropy flux $\Pi_b(k)$. The plots are consistent with Bolgiano-Obukhov scaling. (B): Similar plots for Pr = 1 and Ri = 1.6×10^{-7}. The plots are consistent with passive-scalar turbulence with $E_u(k), E_b(k) \sim k^{-5/3}$. From Kumar and Verma (2015). Reprinted with permission from APS.

find a crossover in $E_u(k)$ from $k^{-11/5}$ to $k^{-5/3}$, which is inconsistent with the predictions of Bolgiano-Obukhov phenomenology.

For Ri $= 1.6 \times 10^{-7}$, which is small, buoyancy is much weaker than the nonlinear term. For such flows, we expect that $E_u(k) \sim k^{-5/3}$ and $E_b(k) \sim k^{-5/3}$, as in passive scalar turbulence. This is what was observed by Kumar and Verma (2015), as exhibited in Fig. 16.4(B.a). The fluxes of kinetic and potential energies ($\Pi_u(k)$ and $\Pi_b(k)$) are constants, as shown in Fig. 16.4(B.b).

Kumar and Verma (2015) also studied the kinetic energy feed by buoyancy, $\mathcal{F}_B(k)$. They observed that $\mathcal{F}_B(k) < 0$ for both the cases, as shown in Fig. 16.3(b). However, $\mathcal{F}_B(k)$ is significant for Ri $= 0.10$ (SST1), but it is quire small for Ri $= 1.6 \times 10^{-7}$ (SST2). The former $\mathcal{F}_B(k)$ affects the kinetic energy flux so as to yield Bolgiano-Obukhov scaling, but the latter $\mathcal{F}_B(k)$ is too insignificant to affect the kinetic energy flux.

In summary, the shell model for SST exhibits behaviour consistent with the theoretical models and numerical simulations discussed in Chapter 12.

16.5   Other shell models of buoyancy-driven flows

Brandenburg (1992) constructed a shell model with the following nonlinearity

$$M_n[u, u] = Ak_n(u_{n-1}^2 - \lambda u_n u_{n+1}) + Bk_n(u_n u_{n-1} - \lambda u_{n+1}) \qquad (16.23)$$

$$N_n[u, \theta] = \bar{A}k_n(u_{n-1}\theta_{n-1} - \lambda u_n \theta_{n+1})$$
$$+\bar{B}k_n[u_n\theta_{n-1} - \lambda u_{n+1}\theta_{n+1}] \qquad (16.24)$$

where λ, A, B, $\bar{A}$, $\bar{B}$ are constants. The constants A, B, $\bar{A}$, and $\bar{B}$ are chosen so as to force either an inverse or a forward cascade of kinetic energy. Brandenburg (1992) reported $E_u(k) \sim k^{-5/3}$ for the forward cascade regime, consistent with the arguments of Sec. 13.1. But for the inverse cascade regime, as in 2D flows, Brandenburg (1992) reported $E_u(k) \sim k^{-11/5}$, consistent with the energetic arguments of Sec. 13.11.

Ching and Cheng (2008) employed Brandenburg (1992)'s shell model and studied intermittency in thermal turbulence. They argued that for a fixed entropy transfer rate, the conditional velocity and temperature structure functions have simple scaling exponents consistent with Bolgiano-Obukhov scaling. However, the intermittency corrections appear in the scaling exponents of the entropy transfer rate.

Mingshun and Shida (1997) constructed a shell model for thermal convection with the following features: (a) Its nonlinear terms are very similar to the shell model described in Sec. 16.1; (b) $d\bar{T}/dz = 0$, thus, the model is applicable to neutral stratification. Mingshun and Shida (1997) analysed the structure function of the velocity and temperature fields, and observed Kolmogorov scaling for the convective turbulence.

Lozhkin and Frick (1998) constructed another shell model for thermal convective turbulence. The nonlinear terms of the shell model are as follows:

$$M_n[u, u] = ik_n \left(u^*_{n+1} u^*_{n+2} - \frac{\eta}{2} u^*_{n-1} u^*_{n+1} - \frac{(1-\eta)}{4} u^*_{n-1} u^*_{n-2} \right), \qquad (16.25)$$

$$N_n[u, \theta] = ik_n [u^*_{n+1} \theta^*_{n+2} + u^*_{n-1} \theta^*_{n+1} - \frac{1}{2} u^*_{n-2} \theta^*_{n-1}$$

$$+ \theta^*_{n+1} u^*_{n+2} + \frac{1}{2} u^*_{n+1} \theta^*_{n-1} - \frac{1}{4} \theta^*_{n-2} u^*_{n-1}]. \qquad (16.26)$$

Here $\lambda = 1/2$. Lozhkin and Frick (1998) reported that in turbulence convection Bolgiano-Obukhov model develops first, but it gives way to passive scalar turbulence in which both velocity and temperature fields yield $k^{-5/3}$ spectra. In 2D thermal turbulence however, Lozhkin and Frick (1998) report Bolgiano-Obukhov's spectrum. As described in Chapter 13, these results are in good agreement with the 3D and 2D turbulent thermal convection.

In summary, the shell models of Brandenburg (1992), Mingshun and Shida (1997), Lozhkin and Frick (1998), and Kumar and Verma (2015) in general indicate Kolmogorov-like scaling for turbulent convection in three dimensions. This is in spite of differences in the nonlinear terms of the shell models. This issue needs further investigation. We also remark that the shell models of Sections 16.1 and 16.2 should be applicable to unstably and stably stratified buoyancy-driven flows respectively.

Further reading

The shell models detailed in Secs. 16.1 and 16.2 were proposed by Kumar and Verma (2015). The other popular shell models of thermal turbulence are by Brandenburg (1992), Mingshun and Shida (1997), and Lozhkin and Frick (1998).

Exercises

(1) Consider the results on turbulent convection obtained by the shell model and by direct numerical simulations (see Chapter 13). What are the main similarities and differences between these results?
(2) Generalize the shell models of Secs. 16.1 and 16.2 to 2D RBC and 2D stably stratified turbulence (SST).
(3) By simulating the shell models of Secs. 16.1 and 16.2, study the scaling of large-scale quantities of RBC and SST, e.g., Reynolds number as a function of Ra.

Chapter 17

Structures and Flow Reversals in RBC

In Chapters 6, 7, and 8 we discussed various structures and patterns in RBC near the onset of thermal instability. These patterns arose due to thermal instability, and nonlinear interactions among the large-scale Fourier modes. In the turbulent regime, one may expect these patterns to be washed out. But this is not the case; turbulent RBC too exhibits coherent structures. In this chapter we will discuss the nature of the structures in turbulent RBC, as well as a well-known phenomena called *flow reversal*.

In the next section we introduce large-scale circulation.

17.1 Structures and Large-scale circulation in RBC

At the onset of convection, thermal instability generates primary convection rolls. As the Rayleigh number Ra is increased, near the onset of convection, secondary rolls and a variety of patterns are generated [Krishnamurti (1970a,b); Getling (1998); Lappa (2010)] (see Sec. 8.4). Interestingly, the primary and several secondary rolls persist even for Ra $\gg$ Ra$_c$, but they appear as variants of the original structures, and they are time-dependent. Such structures in turbulent thermal convection are referred to as *large-scale circulation* or *LSC* [Niemela *et al.* (2000); Niemela and Sreenivasan (2003)]. In Fig. 17.1, we illustrate a LSC of 2D turbulent convection (for Ra $= 2 \times 10^7$) at different times. The large-scale roll structures— Fourier mode (1,1) and (2,2)—are major components of LSC of Fig. 17.1. In the next section we will describe these ideas in more detail.

LSC have been observed in all sorts of geometries—rectangular, cylindrical, and spherical, and for range of parameters—Ra and Pr. Planetary and stellar atmospheres too contain such structures. Experiments and numerical simulations reveal that LSCs are dynamic. In a phenomenon called *flow reversal*, the vertical velocity of a LSC reverses its direction. It was first reported by Niemela *et al.* (2000); they measured the vertical velocity near the lateral wall and observed that it reverses randomly as shown in Fig. 17.2 [Sreenivasan *et al.* (2002)]. Later, Xi *et al.* (2006), Brown and Ahlers (2006), Mishra *et al.* (2011), Chandra and Verma (2011), Gallet *et al.* (2012), Vasiliev *et al.* (2016) reported flow reversals in various

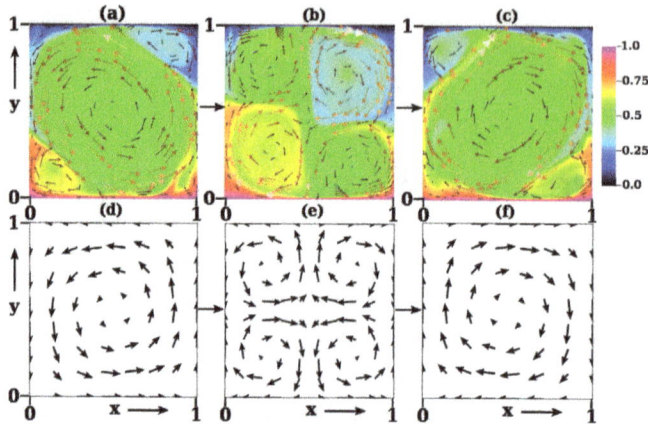

Fig. 17.1   In a 2D RBC, flow profiles (a) before, (b) during, (c) after the reversal. Red and blue regions indicate hot and cold regions. The real space representation of: (d, f) Fourier mode $\mathbf{u}(1,1)$, and (e) Fourier mode $\mathbf{u}(2,2)$. From Chandra and Verma (2011). Reprinted with permission from APS.

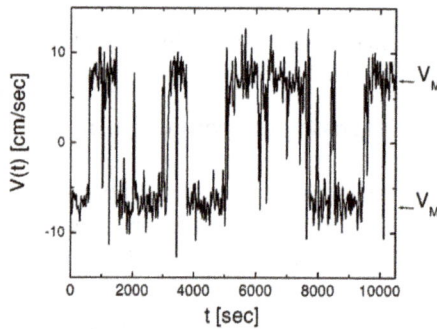

Fig. 17.2   In a RBC experiment with Ra = 1.5×10^{11}, a plot of the time series of u_z measured near the lateral wall exhibits flow reversals. From Sreenivasan *et al.* (2002). Reprinted with permission from APS.

geometries and for a wide range parameters. In the following discussion, we will focus on the dynamics of flow reversal. We start with flow reversals in 2D RBC.

17.2   LSC and its reversal in 2D closed box

Sugiyama *et al.* (2010) performed a RBC experiment in a quasi-2D closed box containing water, and of dimension $(x, y, z) = $ 24.8cm × 7.5cm × 25.4cm. They observed LSC for Ra in the range from 10^7 to 10^{10}. The LSC reversed intermittently for a band of intermediate Ra, but no reversal was observed for very large Ra.

Sugiyama *et al.* (2010), Chandra and Verma (2011), and Chandra and Verma (2013) performed numerical simulations of turbulent RBC in a 2D square box, and observed reversals similar to those observed in Sugiyama *et al.* (2010)'s experiment. Here we detail some of the salient features of flow reversals of 2D RBC observed for Pr $= 1$ and Ra $= 2 \times 10^7$. Specifically, we will discuss the nonlinear interactions among the large-scale flow structures, fluctuations in the Nusselt number, and symmetries of reversing and non-reversing large-scale modes.

17.2.1 *Nonlinear interactions among the large-scale Fourier modes*

In this section we consider one of the flow reversals observed in numerical simulation of Chandra and Verma (2011, 2013) for Pr $= 1$ and Ra $= 2 \times 10^7$. In Fig. 17.1(a,b,c) we depict the flow profiles at three snapshots—(a) before the reversal, (b) during the reversal, and (c) after the reversal[1]. The portraits clearly demonstrates the existence of large-scale structures that can be approximated using large-scale Fourier modes $\hat{\mathbf{u}}(k_x, k_y, t), \hat{\theta}(k_x, k_y, t)$:

$$u_x(\mathbf{r}, t) = \sum_{\mathbf{k}} 4\hat{u}_x(k_x, k_y, t) \sin(\pi n_x x) \cos(\pi n_y y), \tag{17.1}$$

$$u_y(\mathbf{r}, t) = \sum_{\mathbf{k}} 4\hat{u}_y(k_x, k_y, t) \cos(\pi n_x x) \sin(\pi n_y y), \tag{17.2}$$

$$\theta(\mathbf{r}, t) = \sum_{\mathbf{k}} 4\hat{\theta}(k_x, k_y, t) \cos(\pi n_x x) \sin(\pi n_y y), \tag{17.3}$$

where $\mathbf{k} = (k_x, k_y) = (\pi n_x, \pi n_y)$. The box dimension is taken to be unity.

Figure 17.1(d,e,f) exhibit the vector plots of the velocity field corresponding to the dominant Fourier modes $\mathbf{u}(1,1)$ and $\mathbf{u}(2,2)$. Note that the $\hat{\mathbf{u}}(1,1)$ of subfigure (f) has amplitude opposite to that of subfigure (d). Also, the Fourier mode $\hat{\mathbf{u}}(2,2)$ represents the four corner rolls.

It is easy to verify that the flow profiles of the three snapshots of Fig. 17.1(a,b,c) can be approximated by the following superposition of the Fourier modes $\hat{\mathbf{u}}(1,1)$ and $\hat{\mathbf{u}}(2,2)$:

      subfig. (a) $\approx \hat{\mathbf{u}}(1,1)$ of subfig. (d) $+ \hat{\mathbf{u}}(2,2)$ of subfig. (e)

      subfig. (b) $\approx \hat{\mathbf{u}}(2,2)$ of subfig. (e)

      subfig. (c) $\approx \hat{\mathbf{u}}(1,1)$ of subfig. (f) $+ \hat{\mathbf{u}}(2,2)$ of subfig. (e)

In Fig. 17.3(a) we exhibit the time series of the vertical velocity measured at the real space location $(x, y) = (0.25, 0.25)$. The time series clearly demonstrates reversals of $u_y(\mathbf{r})$. In Fig. 17.3(b,c) we plot the time series of Fourier modes $\hat{u}_y(1,1), \hat{u}_y(2,2)$, and their ratio $|\hat{u}_y(2,2)/\hat{u}_y(1,1)|$. Note strong correlation between the time series of the real-space vertical velocity $u_y(0.25, 0.25)$ and the Fourier mode $\hat{u}_y(1,1)$, both of which exhibit flow reversals. Note that in Fig. 17.3, $\hat{u}_y(k_x, k_y)$ is written as $\hat{v}_{k_x, k_y}$,

---

[1]To keep consistency with the published literature, in this chapter, we choose $\hat{y}$ as the vertical direction for two-dimensional flows. Also, we employ hat to represent Fourier transform of a field, e.g., $\hat{f}(\mathbf{k})$ is the Fourier transform of field $f(\mathbf{r})$.

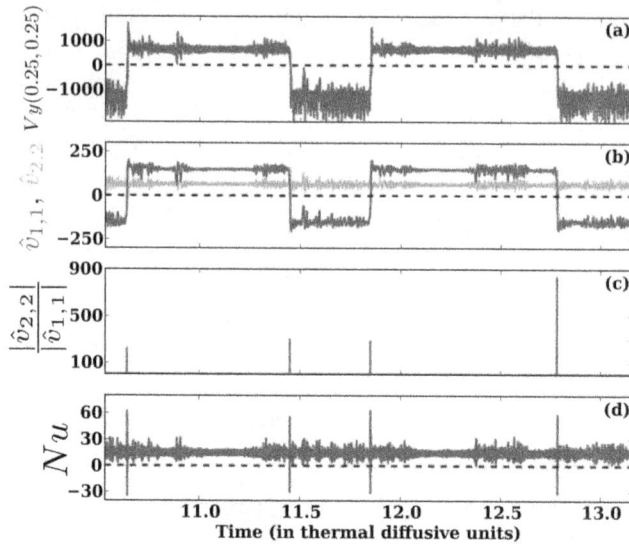

Fig. 17.3   For a flow reversal in 2D RBC, time series of (a) vertical velocity u_y at $(x, y) = (0.25, 025)$, (b) Fourier modes $u_y(1, 1)$ and $u_y(2, 2)$, (c) $|u_y(2, 2)/u_y(1, 1)|$, (d) Nusselt number. In the figure, $\hat{u}_y$ is denoted by $\hat{v}$. From Chandra and Verma (2011). Reprinted with permission from APS.

The Fourier mode $\hat{u}_y(2, 2)$ however does not flip during the reversals (see Fig. 17.3(b)). The amplitude of the Fourier mode $\hat{u}_y(2, 2)$ increases during the reversal, while the amplitude of $u_y(1, 1)$ decreases, crosses zero, and then takes opposite sign. This is the reason why the ratio $|\hat{u}_y(2, 2)/\hat{u}_y(1, 1)|$ peaks during a flow reversal.

Chandra and Verma (2011) observed that $\langle|\hat{\mathbf{u}}(1, 1)|\rangle$ increases with Ra, primarily due to the strengthening of the inverse energy cascade. In comparison, $\langle|\hat{\mathbf{u}}(2, 2)|\rangle$ does not increase as much. As a result, the ratio $\langle|\hat{\mathbf{u}}(2, 2)|\rangle/\langle|\hat{\mathbf{u}}(1, 1)|\rangle$ decreases with Ra. Since $\hat{\mathbf{u}}(2, 2)$ plays an important role in flow reversals, its weakening for large Ra's tends to stop the reversals. This feature has been observed in experiments and simulations [Sugiyama *et al.* (2010); Chandra and Verma (2011)].

Chandra and Verma (2013) examined the flow patterns near a reversal. Fig. 17.4(a–e) exhibits six snapshots that provide important clues. Figure 17.4(b,c) indicates that the corner rolls reconnect (or merge) and become a larger roll, and two new corner rolls emerge along the opposite diagonal. This vortex reconnection is similar to the magnetic reconnection of plasma physics.

Fourier transform of Fig. 17.4(d,e) indicates importance of large-scale Fourier modes other than (1,1) and (2,2). In the following discussion we discuss the role of these modes. Flow reversals are caused by nonlinear interactions among the

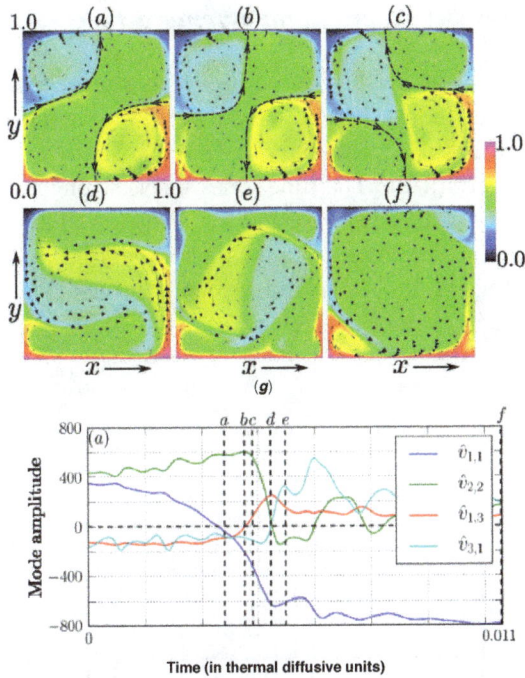

Fig. 17.4   Six snapshots of the flow profiles during a reversal. The two corners rolls of (a,b) reconnect or merge in (c). (d,e) Contains contributions from the Fourier modes $\hat{\mathbf{u}}(1,3)$ and $\hat{\mathbf{u}}(3,1)$. (f) Flow after the reversal. (g) Time series of the dominant Fourier modes during a reversal; the dashed markers in (g) correspond to the five snapshots (a–e). Here $\hat{v} \equiv \hat{u}_y$. From Chandra and Verma (2013). Reprinted with permission from APS.

large-scale Fourier modes, namely,

$$(2,2) \oplus (1,1) = (1,3), \tag{17.4}$$

$$(2,2) \oplus (1,1) = (3,1) \tag{17.5}$$

with $\oplus$ representing the nonlinear interaction. The aforementioned modes form two interacting triads—$\{(1,1),(2,2),(1,3)\}$ and$\{(1,1),(2,2),(3,1)\}$. Here $2 \oplus 1 = 3$ or 1 because of the following identity and other related identities:

$$2 \sin k_1 x \cos k_2 x = \sin(k_1 + k_2)x + \sin(k_1 - k_2)x. \tag{17.6}$$

The above relation indicates that the Fourier modes $(1,3)$ and $(3,1)$ too play significant role during a flow reversal. The modes $(1,3)$ and $(3,1)$ represent three vertical and three horizontal rolls shown in Fig. 17.4(d,e) respectively. These observations are consistent with the time series of the Fourier modes shown in Fig. 17.4(g). The secondary modes (2,2), (1,3), and (3,1) are significant during a reversal.

The aforementioned results indicate that flow reversals in turbulent RBC is strongly influenced by the nonlinear interactions of large-scale Fourier modes. These interactions lead to strong fluctuations in the Nusselt number, as illustrated in Fig. 17.3(d). We will be discuss this issue in the next subsection.

17.2.2   *Nusselt number fluctuations during a reversal*

A flow reversal reorganises the flow structures that causes violent fluctuations in the Nusselt number Nu. In Fig. 17.5(A) we illustrate the time series of Nu(t) during the reversal shown in Fig. 17.4; the markers (a–e) in the time series correspond to the snapshots of Fig. 17.4(a–e). The time series shows strong fluctuations including a negative Nu for a short interval covering snapshot (e). The negative Nu occurs due to descension of a yellow-colored hot structure and ascension of a blue-coloured cold structure (see Fig. 17.4(e)). Such movements are not possible during normal thermal convection, i.e. away from a reversal.

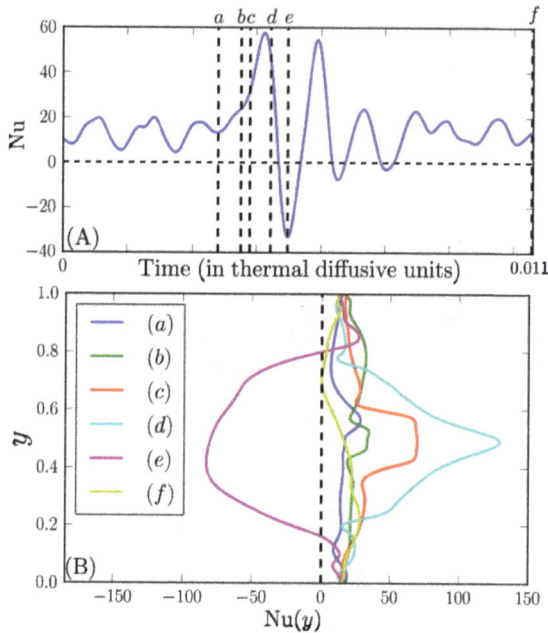

Fig. 17.5   (a) Time series of Nu(t) shows strong fluctuations near the reversals. The vertical dashed lines correspond to the snapshots (a–e) of Fig. 17.4. (b) Nu(y) vs. y for the snapshots (a–f) of Fig. 17.4. Nu < 0 for Snapshot (e). From Chandra and Verma (2013). Reprinted with permission from APS.

Negative Nusselt number indicates a heat transport from colder region to hotter region, and may appear to contradict thermodynamic principles. However, there is no contradiction because the structures within the bulk are driven by dynamics, not thermodynamics. It is the dynamics that causes hot plumes to descend and cold plumes to ascend *temporarily*. The heat transport near the thermal plates is dominated by thermal diffusion, a thermodynamic process, hence it is *always* from hotter region to colder region.

For more detailed analysis, we compute the planar-averaged Nusselt number $\mathrm{Nu}(y)$:

$$\mathrm{Nu}(y) = 1 + \frac{\langle u_z \theta \rangle_x}{\kappa \Delta / d}, \qquad (17.7)$$

with averaging performed along horizontal lines. Refer to Sec. 2.9 for the definitions of Nusselt number. For a steady state, we expect $\langle u_z \theta \rangle_x$ to be independent of y, and hence $\mathrm{Nu}(y) \approx \mathrm{Nu}$. However, the flows illustrated in Fig. 17.4(a–f) are far from steady, hence we expect $\mathrm{Nu}(y)$ to vary strongly with y. In Fig. 17.5(B) we illustrate $\mathrm{Nu}(y)$ for the six snapshots shown in Fig. 17.4(a–f). From the plots we make the following important deductions:

(1) $\mathrm{Nu}(y)$ is always positive near the top and bottom thermal plates. This is consistent because the heat transport in the boundary layer, which is via thermal diffusion, follows thermodynamics.
(2) For the snapshots Fig. 17.4(a,b) and (f), $\mathrm{Nu}(y) \approx$ constant in y, and it equals the average or global Nu. During these phases, the system is away from the reversal process.
(3) For the snapshots (c,d), $\mathrm{Nu}(y)$ in the central region is much larger than elsewhere, which is due to the strong upward (downward) motion of the hot (cold) plumes in the bulk region.
(4) For the snapshot (e), $\mathrm{Nu}(y) < 0$ in the bulk because of the upward (downward) motion of the cold (hot) structure in the bulk. Note that in Fig. 17.4(e), $\mathrm{Nu}(y) > 1$ in the boundary layer, consistent with item (1).
(5) Increase in the amplitudes of secondary modes and Nusselt number may be used as precursors to a flow reversal.

Thus, we show that the heat transport or Nu exhibits unusual behaviour during a flow reversal. Such unpredictable and violent changes in the heat transport during flow restructuring would require special attention in engineering and atmospheric applications.

In the next subsection we will study the symmetry properties of reversing and non-reversing modes.

17.2.3 *Symmetries of Fourier modes*

The reversing modes exhibit interesting patterns that can be inferred from the following evolution equations for $\mathbf{u}(\mathbf{k})$ and $\theta(k)$:

$$\frac{d}{dt}\hat{\mathbf{u}}(\mathbf{k}) = -i\sum_{\mathbf{k}}[\mathbf{k} \cdot \hat{\mathbf{u}}(\mathbf{q})] \cdot \hat{\mathbf{u}}(\mathbf{p}) - ik\hat{\sigma}(\mathbf{k}) + \alpha g\hat{\theta}(\mathbf{k})\hat{z} - \nu k^2 \hat{\mathbf{u}}(\mathbf{k}), \qquad (17.8)$$

$$\frac{d}{dt}\hat{\theta}(\mathbf{k}) = -i\sum_{\mathbf{k}}[\mathbf{k} \cdot \hat{\mathbf{u}}(\mathbf{q})]\cdot\hat{\theta}(\mathbf{p}) - \frac{d\bar{T}}{dz}\hat{u}_z(\mathbf{k}) - \kappa k^2 \hat{\theta}(\mathbf{k}), \qquad (17.9)$$

Table 17.1   In 2D RBC, the rules for nonlinear interactions among the modes. The
elements form Klein four-group $Z_2 \times Z_2$.
Adopted from a table of Verma *et al.*
(2015).

$\times$	E	M_1	M_2	O
E	E	M_1	M_2	O
M_1	M_1	E	O	M_2
M_2	M_2	O	E	M_1
O	O	M_2	M_1	E

Thus, the nonlinear terms are sums of quadratic products of interacting Fourier modes. We denote

$$\mathbf{k} = (k_c n_x, \pi n_z), \tag{17.10}$$
$$\mathbf{p} = (k_c l_x, \pi l_z), \tag{17.11}$$
$$\mathbf{q} = (k_c m_x, \pi m_z), \tag{17.12}$$

where $k_c = 2\pi/L_x$. From the interaction rules—$\mathbf{k} = \mathbf{p} \pm \mathbf{q}$, we deduce that

$$(n_x, n_z) = (\pm l_x \pm m_x, \pm l_z \pm m_z). \tag{17.13}$$

The $\pm$ sign appear because of Eq. (17.6).

We classify the Fourier modes according to the mode parity, that is, we divide the Fourier components, k_x, k_y, k_z of $\mathbf{k}$, according to their parities—even, represented by e, and odd, represented by o. Let P denote the parity function, for example, $P(3) = o$, $P(4) = e$. In this scheme, the Fourier modes of the velocity and temperature fields can be classified into four classes—$E = (ee)$, $M_1 = (eo)$, $M_2 = (oe)$, $O = (oo)$. To illustrate, $u_x(1,1) \in O$, $u_x(2,2) \in E$, $u_x(2,1) \in M_1$, and $u_x(1,2) \in M_2$. In the following discussion, we will derive a group structure for the modes. This mathematical exercise helps us identify the reversing and nonreversing Fourier modes.

Using the rules of addition, even+even = even, even+odd = odd, and odd+odd = even, we obtain the product rules shown in Table 17.1. Here we list $A \oplus B = C$, where the 4 elements of A, (E, M_1, M_2, O) are listed in the first column, while the 4 elements of B are listed in the first row. The product rules indicate that the 4 elements form an abelian group with E as the identity element ($E \oplus X = X$, where X is any element of the group). This is, in fact, Klein-4 $Z_2 \times Z_2$ group. From the multiplication table, we observe that $O \oplus M_1 = M_2$, $O \oplus M_2 = M_1$, and $M_2 \oplus M_1 = O$, hence, M_1 and M_2 are complements of each other. Also, $O \oplus O = E$.

Using the symmetry properties of RBC equations and the product rules of Table 17.1 we deduce several interesting results. If $\{M_1, M_2\} = 0$ (zero), and nonzero modes $\{O\}$, $\{E\}$ are the solution of RBC equations, then $\{-O\}$, $\{E\}$ would also be the solution of the RBC equation. Similar deductions can be made when $\{M_1\}$, $\{E\} \neq 0$, but $\{M_2, O\} = 0$. Note however that in turbulent flows, we expect the modes to take small values ϵ, rather than being zero. Also, if all the entries are nonzero, and $(\{E, M_1, M_2, O\})$ are solution of the RBC equations, then $\{E, -M_1, M_2, -O\}$, or $\{E, M_1, -M_2, -O\}$, or $\{E, -M_1, -M_2, O\}$ are also the solution of RBC equations.

Using the properties of the product table, we can deduce which modes flip during a reversal, and which ones do not. The rules are as follows:

(1) $\{O\} \to \{-O\}$; $\{E\} \to \{E\}$; $\{M_1, M_2\} = \epsilon$
(2) $\{M_1\} \to \{-M_1\}$; $\{E\} \to \{E\}$; $\{O, M_2\} = \epsilon$
(3) $\{M_2\} \to \{-M_2\}$; $\{E\} \to \{E\}$; $\{O, M_1\} = \epsilon$
(4) $\{O\} \to \{-O\}$; $\{M_1\} \to \{-M_1\}$; $\{M_2\} \to \{M_2\}$; $\{E\} \to \{E\}$
(5) $\{O\} \to \{-O\}$; $\{M_2\} \to \{-M_2\}$; $\{M_1\} \to \{M_1\}$; $\{E\} \to \{E\}$
(6) $\{M_1\} \to \{-M_1\}$; $\{M_2\} \to \{-M_2\}$; $\{O\} \to \{O\}$; $\{E\} \to \{E\}$

Here ϵ stands for a small number. Note that the modes in the class $\{E\}$ can never change sign. We remark that the aforementioned discrete group structure corresponds to the discrete symmetry of the system; here the Fourier modes take $+$ or $-$ sign. The flows in cylindrical and spherical geometries have continuous symmetries in which $\mathbf{u}(\mathbf{k})$ takes complex values; in subsequent sections we will discuss reversals in such systems.

The reversals in a 2D box described earlier belong to the set (1) where modes in $\{O\}$ reverse, modes in $\{E\}$ do not reverse, and the modes in the classes $\{M_1, M_2\}$ take small values. Thus, we expect the modes $(1, 3)$ and $(3, 1)$ to switch sign, modes $(2, 2)$ and $(4, 4)$ to not switch sign, but $(2, 1)$ and $(1, 2)$ to remain vanishingly small. Using numerical simulations Chandra and Verma (2011, 2013); Verma *et al.* (2015) verified some of the above symmetry rules.

17.2.4 *Reversals under variations of Prandtl number and geometries*

Researchers studied flow reversals in 2D RBC under variations of Prandtl number, aspect ratio, and boundary conditions. Sugiyama *et al.* (2010), Chandra and Verma (2011), and Chandra and Verma (2013) studied reversals in 2D RBC for different Prandtl numbers and aspect ratios, while Breuer and Hansen (2009) and Verma *et al.* (2015) studied flow reversals in free-slip boxes of aspect ratios 1 and 2 for various Prandtl numbers (including Pr $= \infty$). There are not many data points to make definitive conclusions, but following deductions can be made from the aforementioned experiments and numerical simulations:

(1) Reversals are easier for very small or very large Prandtl numbers. This is possibly because the nonlinear term $\mathbf{u} \cdot \nabla \mathbf{u}$ is very dominant for very small Pr, while the nonlinear term $\mathbf{u} \cdot \nabla \theta$ becomes large for large Pr [Breuer and Hansen (2009); Verma *et al.* (2015)].
(2) In 2D RBC, flow reversals are absent for very large Ra. This is due to relative

strengthening of the primary mode $(1,1)$ with relative to $(2,2)$ because of an inverse cascade of kinetic energy [Chandra and Verma (2011, 2013)].

(3) The properties of reversals for the free-slip and no-slip RBC are, in general, different. As a result, the nonlinear interactions responsible for the reversals could be very different for the two boundary conditions. For example, for aspect ratio of 2, the most dominant and reversing Fourier modes for no-slip and free-slip boundary conditions are $(2,1)$ and $(1,1)$ respectively.

Before we close the discussion on reversals in 2D Cartesian box, we highlight several common features in all 2D reversals. The interval distribution between two consecutive reversal is quite interesting. Mannattil *et al.* (2017) studied the interval distribution between two consecutive reversals in infinite-Pr RBC with free-slip boundary condition and reported a Poisson distribution for Δt:

$$P(\Delta t) \approx \exp\left(-\frac{\Delta t}{\tau_{\text{avg}}}\right), \tag{17.14}$$

where τ_{avg} is the average interval between two consecutive reversals. The Poisson distribution is due to the loss of memory during the rebirth of a new (reversed) structure [Verma *et al.* (2006b)].

In the aforementioned discussion we described importance of the Fourier modes in flow reversals. We remark that the POD (proper orthogonal decomposition) modes are also used for the analysis of flow reversals in RBC. For example, Podvin and Sergent (2015) quantified the reversals in a square box using POD modes and related low-dimensional model. Interestingly, there are strong similarities between the large-scale POD and Fourier modes [Paul and Verma (2015)]. Refer to Appendix C for a detailed comparison.

In the next section, we will describe properties of flow reversal in a 3D Cartesian box.

17.3 LSC and flow reversals in 3D Cartesian box

Several experiments and numerical simulations have been performed to investigate reversals in a 3D Cartesian box. Gallet *et al.* (2012), Vasil'ev and Frick (2011), and Vasiliev *et al.* (2016) performed experiments in cartesian boxes with mercury and water as experimental fluids. These researchers reported that the flow has a dominant roll along one of the diagonals. After a random interval, this dominant roll reorients along the other diagonal. Foroozani *et al.* (2017) performed large-eddy simulations of turbulent RBC in a cube, and observed similar swapping of the diagonal rolls. Gallet *et al.* (2012), Vasil'ev and Frick (2011), Vasiliev *et al.* (2016), and Foroozani *et al.* (2017) attribute the reorientation of the diagonal roll to *lateral or azimuthal rotation of LSC*.

Kumar and Verma (2017) simulated turbulent RBC in a cube for Ra $= 10^8$ and observed a dominant diagonal roll as described above. They filtered the large-scale Fourier modes and ranked them according to the kinetic energy. Their mode analysis shows that the most dominant rolls are $\mathbf{u}(1,0,1)$ and $\mathbf{u}(0,1,1)$, and the subdominant ones are $\mathbf{u}(1,1,2)$, $\mathbf{u}(1,3,3)$, $\mathbf{u}(1,2,2)$. The Fourier modes $(\pm1, \pm1, \pm1)$ corresponding to the diagonal rolls are much weaker than the above modes.

In Fig. 17.6 we illustrate the flow structures at three vertical cross sections of the cube. Clearly, the dominant Fourier modes $\mathbf{u}(1,0,1)$ and $\mathbf{u}(0,1,1)$, shown in Fig. 17.6(b,c), superpose to create a diagonal roll along one of the diagonals, as shown in Fig. 17.6(d). If the mode $\mathbf{u}(1,0,1)$ switches sign, then the dominant roll will flip to the other diagonal. Similar phenomena occurs when the mode $\mathbf{u}(0,1,1)$ reverses. Thus the swapping of the diagonal rolls in a cube is due to the reversal of $\mathbf{u}(1,0,1)$ or $\mathbf{u}(0,1,1)$ mode, not due to a lateral rotation of the LSC, as envisaged by earlier researchers [Gallet *et al.* (2012); Vasil'ev and Frick (2011); Vasiliev *et al.* (2016); Foroozani *et al.* (2017)].

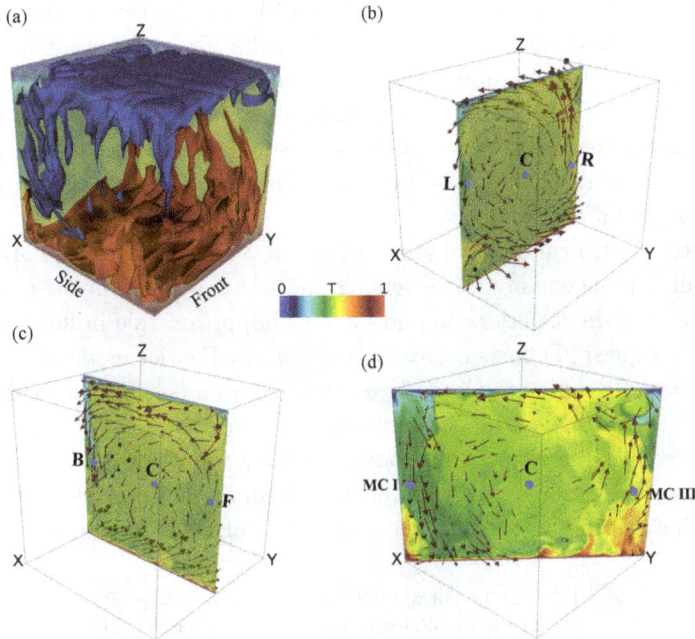

Fig. 17.6 Turbulent convection in a cube with $\mathrm{Pr} = 1, \mathrm{Ra} = 10^8$: (a) Temperature isosurfaces. (b) Convection rolls in xz plane. (c) Convection roll in yz plane. (d) Superposition of rolls of (b,c) yield a diagonal roll. Adopted from a figure of Kumar and Verma (2017).

The group structure of a 2D Cartesian box can be easily extended to 3D Cartesian box. For a 3D RBC, the modes are classified into 8 classes: $E = (eee)$, odd $O = (ooo)$, and mixed modes—$M_1 = (eoo)$, $M_2 = (oeo)$, $M_3 = (ooe)$, $M_4 = (eeo)$, $M_5 = (oee)$, $M_6 = (eoe)$. The multiplication rules for these modes, listed in Table 17.2, indicate that the group is abelian with E as the identity element. In fact, the 8 elements $\{E, O, M_i\}$ form Klein-8 $Z_2 \times Z_2 \times Z_2$ group.

Table 17.2   In 3D RBC, rules of nonlinear interactions among the modes.
The elements form the Klein 8-group $Z_2 \times Z_2 \times Z_2$.

$\times$	E	O	M_1	M_2	M_3	M_4	M_5	M_6
E	E	O	M_1	M_2	M_3	M_4	M_5	M_6
O	O	E	M_5	M_6	M_4	M_3	M_1	M_2
M_1	M_1	M_5	E	M_3	M_2	M_6	O	M_4
M_2	M_2	M_6	M_3	E	M_1	M_5	M_4	O
M_3	M_3	M_4	M_2	M_1	E	O	M_6	M_5
M_4	M_4	M_3	M_6	M_5	O	E	M_2	M_1
M_5	M_5	M_1	O	M_4	M_6	M_2	E	M_3
M_6	M_6	M_2	M_4	O	M_5	M_1	M_3	E

Based on the multiplication table, we derive the rules listed in Table 17.3 for the reversing and non-reversing modes [Bandyopadhyay and Verma (2017)]. For the 3D RBC example described above, the reversing modes are either $(1, 0, 1)$ or $(0, 1, 1)$, which belong to the classes $\{M_2\}$ and $\{M_1\}$ respectively. The other possible reversing modes have not been studied for the cube. From the available limited data we conjecture that for the cube discussed above, (a) $\{M_1\}$ modes reverse, $\{E\}$ modes do not reverse, and other modes take small values; or (b) $\{M_2\}$ modes reverse, $\{E\}$ modes do not reverse, and other modes take small values. Both these rules belong to category (2) of Table 17.3. We need to perform detailed simulations/experiments to come to a definite conclusion.

We need to perform detailed study of flow reversals in Cartesian 3D box. These studies will help us classify the reversing and non-reversing modes in such geometries. Note that the aspect ratio plays a very important role in flow reversals. For example, in a quasi-2D box suppressed along $\hat{y}$, the Fourier mode $(0, 1, 1)$ is excited at a very large Ra due to the large k_c in that direction (see Fig. 6.1 and Eq. (6.20)).

Before closing this section, we remark that there are some recent works on *superstructures* in large aspect-ratio boxes [Getling *et al.* (2013)]. These works are motivated by the granules, mesogranules, and supergranules observed in solar convection. The horizontal extent of the rolls are determined by nonlinear interactions

Table 17.3   For 3D RBC, the reversing modes (R), non-reversing modes (NR) during a reversal. The modes not included in R or NR are small. Adopted from a table of Bandyopadhyay and Verma (2017).

Item	R	NR	classes
0	-	All elements	No reversal
1	O	E	1 class
2	M_i	E	6 classes
3	O and M_i	E and M_i'	6 classes
4	O, M_1, M_2	E, M_3, M_5, M_6	1 class
5	O, M_2, M_3	E, M_1, M_4, M_6	1 class
6	O, M_1, M_3	E, M_2, M_4, M_5	1 class
7	M_1, M_2, M_5, M_6	E, O, M_3, M_4	1 class
8	M_2, M_3, M_4, M_6	E, O, M_1, M_5	1 class
9	M_1, M_3, M_4, M_5	E, O, M_2, M_6	1 class

among the Fourier modes; such analysis may provide valuable insights for understanding superstructures.

In the next section, we report flow reversals in cylindrical annulus, which is analogous to a 2D box with periodicity along the horizontal direction.

17.4 LSC and its reversal in a cylindrical annulus or 2D periodic RBC

In Secs. 17.2 and 17.3 we studied flow reversals in closed boxes. It is interesting to investigate reversals in a 2D box which is open along $\hat{x}$. For such geometries, we employ periodic boundary condition along $\hat{x}$.

In turbulent convection, the convective rolls travel along $\hat{x}$ similar to the travelling motion of convective rolls near the onset of convection [Cross and Hohenberg (1993)], except that the roll movement for large Ra is chaotic. This is the topic of this section. Figure 17.7 illustrates such movements for the free-slip and no-slip RBC simulations [Paul *et al.* (2010)]. Incidentally, the velocity measured by a real-space probe in the mid plane exhibits a flow reversal when the convective roll moves by its horizontal extant.

A cylindrical annulus is geometrically equivalent to the aforementioned Cartesian box with periodicity along $\hat{x}(\hat{x} \equiv \hat{\phi}$ of the cylinder). Hence, the flow properties in an annulus is analogous to that in a periodic box. In an annulus, Nath and Verma (2014) showed that for large Ra, the convection rolls move azimuthally in a chaotic manner.

In RBC with periodicity along $\hat{x}$ or $\hat{\phi}$, a roll can move by any distance. Hence, the flow movement along $\hat{x}$ or $\hat{\phi}$ is governed by *continuous symmetry* or *continuous group*. We are not aware of any group-theoretic description of such interacting Fourier modes (in a manner described in Secs. 17.2 and 17.3). It is obvious, but worth remarking, that each walled direction is governed by discrete symmetry, and each periodic direction by continuous symmetry.

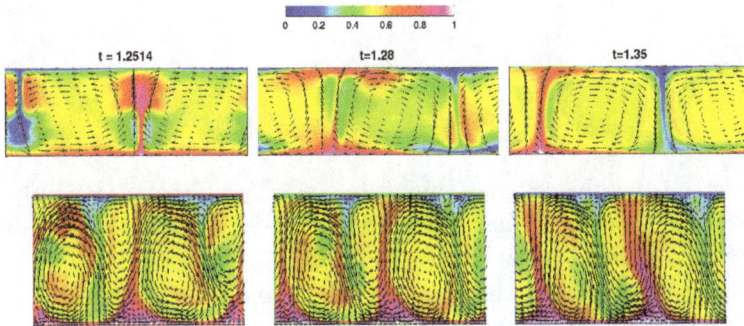

Fig. 17.7 Chaotic travelling rolls in 2D RBC with periodicity along $\hat{x}$: (top panel) free-slip boundary condition along the vertical; (bottom panel) no-slip boundary condition along the vertical. From Paul *et al.* (2010). Reprinted with permission from Indian Academy of Science.

The continuous symmetry would be broken for an elliptical cylinder for which the convective roll would be aligned along the major axis of the cylinder. We expect no azimuthal movement for such system; this conjecture needs to investigated by experiments or simulations. Similarly, the azimuthal symmetry is broken for an inclined cylinder with related to **g**, hence flow reversals are hindered in such systems as well.

A Cartesian box with periodicity in all directions does not have a well defined LSC. This box has continuous symmetry along all the directions. A detailed and comparative group-theoretic analysis of these systems would be very valuable.

In the next section, we will study reversals in a cylinder.

17.5   LSC and flow reversals in a cylinder

Sreenivasan *et al.* (2002), Brown and Ahlers (2006), and Xi *et al.* (2006) performed RBC experiments in a cylindrical geometry for Ra in the range of 10^8 to 10^{13}. They used either water or Helium gas as experimental fluids. All these experiments exhibit flow reversals that will be described below. Mishra *et al.* (2011) performed numerical simulations for a similar setup but at Ra $\sim 10^6$, and observed reversals similar to those in the above experiments.

For brevity, in this section we will focus on a cylinder of unit aspect ratio (height = diameter). For studying the reversals, researchers measured the temperature or the vertical velocity using real-space probes. In Fig. 17.8(a) we illustrate 8 such probes whose angular coordinates are

$$\phi_j = j\pi/4 \quad (j = 0, 1, .., 7). \tag{17.15}$$

We denote the vertical velocity measured by the jth probe at time t by $u_j(t)$, which is expanded as Fourier series:

$$u_j(t) = \sum_{k=-4}^{k=4} \hat{u}_k(t) \exp(ik\phi_j)$$

$$= \hat{u}_0(t) + \sum_{k=1}^{k=4} |\hat{u}_k(t)| \exp[i(k\phi_j + \delta_k(t)] + c.c.$$

$$= 0 + \sum_{k=1}^{k=4} 2|\hat{u}_k(t)| \cos(k\phi_j + \delta_k(t)), \tag{17.16}$$

where *c.c.* stands for the complex conjugate, and $\hat{u}_k = |\hat{u}_k| \exp(i\delta_k)$ with $|\hat{u}_k|$ and δ_k being the amplitude and phase of the Fourier mode. Note that $\hat{u}_0 = 0$ due to the absence of a travelling wave at large Ra. In Fig. 17.8(b,c) we illustrate schematic diagrams of the flow profiles corresponding to $\hat{u}_1$ and $\hat{u}_2$ respectively. In the flow corresponding to $\hat{u}_1$, the hot fluid moves upward at A, and the cold fluid moves downward at B. But for $\hat{u}_2$, the upward and downward movements of the fluid occur at two locations each. The phase δ_k denotes the orientation of the mode k with relative to $\phi = 0$ or x axis, as shown in Fig. 17.8(b,c).

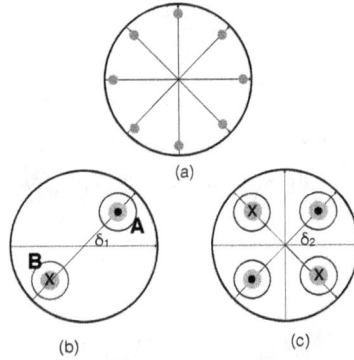

(a)

(b)   (c)

Fig. 17.8   (a) Real space probes located on a horizontal cross section of a cylinder. (b) For $\hat{u}_1$ of Eq. (17.16), the hot fluid (orange coloured) moves upward at A, and the cold fluid (blue coloured) downward at B. (c) For $\hat{u}_2$, the fluid ascend and descend at two locations each.

For turbulent RBC, one may expect the phases δ_k to be random. But this is not the case. Both amplitude and phase of the large-scale modes exhibit very interesting dynamics. Here we present the behaviour of the modes $\hat{u}_1$ and $\hat{u}_2$ that yields valuable insights into the reversal dynamics in a cylinder.

The flow appears to be random, yet, on an average, a hot plume ascends from one side in a slab and a cold plume descends from the other side of the slab. This slab forms the primary convective roll. On a horizontal cross section near the middle of the cylinder, the upward movement of the hot fluid and downward movements of the cold fluid appear as in Fig. 17.8(b), which corresponds to the Fourier mode $\hat{u}_1$. The orientation of the roll is described by δ_1.

In experiments and numerical simulations we observe that the orientation of the primary roll changes abruptly, and sometimes $\delta_1 \sim 180°$. Changes in δ_1, specially by 180° lead to flow reversals in a cylinder. Such reversals are classified in two categories—*rotation-led reversals* and *cessation-led reversals*, which will described below.

17.5.1   *Rotation-led reversals*

We describe the properties of Rotation-led reversals using the modes $\hat{u}_1$ and $\hat{u}_2$. Figure 17.9 exhibits a generic time series of $|\hat{u}_1|$, $|\hat{u}_2|$, and the phase δ_1 during a roll reorientation. The time series shows that $|\hat{u}_1|$ fluctuates around a mean value, but $|\hat{u}_2| \ll |\hat{u}_1|$, thus the flow is dominated by a primary roll structure [Mishra *et al.* (2011)]. The phase δ_1 (or the orientation of the roll) remains steady for a while, and then it jumps abruptly. This leads to a rotation-led reorientation of the roll. The flow would appear to have reversed when $\delta_1 \approx \pi$. Note that the reorientations in a cylinder, in a cylindrical annulus, and in a 2D-periodic RBC discussed in the previous section has a very similar behaviour.

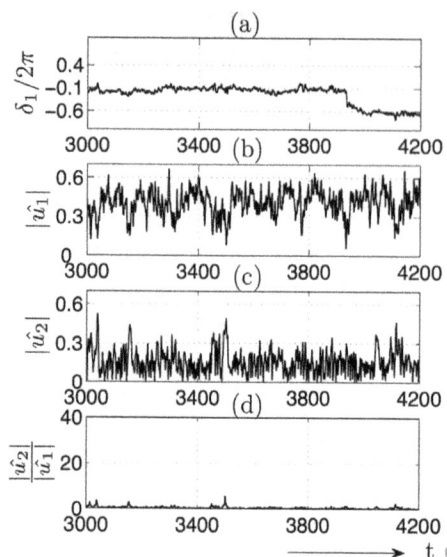

Fig. 17.9   For rotation-led reversal, time series of (a) phase δ_1, (b) magnitude $|\hat{u}_1|$, (c) magnitude $|\hat{u}_2|$, and (d) $|\hat{u}_2/\hat{u}_1|$. Reversal occurs near $t = 3900$. From Mishra *et al.* (2011). Reprinted with permission from Cambridge University Press.

Since the rotation-led reversals occurs due to a continuous rotation of the primary roll, their properties are to be described using continuous groups. This is contrary to the flow reversals in a closed Cartesian geometry that are described by discrete groups. Till date, there has not been any detailed study on the continuous group structure of the flow reversals.

Brown and Ahlers (2006) studied the probability distribution function (PDF) of the interval between two consecutive reorientations. They argue this to be a power law:

$$P(\Delta t) \approx (\Delta t)^{-\alpha}, \qquad (17.17)$$

where α is a positive constant. This is reasonable since the two consecutive reorientations are temporally correlated, similar to those in earthquakes [Verma *et al.* (2006b)]. The time interval between two consecutive flow reversals does not have a time scale, hence power law distribution $P(\Delta t)$ of Eq. (17.17) is expected [Verma *et al.* (2006b)].

17.5.2   *Cessation-led reversals*

In a cylinder we observe another class of reversals, called cessation-led reversals. In Fig. 17.10, we illustrate a cessation-led reversal using the time series of $|\hat{u}_1|$, $|\hat{u}_2|$, and δ_1. During a reversal, near $t \approx 1175$, $|\hat{u}_1| \to 0$, while $|\hat{u}_2|$ shows a slight increase. As a result, the ratio $|\hat{u}_2|/|\hat{u}_1|$ shows a sharp spike. Note that $|\hat{u}_1|$ takes

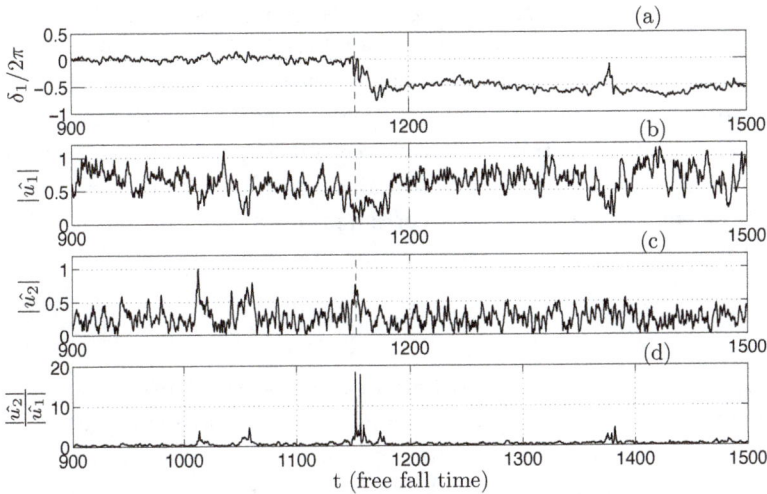

Fig. 17.10 For cessation-led reversal, time series of (a) phase δ_1, (b) magnitude $|\hat{u}_1|$, (c) magnitude $|\hat{u}_2|$, and (d) $|\hat{u}_2/\hat{u}_1|$. Reversal occurs near $t = 1175$ (denoted by vertical dashed marker). There is a failed reversal near $t = 1400$. From Mishra *et al.* (2011). Reprinted with permission from Cambridge University Press.

finite values before and after the reversal. The phase δ_1 before and after the reversal are very different.

In Fig. 17.11 we illustrate the cross-sectional flow profiles during a cessation-led reversal. We start with a primary roll structure of Fig. 17.11(a). During a reversal, the primary roll disappears in preference to two short-lived secondary rolls, as shown in Fig. 17.11(b). Here $|\hat{u}_1| \to 0$, and $|\hat{u}_2|$ increases significantly. The secondary roll $\hat{u}_2$ however vanishes quickly, and the primary roll $\hat{u}_1$ is reborn, as in Fig. 17.11(c). The orientation of the new primary roll however is uncorrelated to the previous one.

As shown in Fig. 17.10, during a reversal $|\hat{u}_1| \to 0$, but the secondary modes, primarily $|\hat{u}_2|$, grows in time. Thus, only the primary roll disappears, not the overall

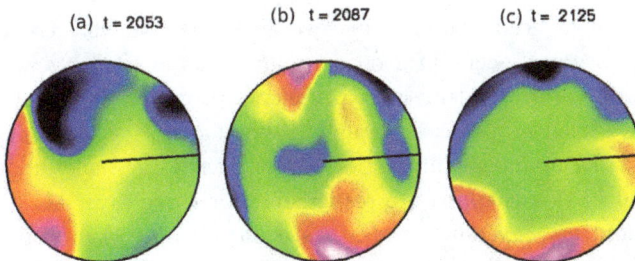

Fig. 17.11 During a cessation-led reversal, flow profiles on a horizontal cross section (a) before the reversal, (b) during the reversal, and (c) after the reversal. Adopted from a figure of Mishra *et al.* (2011).

flow. Hence, the cessation word of the phrase "cessation-led reversals" is meant only for the primary roll. Also, in an experimental study, Xi *et al.* (2016) showed that the higher modes—$\hat{u}_3$ and $\hat{u}_4$—are negligible during normal convection, but they too become significant during a cessation-led reversal. Thus, the higher secondary modes too play significant role in the flow reversals in a cylinder, as in 2D geometry.

The new primary roll does not have any memory of its earlier manifestation. Due to this memoryless reorientations of the primary rolls, the interval distribution of consecutive reversals is Poissonian [Verma *et al.* (2006b)], i.e.,

$$P(\Delta t) \approx \exp\left(-\frac{\Delta t}{\tau_{\text{avg}}}\right), \tag{17.18}$$

where τ_{avg} is the average interval between two consecutive reversals.

As described in Sec. 17.2, during a flow reversal in a square, the primary roll $(1,1)$ vanishes, while the secondary roll $(2,2)$ increases in magnitude. The primary roll $(1,1)$ is reborn after the reversal, but with an opposite sign. Thus, there are strong similarities between the cessation-led flow reversals in a cylinder and the flow reversals in a square. The cessation-led reversals are related to the discrete group structure, which was described in Secs. 17.2.3 and 17.3.

The flow reversals in a cylindrical geometry have several more important properties that are described below:

(1) There are failed reversals in a cylindrical geometry. For example, near $t = 1400$ of Fig. 17.10. Here $|\hat{u}_1|$ decreases significantly, but it recovers to its original amplitude. Thus flow does not reverse.
(2) The properties of flow reversals depend on Prandtl number and aspect ratio.
(3) The reversals would be hindered in an inclined cylinder with relative to **g** and in elliptic cylinder. If the eccentricity of the ellipse is significant, then the flow may show similarities with those in 2D Cartesian geometry discussed in Sec. 17.2.

In the next section we will describe reversals in dynamo and in Kolmogorov flow.

17.6   Reversals in dynamo and Kolmogorov flow

Field reversals have been observed in several other systems. For example, the magnetic field reversals occur in geo and solar dynamo [Roberts and Glatzmaier (2000)], and in several liquid metal experiments [Berhanu *et al.* (2007)]. Several mechanism have been proposed to explain a dynamo reversal. Gallet *et al.* (2012) emphasize that the nonlinear interactions among the large-scale modes, similar to those described in this chapter, are responsible for these reversals. They identify the reversing and non-reversing modes based on mirror and rotation symmetries; the symmetry arguments of Gallet *et al.* (2012) are similar to the group-theoretic arguments presented in Secs. 17.2 and 17.3. However, the algebraic structure of the Klein group provides a simpler and more general classification of the reversing and non-reversing modes. Another remark is in order. Astrophysical dynamos are spherical, unlike majority of convection experiments that are in Cartesian and cylindrical geometries.

When a strong external magnetic field is employed to liquid metals, the flow becomes quasi-2D [Verma (2017)]. When forced at the intermediate length scales, this flow, called *Kolmogorov flow*, exhibits interesting features—patterns, chaos, reversals, and condensate state. As shown by Mishra *et al.* (2015), the properties of reversals in Kolmogorov flow are quite similar to those in 2D RBC. For weak forcing and strong nonlinearity, Mishra *et al.* (2015) observed large-scale structures, called *condensate*, that do not reverse. This non-reversal of condensate is essentially due to the relative weakening of the secondary modes compared to the primary mode, a feature similar to that observed in 2D RBC in which the ratio $|u_y(2,2)/u_y(1,1)|$ becomes weak for large Ra (see Sec. 17.2.1).

At the outset, the flow restructuring due to the interactions of large-scale modes appear generic, and one would expect them to occur in nonlinear optics, atmospheric flows, and other nonlinear systems. Gallet *et al.* (2012) argued that the quasi-biennial oscillation (QBO) in Earth's atmosphere may arise due to such interactions.

In the next section we will discuss whether flow reversals are low-dimensional or stochastic.

17.7 Flow reversals: Low-dimensional or stochastic?

As shown in Sec. 17.2, large-scale Fourier modes play an important role in the reversal dynamics. Motivated by such observations, researchers have constructed low-dimensional models with large-scale modes as variables. Lorenz equations, discussed in Sec. 7.1, is one such model with $\hat{u}(1,1)$, $\hat{\theta}(1,1)$, $\hat{\theta}(0,2)$ as dynamical variables. Though Lorenz model captures features of RBC near the onset of convection, it falls short in capturing other features of RBC. One reason for this failure is absence of important modes such as $(2,2)$, $(1,3)$, $(3,1)$ in the Lorenz model. Earlier, Araujo and Grossmann (2005) constructed a model containing three Fourier modes, while Podvin and Sergent (2015) constructed a model with the three most energetic POD modes. Such models, called *low-dimensional models*, have had limited success in explaining several features of flow reversals.

The discussion of Sec. 17.2 deals with large-scale modes only. The role of inertial-range Fourier modes in flow reversals is still uncertain. Some researchers argue that the modes in the inertial range provide the necessary stochasticity to the reversals. For example, Benzi (2005) explained flow reversals using a shell model whose first mode captures the LSC, while the higher modes provide stochasticity to the LSC reversals.

Several researchers argue that the dynamics of two stable states of LSC (positive and negative) is similar to the dynamics of a bistable system. The system noise induces jumps from one stable to the other state resulting in a flow reversal. Sreenivasan *et al.* (2002) attempted to connect the flow reversal to noisy over-damped bistable oscillator. Researcher have also attempted to relate flow reversals to self-organised criticality and stochastic resonance.

Mannattil *et al.* (2017) examined whether the flow reversals in RBC are low-dimensional or stochastic. Using tools of nonlinear dynamics [Strogatz (2014)], they analysed the reversal time series of a 3D RBC with $Pr = 0$, and of 2D RBC with $Pr = \infty$, and showed that zero-Pr RBC is low-dimensional, while infinite-Pr RBC is stochastic. Hence, to reconstruct the phase-space attractor from a numerical time series, it may be possible to make a low-dimensional model for $Pr = 0$, but not for $Pr = \infty$. Note however that a carefully constructed low-dimensional model for RBC with $Pr = \infty$ could capture some important features of the flow reversal, but this model cannot reconstruct the phase space attractor.

Numerical simulations of the low-dimensional models are much less expensive than the DNS, hence they are very valuable for predicting system behaviour. Therefore, accurate low-dimensional models of flow reversals and dynamo reversals are very valuable to the community. Though several presently-available models capture several features of the above phenomena, we can safely say that none of the present models capture all the features of flow reversals, e.g., Prandtl and Rayleigh number dependence of 2D flow reversals, time scale of reversals, etc.

So far we have covered a large number of topics on reversals. We summarise them in the next section.

17.8   Summary of flow reversal studies

Flow reversals in a Cartesian box are primarily caused by the nonlinear interactions among the large-scale Fourier modes. Some of these modes switch sign during a reversal. The reversing and non-reversing modes can be identified using their group structures. During a reversal, the heat transport or the Nusselt number exhibits strong fluctuations.

In a cylinder, one class of flow reversals have behaviour similar to those in Cartesian box. Such reversals are called cessation-led reversal. However, the other class of reversals, called rotation-led reversals, are due to azimuthal rotation of the convection rolls, and they are governed by continuous symmetry.

The properties of flow reversals depend quite crucially on the Prandtl and Rayleigh numbers, box geometry, aspect ratio, and boundary condition. In fact, the nature of reversals, whether it is low-dimensional or stochastic depends on the aforementioned parameters [Mannattil *et al.* (2017)]. To some degree we understand the patterns of reversals under the variation of the above parameters, but much more work is required for definitive answers. Also, the flow reversals would be suppressed in elliptic and inclined cylinders.

So far we confined our discussion to LSC and flow reversals. In the last two sections of the chapter, we make some general observations on flow structures in turbulence with or without walls, and in large-Pr RBC.

17.9 Structures in turbulence with and without walls

Structures are observed in turbulent flows with and without walls. Hydrodynamic turbulence without walls (free turbulence) appear to be isotropic and homogeneous, yet it generates hierarchical flow structures, e.g., vortex within vortex. Turbulent flows with walls, as in RBC and channel flow, too exhibit mutliscale structures. Comparatively, walled turbulence have more pronounced structures than free turbulence. For example, well pronounced LSCs are observed in RBC with walls, but not in RBC with periodic boundary condition. As a consequence, real space probes embedded in a periodic box exhibits random $\mathbf{u}(\mathbf{r})$, contrary to flow reversals $(u_z(\mathbf{r}) \to -u_z(\mathbf{r}))$ in RBC in a closed box.

LSC in a 2D closed box of Sec. 17.2 is more structured than the ones in cylindrical annulus or in cylinder. A Cartesian box with periodicity in all directions does not have a very well defined LSC. Thus, inclusion of walls makes the flow more structured. The walled direction is governed by discrete symmetry, while the periodic direction by continuous symmetry. Therefore, the walled structures are less symmetric than those in periodic geometry.

The energy cascade rates or energy flux also shows interesting differences between the free and walled turbulence. Some Fourier modes are absent in flows with walls. For example, in 2D RBC, $\hat{\mathbf{u}}(0, n) = 0; \hat{\mathbf{u}}(n, 0) = 0$, hence an important nonlinear triad $\{(0, 1), (1, 0), (1, 1)\}$ of a homogeneous isotropic turbulence is absent in 2D RBC. Consequently, several channels of nonlinear interactions and energy cascades are blocked in walled turbulence. This feature is one of the reasons for the suppression of nonlinearity and energy flux in RBC compared to free turbulence (see Secs. 11.2 and 11.5).

Coherent structures are of major interest to RBC community. We believe that more refined experiments and simulations at very high Ra will provide further insights into the dynamics of convective structures. For example, an understanding of the RBC structures in the ultimate regime would be valuable for modelling solar convection that has very large Ra.

Before we close this chapter, we revisit large-Pr RBC because they exhibit interesting set of patterns.

17.10 Two-dimensionalization of large-Pr RBC

For large Rayleigh numbers, thermal convection is three-dimensional, except when $\mathrm{Pr} \gg 1$. Pandey *et al.* (2014) solved RBC equations for $\mathrm{Pr} = 100, 1000$, and ∞ with $L_x = L_y = 2\sqrt{2}$ and $L_z = 1$. Figure 17.12 illustrates temperature isosurfaces for these there cases. As shown in the figure, the xy symmetry is reduced as Pr is increased. The flow is quasi-2D for $\mathrm{Pr} = \infty$.

The xy symmetry is broken for $\mathrm{Pr} = \infty$. Note that for $\mathrm{Pr} = \infty$, the nonlinear term of the momentum equation, $\mathbf{N}_u = 0$. Hence, the secondary modes are generated and driven by the nonlinear term of the temperature equation. Due to the

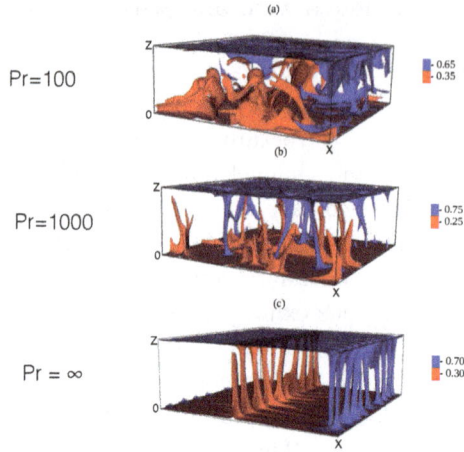

Fig. 17.12 Temperature isosurfaces for (a) Pr = 100 and Ra = 10^7, (b) Pr = 1000 and Ra = 6×10^6, and (c) Pr = ∞ and Ra = 6.6×10^6. Here the red and blue colours represent the hot and cold plumes. From Pandey *et al.* (2014). Reprinted with permission from APS.

intense dissipation, the system possibly finds it efficient to transfer energy either in xz plane or in yz plane and maintain the steady state.

Thus, for Pr = ∞, the system gets locked into one of the above configurations due to very large potential barrier for a crossover. This is a unique symmetry breaking in which a system parameter, here Pr, breaks the xy symmetry in a rectangular geometry with $L_x = L_y$. Typical symmetry breaking discussed in literature involves either external field (e.g., external magnetic field for Ising spin) or asymmetric geometry (e.g., $L_x \neq L_y$). It is important to analyse the bifurcation diagram and energy transfers for this scenario.

To get further insights into quasi-2D nature of large-Pr RBC, Pandey *et al.* (2016b) performed numerical simulations of RBC in 2D and 3D rectangular boxes and analysed the dominant Fourier modes. The ten most dominant $\theta(\mathbf{k})$ modes in both 2D and 3D RBC are $(0, 0, 2)$, $(0, 0, 4)$, $(0, 0, 6)$, $(0, 0, 8)$, $(0, 0, 10)$, $(0, 0, 12)$, $(1, 0, 1)$, $(3, 0, 1)$, $(1, 0, 3)$, and $(3, 0, 3)$. Note that $k_y = 0$ for these modes thus affirming quasi-2D behaviour of the flow. The origin of the modes $(0, 0, 2n)$ have been discussed in Chapter 9. In fact, they are most dominant for Pr = ∞. The primary mode $(1, 0, 1)$ interacts with $(0, 0, 2)$ to yield $(1, 0, 3)$. Then, $(1, 0, -1) \oplus (1, 0, 3) = (2, 0, 2)$ and $(2, 0, -2) \oplus (1, 0, 3) = (3, 0, 1)$.

Also note that the plumes get sharper with the increase of Prandtl number. One way to estimate the finger width, δ, is by setting Ra$_\delta$ = 1 and $\alpha = 1/T_m$, where T_m is the mean temperature, or [Turner (2009); Yoshida and Nagashima (2003); Sadhukhan *et al.* (2017)]

$$\text{Ra}_\delta = \frac{g\delta^4 d\bar{T}/dz}{T_m \nu \kappa} \sim 1. \tag{17.19}$$

Hence,

$$\delta \sim \left[\frac{T_m \nu \kappa}{g d \bar{T}/dz} \right]^{1/4} \sim d\text{Ra}^{-1/4}. \tag{17.20}$$

In this formula, δ scales as the width of the viscous boundary layer.

The formula of Eq. (17.20) appears to have a flaw. Numerical simulations reveal that the fingers become thinner with the increase of Pr. For a given Ra, low-Pr fluids is expected to have wider fingers compared to large-Pr fluid. Therefore, the thickness of the thermal boundary layer, $d/(2\text{Nu})$, or the diffusion length scale κ/U, or their combinations, could be other possible candidates for the finger width. We believe that more careful scaling and numerical analysis is required to settle this issue.

With this, we conclude our discussion on flow structures in turbulent RBC.

Further reading

Detailed study of large-scale circulation was first performed by Niemela *et al.* (2000) and Niemela and Sreenivasan (2003). For details on flow reversals in cylindrical geometry, refer to Niemela and Sreenivasan (2003), Xi *et al.* (2006), and Brown and Ahlers (2006). Refer to Sugiyama *et al.* (2010), Chandra and Verma (2011), and Chandra and Verma (2013) for reversals in 2D square, and to Gallet *et al.* (2012), Vasiliev *et al.* (2016), and Foroozani *et al.* (2017) for those in 3D geometry. Verma *et al.* (2015) studied reversals in 2D box with free-slip boundary condition at all the walls, and Paul *et al.* (2010) studied in 2D box with periodic boundary condition along the horizontal direction. Mannattil *et al.* (2017) examined applicability of low-dimensional models to convective flow reversals. Gallet *et al.* (2012) and references therein cover magnetic field reversals in dynamo.

Exercises

(1) Consider RBC experiments in a 2D square and in a 3D cube. What are the known differences between the properties of flow reversals in these two systems?
(2) Contrast the properties of flow reversals in a cube, a cylinder, and a sphere.
(3) What are the known theories on the probability distribution of interval distribution between two consecutive flow reversals?
(4) Do you expect flow reversals in a RBC setup with periodic boundary condition along all the directions?

PART 3

Buoyant Flows with Rotation and Magnetic Field

This part covers

(1) Buoyant and rotating flows
(2) Buoyant flows under magnetic field
(3) Buoyant and rotating flows under magnetic field
(4) Double diffusive systems

Chapter 18

Rotating Flows

Before we embark on our discussion on rotating RBC and rotating stratified flows, it is prudent to understand the basic properties of rotating flows. Here we will focus on the linear regime, and introduce Taylor-Proudman theorem and inertial waves. More complex topics such as rotating turbulence are beyond the scope of this book. Reader is referred to textbooks, e.g., Chandrasekhar (2013), Greenspan (1968) and Davidson (2013) for a detailed discussion.

Without loss of generality, we take the axis of rotation to be along $\hat{z}$. For simplicity, we assume the rotation frequency Ω to be a constant in space and time, and it to be aligned along $\hat{z}$. Hence,

$$\mathbf{\Omega} = \Omega\hat{z}. \tag{18.1}$$

We will work out the solution in rotating reference frame in which the equations of motion for an incompressible fluid are

$$\frac{\partial \mathbf{u}}{\partial t} + (\mathbf{u} \cdot \nabla)\mathbf{u} = -\nabla\sigma - 2\mathbf{\Omega} \times \mathbf{u} + \nu\nabla^2\mathbf{u}, \tag{18.2}$$

$$\nabla \cdot \mathbf{u} = 0, \tag{18.3}$$

where $-2\mathbf{\Omega} \times \mathbf{u}$ is the Coriolis acceleration. The centrifugal force has been absorbed in the pressure.

In the next section we describe Taylor-Proudman theorem.

18.1 Taylor-Proudman theorem

Under linear and inviscid limit, Eq. (18.2) becomes

$$\frac{\partial \mathbf{u}}{\partial t} = -\nabla\sigma - 2\mathbf{\Omega} \times \mathbf{u}. \tag{18.4}$$

For slow and steady motion, $\partial \mathbf{u}/\partial t \approx 0$. In this limit, taking curl of Eq. (18.4) yields

$$\nabla \times (\mathbf{\Omega} \times \mathbf{u}) = 0. \tag{18.5}$$

Under the assumption of $\mathbf{\Omega} = $ constant and $\nabla \times \mathbf{u} = 0$, the above equation becomes

$$\mathbf{\Omega} \cdot \nabla\mathbf{u} = \Omega\frac{\partial}{\partial z}\mathbf{u} = 0, \tag{18.6}$$

which implies that the velocity field is invariant along $\hat{z}$. As a consequence, rotating flows have strong vortical structures that do not vary along $\hat{z}$. This is the *Taylor-Proudman theorem* [Chandrasekhar (2013)]

Taylor-Proudman theorem describes time-independent solution of Eq. (18.4). In the linear limit, the time-dependent equation (18.2) admits waves that are called *inertial waves*, which will be described in the next section.

18.2 Inertial wave

In Fourier space, the linearised version of Eqs. (18.2, 18.3) yields [Chandrasekhar (2013)]

$$\frac{d}{dt}\mathbf{u}(\mathbf{k}) = -ik\sigma(\mathbf{k}) - 2\mathbf{\Omega} \times \mathbf{u}(\mathbf{k}),$$ (18.7)

$$\mathbf{k} \cdot \mathbf{u}(\mathbf{k}) = 0.$$ (18.8)

It is convenient to use Craya-Herring decomposition for deriving inertial waves. Refer to Figs. 3.5 and 3.6 for illustrations of Craya-Herring basis. The Coriolis force in this basis is

$$\mathbf{F}(\mathbf{k}) = -2\mathbf{\Omega} \times \mathbf{u}(\mathbf{k})$$
$$= -2\Omega\{\hat{e}_3 \cos\zeta - \hat{e}_2 \sin\zeta\} \times \{u_1\hat{e}_1 + u_2\hat{e}_2\}$$
$$= 2\Omega\{\hat{e}_1 u_2 \cos\zeta - \hat{e}_2 u_1 \cos\zeta - \hat{e}_3 u_1 \sin\zeta\},$$ (18.9)

where ζ is the angle between $\mathbf{k}$ and $\hat{z}$. Hence the dynamical equations along $\hat{e}_1, \hat{e}_2$ are

$$\dot{u}_1(\mathbf{k}) = \omega u_2(\mathbf{k}),$$ (18.10)
$$\dot{u}_2(\mathbf{k}) = -\omega u_1(\mathbf{k}),$$ (18.11)

where

$$\omega = 2\Omega \cos\zeta.$$ (18.12)

Along $\hat{e}_3$, the equation is

$$-ik\sigma(\mathbf{k}) - 2\Omega u_1(\mathbf{k}) \sin\zeta = 0$$ (18.13)

that yields the pressure σ as

$$\sigma(\mathbf{k}) = \frac{i}{k} 2\Omega u_1(\mathbf{k}) \sin\zeta.$$ (18.14)

The equations for u_1 and u_2 are coupled, but they can be decoupled in the helical basis (see Sec. 3.9), in which the equations are

$$\dot{u}_+(\mathbf{k}) = \frac{1}{2}(\dot{u}_2(\mathbf{k}) + i\dot{u}_1(\mathbf{k}))$$ (18.15)
$$= \frac{1}{2}[-\omega u_1(\mathbf{k}) + i\omega u_2(\mathbf{k})]$$ (18.16)
$$= i\omega u_+(\mathbf{k}),$$ (18.17)

and

$$\dot{u}_-(\mathbf{k}) = -i\omega u_-(\mathbf{k}).$$ (18.18)

The above equations admit wave solutions:

$$u_+(\mathbf{k}, t) = u_+(\mathbf{k}, 0)\exp(i\omega t),$$ (18.19)
$$u_-(\mathbf{k}, t) = u_-(\mathbf{k}, 0)\exp(-i\omega t)$$ (18.20)

with Eq. (18.12) as the dispersion relation. Note that $\mathbf{u}_+$ mode has maximal helicity with $H_K(k)/(kE(k)) = 1$, while $\mathbf{u}_-$ mode has minimum helicity with $H_K(k)/(kE(k)) = -1$ (see Sec. 3.9). These waves are called *inertial waves* [Chandrasekhar (2013)].

In real space, following similar algebra as in Sec. 3.9, we derive that real-space velocity vector $\mathbf{u}_-(\mathbf{r}, t)$ as

$$\begin{aligned}
\mathbf{u}_-(\mathbf{r}, t) &= \Re[\hat{e}_- u_- \exp i(\mathbf{k} \cdot \mathbf{r} - \omega t)] \\
&= |u_-|\Re[(\hat{e}_2 + i\hat{e}_1)\exp i(kz' - \omega t + \phi_{k-})] \\
&= |u_-|\begin{pmatrix} -\sin(kz' - \omega t + \phi_{k-}) \\ \cos(kz' - \omega t + \phi_{k-}) \end{pmatrix},
\end{aligned}$$ (18.21)

where ω is the frequency of the wave, $u_- = |u_-|\exp(i\phi_{k-})$ with ϕ_{k-} as the phase of the wave, and z' is the coordinate along $\hat{e}_3$. This wave is exactly same as the helical wave $\mathbf{u}_-$ of Sec. 3.10 (see Eq. (3.127)). Hence, the inertial helical wave $\mathbf{u}_-(\mathbf{r}, t)$ given above is right-handed with clockwise or left circular polarization (LCP). See Fig. 3.11 for an illustration.

Analogously, we can work out the motion of $\mathbf{u}_+$ wave:

$$\begin{aligned}
\mathbf{u}_+(\mathbf{r}, t) &= \Re[\hat{e}_+ u_+ \exp i(\mathbf{k} \cdot \mathbf{r} + \omega t)] \\
&= |u_+|\Re[(\hat{e}_2 - i\hat{e}_1)\exp i(kz' + \omega t + \phi_{k+})] \\
&= |u_+|\begin{pmatrix} \sin(kz' + \omega t + \phi_{k+}) \\ \cos(kz' + \omega t + \phi_{k+}) \end{pmatrix},
\end{aligned}$$ (18.22)

where $u_+ = |u_+|\exp(i\phi_{k+})$ with ϕ_{k+} as the phase of the wave. This wave appears to be same as $\mathbf{u}_+$ helical wave of Sec. 3.10. There is however a crucial difference. The wave of Eq. (18.22) is travelling along $-\mathbf{k}$ direction.

A common feature between the above $\mathbf{u}_+$ wave and that of Sec. 3.10 is that the frozen-in configurations of both the fields is left-handed. But temporal evolution of $\mathbf{u}_+$ of Eq. (18.22) for a given z', say at $z' = 0$, is

$$\mathbf{u}_+(z' = 0, t) = \begin{pmatrix} \sin(\omega t + \phi_{k+}) \\ \cos(\omega t + \phi_{k+}) \end{pmatrix}.$$ (18.23)

Therefore, the tip of the velocity vector rotates in the clockwise direction when viewed from $\hat{z}'$ direction. Note however that the wave is travelling along $-\mathbf{k}$ or $-\hat{z}'$ direction, hence we should report polarisation while viewing from $-\mathbf{k}$ or $-\hat{z}'$ direction. In this direction, the tip of $\mathbf{u}_+(t)$ rotates in the anti-clockwise direction, similar to Fig. 3.10(c). Therefore, the inertial helical wave $\mathbf{u}_+(\mathbf{r}, t)$ of Eq. (18.22) is left-handed with anti-clockwise or right circular polarisation (RCP). See Fig. 3.10 for an illustration.

The phase velocity of the wave $\mathbf{u}_+(\mathbf{r}, t)$ is $\mathbf{c}_+ = -(\omega/k)\hat{k}$, while that of $\mathbf{u}_-(\mathbf{r}, t)$ is $\mathbf{c}_- = (\omega/k)\hat{k}$. The group velocities of the waves are (see Appendix B)

$$\mathbf{c}_{g\pm} = \mp \nabla_\mathbf{k}\omega = \mp \hat{\zeta}\frac{1}{k}\frac{\partial\omega}{\partial\zeta} = \pm\hat{\zeta}\frac{2\Omega}{k}\sin\zeta, \qquad (18.24)$$

where ζ, the polar angle, is the angle between $\mathbf{k}$ and $\hat{z}$, and $\hat{\zeta}$ is the associated unit vector. Note that for the present configuration, $\hat{\zeta}$ is along the Craya-Herring basis vector $\hat{e}_2$. Hence, for both $\mathbf{u}_\pm(\mathbf{r}, t)$ waves, the phase and group velocities are perpendicular to each other as shown in Fig. 18.1. Also, for the waves travelling along $\mathbf{k}$ (away from the source), the wave energy propagates towards the polar regions. On the contrary, for the waves travelling towards the source, the wave energy propagates towards the equator. It is easy to show that

$$\mathbf{c}_\pm + \mathbf{c}_{g\pm} = \mp \frac{2\Omega}{k}\hat{z}, \qquad (18.25)$$

which is a constant vector.

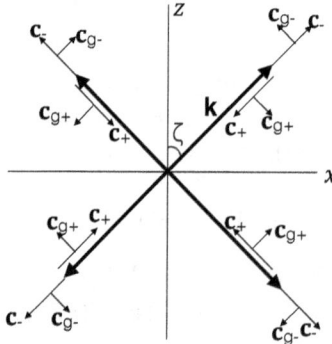

Fig. 18.1   A wave travelling along $\mathbf{k}$ has phase velocity $\mathbf{c}_-$ and group velocity $\mathbf{c}_{g-}$, while a wave travelling along $-\mathbf{k}$ has phase velocity $\mathbf{c}_+$ and group velocity $\mathbf{c}_{g+}$.

We also remark that Eqs. (18.10, 18.11) could be written in matrix form as

$$\frac{d}{dt}\begin{pmatrix} u_1(\mathbf{k}) \\ u_2(\mathbf{k}) \end{pmatrix} = \begin{pmatrix} 0 & \omega \\ -\omega & 0 \end{pmatrix}\begin{pmatrix} u_1(\mathbf{k}) \\ u_2(\mathbf{k}) \end{pmatrix} = A\begin{pmatrix} u_1(\mathbf{k}) \\ u_2(\mathbf{k}) \end{pmatrix} \qquad (18.26)$$

that yield the same solution as before. The eigenvalues are $\pm i\omega$, and the corresponding eigenvectors are

$$\begin{pmatrix} -i \\ 1 \end{pmatrix} u_+(\mathbf{k}); \quad \begin{pmatrix} i \\ 1 \end{pmatrix} u_-(\mathbf{k}). \qquad (18.27)$$

These solution are same as $\mathbf{u}_\pm(\mathbf{k})$ discussed above.

Let us consider waves close to the $k_x k_y$ plane for which $\zeta \to \pi/2$. Such waves have $\omega \to 0$, hence they evolve very slowly, or

$$\mathbf{u}(\mathbf{k}, t) = u_1(\mathbf{k})\hat{e}_1(\mathbf{k}) + u_2(\mathbf{k})\hat{e}_2(\mathbf{k}) \qquad (18.28)$$

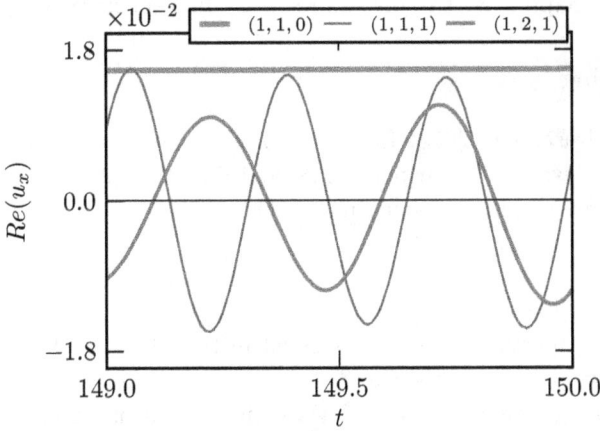

Fig. 18.2 For a numerical simulation of strongly rotating decaying turbulence with $\Omega = 16$, Ro $= U/(\Omega d) \approx 0.003$, and Re ≈ 1900, time series of the large-scale modes $\mathbf{u}(1,1,0)$, $\mathbf{u}(1,1,1)$, and $\mathbf{u}(1,2,1)$ in the asymptotic regime when the flow decays very slowly.

is nearly constant in time with $u_1 \gg u_2$. This feature may be the cause for the near stationary asymptotic flow patterns in rotating turbulence. In the other limiting case—$\zeta \to 0, \pi$, the dispersion relation yields waves with $\omega \approx \pm 2\Omega$, maximal phase velocity, and vanishing group velocity. Hence, such waves exhibit pure oscillations with maximum phase velocity, but with no energy propagation. These types of waves, called *near-inertial oscillations*, are important in oceanic flows.

We performed a numerical simulation of strongly rotating and decaying turbulence with $\Omega = 16$, Ro $= U/(\Omega d) \approx 0.003$, and Re ≈ 1900. In the asymptotic regime, as shown in Fig. 18.2, the time series of the large-scale modes $\mathbf{u}(1,1,1)$ and $\mathbf{u}(1,2,1)$ exhibit oscillatory behaviour with frequencies given by the dispersion relation of Eq. (18.12). The mode $\mathbf{u}(1,1,0)$, the most dominant one of the flow, is constant in time consistent with the fact that $\omega = 0$ for this mode. This is the reason for nearly stationary large-scale flow patterns observed in our simulation. The aforementioned wave nature observed in the turbulent regime occurs due to the dominance of the Coriolis force over the nonlinear term (note that Ro ≈ 0.003).

Davidson (2014) and Ranjan (2017) studied the inertial waves in the context of Earth's outer core and concluded dynamic separation of $\mathbf{u}_+$ and $\mathbf{u}_-$. They (earlier, Chandrasekhar (2013) and Greenspan (1968)) observed that the energy of these waves travel along $\pm\hat{z}$ because the group velocity of such waves is parallel to $\pm\hat{z}$. The inertial waves have been observed in experiments, e.g. by Messio *et al.* (2007). For details, refer to Greenspan (1968) and Davidson (2013).

We can construct standing inertial waves by appropriate superposition of $\mathbf{u}_+(\mathbf{r}, t)$ and $\mathbf{u}_-(\mathbf{r}, t)$. For example, if we take $|u_+| = |u_-|$ and $\phi_{k+} = \phi_{k-} = 0$, the net velocity is

$$\mathbf{u}(\mathbf{r}, t) = |u_+|(\hat{e}_1 \cos kz' \sin \omega t + \hat{e}_2 \cos kz' \cos \omega t). \qquad (18.29)$$

In the next chapter, we will discuss rotating buoyant flows.

Further reading

Refer to Chandrasekhar (2013) for an introduction on linear analysis of rotating flows and inertial waves. Greenspan (1968) and Davidson (2013) cover the nonlinear treatment, as well as recent results in the field.

Exercises

(1) Consider the inertial wave $\mathbf{u}_+$ discussed in this chapter. What are the phase and group velocities of this wave?
(2) A point source generates inertial waves in a rotating cylinder. Describe the properties of the generated waves.

Chapter 19

Buoyant and Rotating Flows

In this chapter we discuss the effects of rotation on buoyancy-driven flows. First we will describe rotating stably stratified flow, and then rotating convection.

19.1 Stably stratified flow with rotation

We start with Eqs. (2.38, 2.39), add Coriolis force to it, and then drop the nonlinear, viscous, and diffusion terms. As a result we obtain the following linear equations:

$$\frac{\partial \mathbf{u}}{\partial t} = -\nabla \sigma - Nb\hat{z} - 2\boldsymbol{\Omega} \times \mathbf{u}, \tag{19.1}$$

$$\frac{\partial b}{\partial t} = Nu_z, \tag{19.2}$$

$$\nabla \cdot \mathbf{u} = 0. \tag{19.3}$$

Fourier transformation of the above equations yields

$$\frac{d}{dt}\mathbf{u}(\mathbf{k}) = -ik\sigma(\mathbf{k}) - Nb(\mathbf{k})\hat{z} - 2\boldsymbol{\Omega} \times \mathbf{u}(\mathbf{k}), \tag{19.4}$$

$$\frac{d}{dt}b(\mathbf{k}) = Nu_z(\mathbf{k}), \tag{19.5}$$

$$\mathbf{k} \cdot \mathbf{u}(\mathbf{k}) = 0. \tag{19.6}$$

We employ Craya-Herring basis vectors in which the net external force (other than the pressure gradient) is

$$\begin{aligned}
\mathbf{F}(\mathbf{k}) &= -2\boldsymbol{\Omega} \times \mathbf{u}(\mathbf{k}) - Nb(\mathbf{k})\hat{z} \\
&= 2\Omega\{\hat{e}_1 u_2 \cos\zeta - \hat{e}_2 u_1 \cos\zeta - \hat{e}_3 u_1 \sin\zeta\} \\
&\quad + \hat{e}_2 Nb \sin\zeta - \hat{e}_3 Nb \cos\zeta
\end{aligned} \tag{19.7}$$

where ζ is the angle between $\mathbf{k}$ and $\hat{z}$. In this basis, the dynamical equations are

$$\dot{u}_1(\mathbf{k}) = \omega_1 u_2(\mathbf{k}), \tag{19.8}$$

$$\dot{u}_2(\mathbf{k}) = -\omega_1 u_1(\mathbf{k}) + \omega_2 b(\mathbf{k}), \tag{19.9}$$

$$\dot{b}(\mathbf{k}) = -\omega_2 u_2(\mathbf{k}) \tag{19.10}$$

where

$$\omega_1 = 2\Omega \cos\zeta; \quad \omega_2 = N \sin\zeta. \tag{19.11}$$

The above equations can be rewritten in a matrix form as

$$\frac{d}{dt}\begin{pmatrix} u_1(\mathbf{k}) \\ u_2(\mathbf{k}) \\ b(\mathbf{k}) \end{pmatrix} = \begin{pmatrix} 0 & \omega_1 & 0 \\ -\omega_1 & 0 & \omega_2 \\ 0 & -\omega_2 & 0 \end{pmatrix}\begin{pmatrix} u_1(\mathbf{k}) \\ u_2(\mathbf{k}) \\ b(\mathbf{k}) \end{pmatrix} = A\begin{pmatrix} u_1(\mathbf{k}) \\ u_2(\mathbf{k}) \\ b(\mathbf{k}) \end{pmatrix}. \tag{19.12}$$

The aforementioned matrix A has eigenvalues

$$\lambda_0 = 0, \tag{19.13}$$

$$\lambda_\pm = \pm i\sqrt{\omega_1^2 + \omega_2^2}. \tag{19.14}$$

The zero eigenvalue yields a stationary solution, and its corresponding eigenvector is

$$\begin{pmatrix} (\omega_2/\omega_1) \\ 0 \\ 1 \end{pmatrix}, \tag{19.15}$$

while the eigenvalues $\lambda_\pm$ have the eigenvectors

$$\begin{pmatrix} -(\omega_1/\omega_2) \\ -i(\sqrt{\omega_1^2 + \omega_2^2})/\omega_2 \\ 1 \end{pmatrix}; \quad \begin{pmatrix} -(\omega_1/\omega_2) \\ i(\sqrt{\omega_1^2 + \omega_2^2})/\omega_2 \\ 1 \end{pmatrix} \tag{19.16}$$

that correspond to waves with elliptic polarisation. See Sec. 3.9 for discussion on such flows. Note that we recover inertial and gravity waves for $\omega_2 = 0$ and $\omega_1 = 0$ respectively.

Thus, in the linear limit, rotating stably stratified flows admit wave solution. This is natural since the system is stable. In the next section we will discuss rotating convection.

19.2 Rotating convection

The setup for rotating convection is similar to that of Sec. 6.1. Here, we assume free-slip and conducting boundary conditions for the top and bottom thermal plates. To this system we add rotation along $\hat{z}$. In the rotating reference frame, the non-dimensional equations of this system are

$$\frac{\partial \mathbf{u}}{\partial t} = -\nabla \sigma + \mathrm{RaPr}\theta\hat{z} + \mathbf{F}_c + \mathrm{Pr}\nabla^2\mathbf{u}, \tag{19.17}$$

$$\frac{\partial \theta}{\partial t} = u_z + \nabla^2\theta. \tag{19.18}$$

Here κ/d is the velocity scale, d is the length scale, and the non-dimensional Coriolis force is

$$\mathbf{F}_c = -\frac{2\mathbf{\Omega} \times (\kappa/d)\mathbf{u}}{\kappa^2/d^3} = -\frac{2\Omega d^2}{\kappa}\hat{z} \times \mathbf{u} \tag{19.19}$$

$$= -\frac{\mathrm{Pr}}{E}\hat{z} \times \mathbf{u} \quad \text{or} \quad -\mathrm{Pr}\sqrt{\mathrm{Ta}}\hat{z} \times \mathbf{u}, \tag{19.20}$$

where

$$\text{Ta} = \frac{4\Omega^2 d^4}{\nu^2}, \tag{19.21}$$

$$E = \frac{\nu}{2\Omega d^2} \tag{19.22}$$

are Taylor number and Ekman number respectively. Ekman number is the ratio of the viscous force and the Coriolis force, and it equals $\text{Ta}^{-1/2}$. Another related quantity is the Rossby number, which is the ratio of the nonlinear term and the Coriolis force:

$$\text{Ro} = \frac{U}{\Omega d}. \tag{19.23}$$

It is easy to verify that

$$\text{Ro} = 2\text{Re}E. \tag{19.24}$$

In Fourier space, the non-dimensionalized equations of rotating RBC are

$$\frac{d}{dt}\mathbf{u}(\mathbf{k}) = -i\mathbf{k}\sigma(\mathbf{k}) + \text{RaPr}\theta(\mathbf{k})\hat{z} - (\text{Pr}\sqrt{\text{Ta}})\hat{z} \times \mathbf{u}(\mathbf{k}) + \text{Pr}k^2\mathbf{u}(\mathbf{k}), \tag{19.25}$$

$$\frac{d}{dt}\theta(\mathbf{k}) = u_z(\mathbf{k}) + k^2\theta(\mathbf{k}), \tag{19.26}$$

$$\mathbf{k} \cdot \mathbf{u}(\mathbf{k}) = 0. \tag{19.27}$$

In the Craya-Herring basis, the above equations in matrix form are

$$\frac{d}{dt}\begin{pmatrix} u_1(\mathbf{k}) \\ u_2(\mathbf{k}) \\ \theta(\mathbf{k}) \end{pmatrix} = A \begin{pmatrix} u_1(\mathbf{k}) \\ u_2(\mathbf{k}) \\ \theta(\mathbf{k}) \end{pmatrix}, \tag{19.28}$$

where

$$A = \begin{pmatrix} -\text{Pr}k^2 & \text{Pr}\sqrt{\text{Ta}}\cos\zeta & 0 \\ -\text{Pr}\sqrt{\text{Ta}}\cos\zeta & -\text{Pr}k^2 & -\text{RaPr}\sin\zeta \\ 0 & -\sin\zeta & -k^2 \end{pmatrix} \tag{19.29}$$

with ζ as the angle between $\mathbf{k}$ and $\hat{z}$, and $\sin\zeta = k_\perp/k$. The eigenvalues and eigenvectors of the matrix A are quite complex, hence, we only discuss the eigenvalues and eigenvectors for neutral or marginal instability, i.e., for $\lambda = 0$. The condition for the eigenvalue to be zero is

$$\text{Ra}_c k^2 \sin^2\zeta - \text{Ta}k^2 \cos^2\zeta - k^6 = 0, \tag{19.30}$$

where Ra_c is the critical Rayleigh number. From the above equation, we derive

$$\text{Ra}_c = \text{Ta}\cot^2\zeta + \frac{k^6}{k_\perp^2}$$

$$= \text{Ta}\left(\frac{k_z}{k_\perp}\right)^2 + \frac{k^6}{k_\perp^2}$$

$$= \text{Ta}\left(\frac{n\pi}{k_\perp}\right)^2 + \frac{(n^2\pi^2 + k_\perp^2)^3}{k_\perp^2} \tag{19.31}$$

$$= \frac{\text{Ta}}{s} + (n\pi)^4\frac{(1+s)^3}{s}, \tag{19.32}$$

where $s = (k_\perp/n\pi)^2$. The minimum Rayleigh number for the onset of instability, $\mathrm{Ra}_{c,min}$, is derived by minimising Ra_c with relative to k_c or s for $n = 1$ [Chandrasekhar (2013)] that yields the following condition:

$$\mathrm{Ta} + (1+s)^3\pi^4 = 3s(1+s)^2\pi^4. \tag{19.33}$$

The solution of the above cubic equation is reasonably complex, so we focus on the limiting cases—$\mathrm{Ta} = 0$ and $\mathrm{Ta} \gg 1$. When $\mathrm{Ta} = 0$, the flow is same as RBC. But for very large Ta, $s \approx \mathrm{Ta}^{1/3}$, and hence

$$\mathrm{Ra}_{c,min} \approx \frac{\mathrm{Ta}}{s} + (n\pi)^4 s^2 \sim \mathrm{Ta}^{2/3}. \tag{19.34}$$

The eigenvector for $\lambda = 0$ yields the neutral mode:

$$\begin{pmatrix} (\sqrt{\mathrm{Ta}}\cos\varsigma)/k^2 \\ 1 \\ -(\sin\varsigma)/k^2 \end{pmatrix} = \begin{pmatrix} (k_z\sqrt{\mathrm{Ta}})/k^3 \\ 1 \\ -k_\perp/k^3 \end{pmatrix}, \tag{19.35}$$

or $u_1(\mathbf{k}) = (k_z\sqrt{\mathrm{Ta}})/k^3$, $u_2(\mathbf{k}) = 1$, and $\theta(\mathbf{k}) = -k_\perp/k^3$. We employ Eq. (3.64) to construct the Cartesian components of the velocity field. In real space, for the neutral mode, the velocity vector is $\Re(\mathbf{u}(\mathbf{k})\exp(i\mathbf{k}\cdot\mathbf{r}))$, and the temperature field is $\Re(\theta(\mathbf{k})\exp(i\mathbf{k}\cdot\mathbf{r}))$.

For a physical picture of the velocity field, we first compute the field for $\mathrm{Ta} = 0$, for which $u_1 = 0$. Without any loss of generality, in Eq. (3.64), we choose the azimuthal angle $\phi = 0$, hence $k_y = 0$ and $k_\perp = k_x = k_c$. Also, $k_z = n\pi$. Hence,

$$\mathbf{u}(\mathbf{r}) = \begin{pmatrix} u_x(\mathbf{r}) \\ u_y(\mathbf{r}) \\ u_z(\mathbf{r}) \end{pmatrix} = \begin{pmatrix} (n\pi/k)\sin(k_x x + \phi_{k2})\cos(n\pi z) \\ 0 \\ -(k_x/k)\cos(k_x x + \phi_{k2})\sin(n\pi z) \end{pmatrix}, \tag{19.36}$$

$$\theta(\mathbf{r}) = -(k_x/k^3)\cos(k_x x + \phi_{k2})\sin(n\pi z), \tag{19.37}$$

where ϕ_{k2} is the phase of $u_2(\mathbf{k})$ and $\theta(\mathbf{k})$. The aforementioned 2D velocity field with $\phi_{k2} = 0$ is depicted in Fig. 6.2 of Chapter 6.

For nonzero Ta, the velocity field is three-dimensional because $u_1 = (n\pi\sqrt{\mathrm{Ta}})/k^3 \neq 0$. Note that $u_2 = 1$ and $k_y \neq 0$. Therefore, using Eq. (3.64, 19.35), we obtain

$$\mathbf{u}(\mathbf{r}) = \begin{pmatrix} [u_1(k_y/k_\perp)\cos(\mathbf{k}_\perp\cdot\mathbf{r}_\perp + \phi_{k1})]\cos(n\pi z) \\ [-u_1(k_x/k_\perp)\cos(\mathbf{k}_\perp\cdot\mathbf{r}_\perp + \phi_{k1})]\cos(n\pi z) \\ 0 \end{pmatrix}$$

$$+ \begin{pmatrix} [((n\pi k_x)/(kk_\perp))\cos(\mathbf{k}_\perp\cdot\mathbf{r}_\perp + \phi_{k2})]\cos(n\pi z) \\ [((n\pi k_y)/(kk_\perp))\cos(\mathbf{k}_\perp\cdot\mathbf{r}_\perp + \phi_{k2})]\cos(n\pi z) \\ [(k_\perp/k)\sin(\mathbf{k}_\perp\cdot\mathbf{r}_\perp + \phi_{k2})]\sin(n\pi z) \end{pmatrix}, \tag{19.38}$$

$$\theta(\mathbf{r}) = (k_\perp/k^3)\cos(\mathbf{k}_\perp\cdot\mathbf{r}_\perp + \phi_{k2})\sin(n\pi z), \tag{19.39}$$

where $\mathbf{k}_\perp\cdot\mathbf{r}_\perp = k_x x + k_y y$, and ϕ_{k1}, ϕ_{k2} are the phases of $u_1(\mathbf{k})$ and $u_2(\mathbf{k})$ respectively.

We illustrate the velocity profile for a standing wave obtained by superposing two solutions with $\mathbf{k}_{\perp,1} = k_x \hat{x} + k_y \hat{y}$ and $\mathbf{k}_{\perp,2} = k_x \hat{x} - k_y \hat{y}$. We also choose $\phi_{k1} = \phi_{k2} = \pi/2$ for both the waves. A superposition of these modes yields

$$\mathbf{u}(\mathbf{r}) = \begin{pmatrix} -u_1(k_y/k_\perp) \cos k_x x \sin k_y y \cos(n\pi z) \\ u_1(k_x/k_\perp) \sin k_x x \cos k_y y \cos(n\pi z) \\ 0 \end{pmatrix}$$

$$+ \begin{pmatrix} -((n\pi k_x)/(kk_\perp)) \sin k_x x \cos k_y y \cos(n\pi z) \\ -((n\pi k_y)/(kk_\perp)) \cos k_x x \sin k_y y \cos(n\pi z) \\ (k_\perp/k) \cos k_x x \cos k_y y \sin(n\pi z) \end{pmatrix}, \qquad (19.40)$$

$$\theta(\mathbf{r}) = (k_\perp/k^3) \cos k_x x \cos k_y y \sin(n\pi z). \qquad (19.41)$$

The above 3D field configuration, exhibited in Fig. 19.1, is same as that derived by Chandrasekhar (2013) (see Eqs. (196, 197) in his book). Also, it is easy to verify that the aforementioned solution $\mathbf{u}(\mathbf{r})$ and $\theta(\mathbf{r})$ satisfy the linearised equations equations with $\partial_t = 0$. Note that the rolls in rotating RBC are three dimensional due to the Coriolis force.

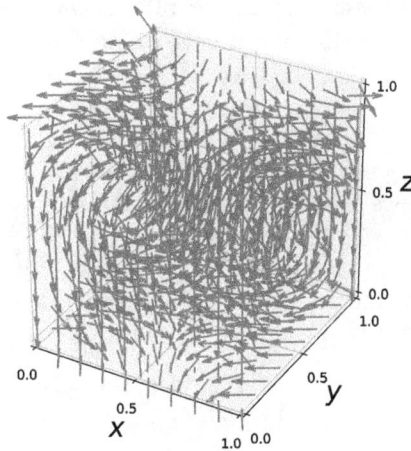

Fig. 19.1 For the neutral mode of rotating convection, the vector plot of the velocity field constructed using Eq. (19.40).

In rotating RBC, the ratio of the Coriolis term and the buoyancy is

$$\frac{\text{Coriolis term}}{\text{Buoyancy term}} = \frac{\sqrt{\text{Ta}}\,\text{Pr}\,u(\mathbf{k})}{\text{Ra}\,\text{Pr}\,\theta(\mathbf{k})}. \qquad (19.42)$$

For large Ta and near the onset of convection, an application of Eq. (19.35) simplifies the above ratio to

$$\frac{\text{Coriolis term}}{\text{Buoyancy term}} \approx \frac{n\pi \text{Ta}}{k_\perp \text{Ra}} \approx \frac{\text{Ta}}{\text{Ra}}. \qquad (19.43)$$

We will revisit the aforementioned ratio in the next section in which we briefly review some of the important results of rotating thermal convection.

19.3 Some important results of rotating convection

Rotating thermal convection is a vast area of research with wide applications in planetary and stellar physics. Both thermal convection and rotation are present in these systems. Many engineering applications too involve rotating thermal convection.

For rotating convection too, we can study instability, patterns, chaos, large-scale quantities, energy spectrum and fluxes, etc. Instabilities, patterns and chaos of rotating convection have been studied is some detail. For details and references, refer to Chandrasekhar (2013) and Lappa (2012). But turbulence in rotating convection is not well understood. For example, the kinetic energy and entropy spectra of rotating convection is not fully understood; the scaling laws for Re and Nu are not well founded. Detailed discussion of these topics would take as much as space as the present book, hence we will not delve on these topics, and only state some of the main results.

It is reasonable to expect that in buoyancy-dominated flows (Ta $\ll$ 1), the kinetic energy and entropy spectra and fluxes would be similar to those for RBC without rotation (as discussed in Chapter 13). However, for strongly rotating flows (Ta $\gg$ 1), turbulent convection would have properties similar to those of strongly-rotating flows. The transition Rayleigh number separating the rotation-dominated and boyancy-dominated flows is [King *et al.* (2012)]

$$\mathrm{Ra_{trans}} \approx \mathrm{Ta}. \tag{19.44}$$

In the range Ra $\gg$ Ta, buoyancy dominates rotation, and converse for Ra $\ll$ Ta. Interestingly, the above equation resembles Eq. (19.43) which was derived for the neutral mode.

There is significant interest in quantifying the heat transport in rotating convection. King *et al.* (2012) argued that

$$\mathrm{Nu} = \begin{cases} \mathrm{Ra}^3 & \text{for } \mathrm{Ra} < \mathrm{Ta} \\ \mathrm{Ra}^{2/7} & \text{for } \mathrm{Ra} > \mathrm{Ta}. \end{cases} \tag{19.45}$$

Note that the behaviour of Nu for the buoyancy-dominated regime (Ra > Ta) is closer to $\mathrm{Ra}^{2/7}$, which is the Nu scaling for RBC (see Sec. 11.4). In the following discussion we reproduce King *et al.* (2012)'s formula that Nu $\sim \mathrm{Ra}^3$ for the rotation-dominated regime (Ra < Ta).

Let us estimate the thermal boundary layer thickness (δ) of rotating convection. In this layer, the temperature drops by $\Delta/2$, hence the Rayleigh number based on these parameters is

$$\mathrm{Ra}_\delta = \frac{\alpha g \Delta \delta^3}{2\nu\kappa}. \tag{19.46}$$

In $0 < z < \delta$, there is a balance between the destabilising force, buoyancy, and the stabilising forces, viscous and Coriolis. Hence, using Eq. (19.34), we obtain

$$\mathrm{Ra}_\delta = \mathrm{Ra}_{c,\min} \approx \mathrm{Ta}^{2/3} \tag{19.47}$$

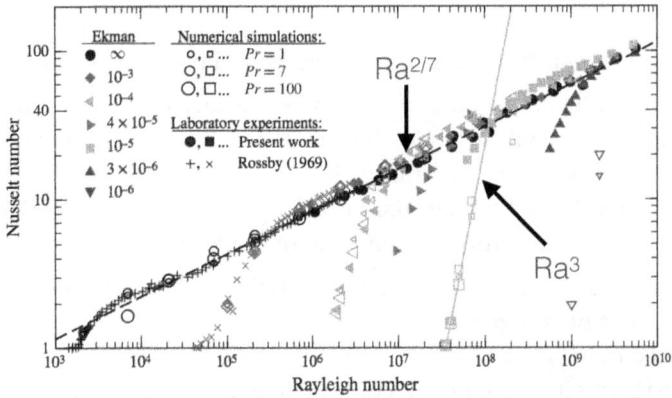

Fig. 19.2   For rotating convection, plot of Nusselt number as a function of Ra and Ekman number E. In the plot the experimental data are represented by solid symbols, and simulation data by open symbols. Note that $Ta = 1/E^2$. $Nu \sim Ra^3$ for $Ra < Ta$, but $Nu \sim Ra^{2/7}$ for $Ra > Ta$. From King *et al.* (2012). Reprinted with permission from Cambridge University Press.

or,

$$\frac{\alpha g \Delta \delta^3}{2\nu\kappa} \approx Ta^{2/3} \approx \left(\frac{2\Omega d^2}{\nu}\right)^{4/3} \implies (d/\delta)^{1/3} \sim Ra/Ra_{c,min}, \qquad (19.48)$$

where

$$\delta \sim \left(\frac{2\Omega}{\nu}\right)^4 \left(\frac{\nu\kappa}{\alpha g \Delta}\right)^3 \qquad (19.49)$$

is the thickness of the boundary layer. Now, using $Nu = d/(2\delta)$, we deduce that for the rotation-dominated regime,

$$Nu \sim \left(\frac{Ra}{Ra_{c,min}}\right)^3. \qquad (19.50)$$

King *et al.* (2012) compiled the numerical and experimental results on Nu scaling of rotating RBC and showed them to be consistent with Eq. (19.45). See Fig. 19.2 for an illustration. Also refer to King *et al.* (2008), Kunnen *et al.* (2010), Davidson (2013), and Pharasi *et al.* (2011) for further details. Kunnen *et al.* (2006) showed that Nusselt number is enhanced in rotating convection.

With this, we close our discussion on rotating convection.

Further reading

Refer to Chandrasekhar (2013) for introduction on linear stability analysis of rotating convection. King *et al.* (2008) cover recent results in the field.

Exercises

(1) Consider the stability matrix of rotating RBC. Perform the following analysis.

    (a) What are the eigenvalues of the matrix? Study them near $Ra = Ra_c$.

    (b) Compute the neutral mode. Interpret it physically. Draw the flow profile corresponding to the neutral mode.

    (c) Study how $\lambda = 0$ becomes nonzero. Is the transition pitchfork of Hopf?

    (d) Study variation of Ra_c as a function of Taylor number.

(2) For rotating convection, what are the vorticity and kinetic helicity of the flow near the onset of convection?

(3) Decompose the neutral mode in helical basis.

(4) Contrast thermally-driven convection with and without rotation.

(5) Compute Rossby, Rayleigh, and Taylor numbers for the Earth's outer core, the Earth's mantle, and solar convection zone.

Chapter 20

Magnetohydrodynamics and Rotating Flows

The magnetic field and rotation are present in many buoyant flows, e.g., in Earth's outer core and in Sun's convection zone. We discuss this topic in the next chapter after setting up the equations for magnetofluids and rotating magnetofluids, which are the subjects of the present chapter.

The equations for magnetohydrodynamics (MHD) are

$$\frac{\partial \mathbf{u}}{\partial t} + \mathbf{u} \cdot \nabla \mathbf{u} = -\frac{1}{\rho} \nabla \sigma + \frac{1}{4\pi\rho} \mathbf{B} \cdot \nabla \mathbf{B} + \nu \nabla^2 \mathbf{u}, \tag{20.1}$$

$$\frac{\partial \mathbf{B}}{\partial t} + \mathbf{u} \cdot \nabla \mathbf{B} = \mathbf{B} \cdot \nabla \mathbf{u} + \eta \nabla^2 \mathbf{B}, \tag{20.2}$$

$$\nabla \cdot \mathbf{u} = \nabla \cdot \mathbf{B} = 0, \tag{20.3}$$

where $\mathbf{u}, \mathbf{B}$ are the velocity and magnetic fields, σ is the total pressure (thermodynamic + magnetic), ρ is the density, and ν, η are the kinematic viscosity and magnetic diffusivity respectively. These equations are in CGS units in which the magnetic field is in Gauss. It is convenient to express $\mathbf{B}$ in velocity units by a transformation

$$\mathbf{B} = \frac{\mathbf{B}_{\mathrm{CGS}}}{\sqrt{4\pi\rho}}, \tag{20.4}$$

that converts Eqs. (20.1–20.3) to

$$\frac{\partial \mathbf{u}}{\partial t} + \mathbf{u} \cdot \nabla \mathbf{u} = -\nabla \sigma + \mathbf{B} \cdot \nabla \mathbf{B} + \nu \nabla^2 \mathbf{u}, \tag{20.5}$$

$$\frac{\partial \mathbf{B}}{\partial t} + \mathbf{u} \cdot \nabla \mathbf{B} = \mathbf{B} \cdot \nabla \mathbf{u} + \eta \nabla^2 \mathbf{B}, \tag{20.6}$$

$$\nabla \cdot \mathbf{u} = \nabla \cdot \mathbf{B} = 0. \tag{20.7}$$

Often magnetofluids are embedded in a constant magnetic field, denoted by $\mathbf{B}_0$, hence

$$\mathbf{B} = \mathbf{B}_0 + \mathbf{b}, \tag{20.8}$$

where $\mathbf{b}$ is the magnetic fluctuation. In the linear limit, we assume that $u \ll B_0$ and $b \ll B_0$, and hence we can drop the nonlinear terms. In this chapter, we will discuss the linear modes of incompressible magnetohydrodynamics (MHD) and incompressible rotating MHD.

20.1 Alfvén waves

In the linear limit, incompressible and inviscid MHD generate waves, which are called *Alfvén waves*. These waves will be discussed in this section. The linearized incompressible and inviscid MHD equations are

$$\frac{\partial \mathbf{u}}{\partial t} = -\nabla \sigma + \mathbf{B}_0 \cdot \nabla \mathbf{b}, \tag{20.9}$$

$$\frac{\partial \mathbf{b}}{\partial t} = \mathbf{B}_0 \cdot \nabla \mathbf{u}, \tag{20.10}$$

$$\nabla \cdot \mathbf{u} = \nabla \cdot \mathbf{b} = 0, \tag{20.11}$$

where the external magnetic field $\mathbf{B}_0 = B_0 \hat{z}$ is assumed to be constant in time and space. In Fourier space, the above equations become

$$\frac{d}{dt}\mathbf{u}(\mathbf{k}) = -ik\sigma(\mathbf{k}) + i[\mathbf{B}_0 \cdot \mathbf{k}]\mathbf{b}(\mathbf{k}), \tag{20.12}$$

$$\frac{d}{dt}\mathbf{b}(\mathbf{k}) = i[\mathbf{B}_0 \cdot \mathbf{k}]\mathbf{u}(\mathbf{k}), \tag{20.13}$$

$$\mathbf{k} \cdot \mathbf{u}(\mathbf{k}) = \mathbf{k} \cdot \mathbf{b}(\mathbf{k}) = 0. \tag{20.14}$$

As usual, we work in Craya-Herring basis that eliminates pressure and yields relatively simpler set of equations. In Craya-Herring basis, the equations are

$$\frac{d}{dt}\begin{pmatrix} u_1(\mathbf{k}) \\ u_2(\mathbf{k}) \\ b_1(\mathbf{k}) \\ b_2(\mathbf{k}) \end{pmatrix} = \begin{pmatrix} 0 & 0 & i\omega & 0 \\ 0 & 0 & 0 & i\omega \\ i\omega & 0 & 0 & 0 \\ 0 & i\omega & 0 & 0 \end{pmatrix} \begin{pmatrix} u_1(\mathbf{k}) \\ u_2(\mathbf{k}) \\ b_1(\mathbf{k}) \\ b_2(\mathbf{k}) \end{pmatrix} = A \begin{pmatrix} u_1(\mathbf{k}) \\ u_2(\mathbf{k}) \\ b_1(\mathbf{k}) \\ b_2(\mathbf{k}) \end{pmatrix}, \tag{20.15}$$

where

$$\omega = \mathbf{B}_0 \cdot \mathbf{k} \tag{20.16}$$

is the dispersion relation. The matrix A has eigenvalues $-i\omega, -i\omega, i\omega, i\omega$, and the corresponding eigenvectors are

$$\begin{pmatrix} -1 \\ 0 \\ 1 \\ 0 \end{pmatrix} ; \begin{pmatrix} 0 \\ -1 \\ 0 \\ 1 \end{pmatrix} ; \begin{pmatrix} 1 \\ 0 \\ 1 \\ 0 \end{pmatrix} ; \begin{pmatrix} 0 \\ 1 \\ 0 \\ 1 \end{pmatrix}. \tag{20.17}$$

In real space, these eigenvectors yield wave solutions:

$$\begin{pmatrix} -1 \\ 0 \\ 1 \\ 0 \end{pmatrix} \cos(kz' - \omega t + \phi_{k1}); \quad \begin{pmatrix} 0 \\ -1 \\ 0 \\ 1 \end{pmatrix} \cos(kz' - \omega t + \phi_{k2});$$

$$\begin{pmatrix} 1 \\ 0 \\ 1 \\ 0 \end{pmatrix} \cos(kz' + \omega t + \phi_{k3}); \quad \begin{pmatrix} 0 \\ 1 \\ 0 \\ 1 \end{pmatrix} \cos(kz' + \omega t + \phi_{k4}), \tag{20.18}$$

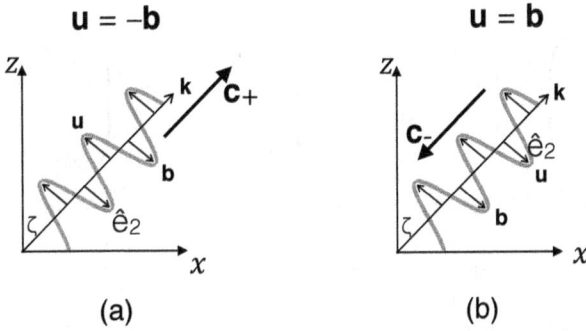

Fig. 20.1 (a) An Alfvén wave with $\mathbf{u} = -\mathbf{b}$ travels along $\mathbf{k}$ with phase velocity of $c_+ = B_0 \cos \zeta$. (b) An Alfvén wave with $\mathbf{u} = \mathbf{b}$ travels along $-\mathbf{k}$ with phase velocity of $c_- = -B_0 \cos \zeta$.

where z' is the coordinate along $\mathbf{k}$. The first two solutions correspond to the waves propagating along $\mathbf{k}$, while the next two are the waves propagating along $-\mathbf{k}$. These waves are called *Alfvén waves*, and they are depicted in Fig. 20.1. Note that the first two waves have $\mathbf{u} = -\mathbf{b}$, but the next two have $\mathbf{u} = \mathbf{b}$.

The phase velocity and group velocity of an Alfvén wave are equal. For the Alfvén waves travelling along $\pm\mathbf{k}$, these velocities are

$$\mathbf{c}_\pm = \mathbf{c}_{\pm g} = \pm\frac{\omega}{k}\hat{\mathbf{k}} = \pm B_0 \cos \zeta \hat{\mathbf{k}}, \tag{20.19}$$

where ζ is angle between $\mathbf{k}$ and $\mathbf{B}_0$. Hence the Alfvén waves are nondispersive. Also note that the $\omega = 0$ for $\mathbf{k} = \mathbf{k}_\perp = k_x\hat{x} + k_y\hat{y}$ (since $\zeta = \pi/2$), hence the fluctuations in the $k_z = 0$ plane are not waves.

We can easily construct helical waves by combining the aforementioned waves. An example of helical Alfvén wave is

$$\begin{pmatrix} -1 \\ 0 \\ 1 \\ 0 \end{pmatrix} \cos(kz' - \omega t) + \begin{pmatrix} 0 \\ -1 \\ 0 \\ 1 \end{pmatrix} \sin(kz' - \omega t). \tag{20.20}$$

For this wave, u_1 and u_2 differ by a phase of $\pi/2$.

In the next section we will describe the effects of rotation on magnetofluids.

20.2 Alfvén wave in rotating fluids

Now let us consider a magnetofluid that is rotated with a constant angular velocity $\mathbf{\Omega} = \Omega\hat{z}$. The linearized equations for this system are

$$\frac{d}{dt}\mathbf{u}(\mathbf{k}) = -ik\sigma(\mathbf{k}) + i[\mathbf{B}_0 \cdot \mathbf{k}]\mathbf{b}(\mathbf{k}) - 2\mathbf{\Omega} \times \mathbf{u}, \tag{20.21}$$

$$\frac{d}{dt}\mathbf{b}(\mathbf{k}) = i[\mathbf{B}_0 \cdot \mathbf{k}]\mathbf{u}(\mathbf{k}), \tag{20.22}$$

$$\mathbf{k} \cdot \mathbf{u}(\mathbf{k}) = 0. \tag{20.23}$$

In Craya-Herring basis, the equations are

$$
\frac{d}{dt}\begin{pmatrix} u_1(\mathbf{k}) \\ u_2(\mathbf{k}) \\ b_1(\mathbf{k}) \\ b_2(\mathbf{k}) \end{pmatrix} = \begin{pmatrix} 0 & \omega_2 & i\omega & 0 \\ -\omega_2 & 0 & 0 & i\omega \\ i\omega & 0 & 0 & 0 \\ 0 & i\omega & 0 & 0 \end{pmatrix} \begin{pmatrix} u_1(\mathbf{k}) \\ u_2(\mathbf{k}) \\ b_1(\mathbf{k}) \\ b_2(\mathbf{k}) \end{pmatrix} = A \begin{pmatrix} u_1(\mathbf{k}) \\ u_2(\mathbf{k}) \\ b_1(\mathbf{k}) \\ b_2(\mathbf{k}) \end{pmatrix},
\tag{20.24}
$$

where $\omega = \mathbf{B}_0 \cdot \mathbf{k}$ and $\omega_2 = 2\Omega \cos \zeta$. The eigenvalues for the matrix A are

$$
\lambda = \pm \left[-\omega^2 - \omega_2^2/2 \pm \omega_2 \sqrt{\omega^2 + \omega_2^2/4} \right]^{1/2}
\tag{20.25}
$$

It is easy to show that all the eigenvalues are pure imaginary, hence the system admits wave solutions. The eigenvectors of the matrix are too complex to be presented here.

In the next chapter we will describe effects of magnetic field on buoyancy-driven flows.

Further reading

Refer to Chandrasekhar (2013) for introduction on linear stability analysis of rotating MHD flows. Davidson (2017), Davidson (2013), and Verma (2004) cover the nonlinear treatment. Rotating MHD is quite important in astrophysical applications, but this topic is beyond the scope of this book.

Exercises

(1) Consider the Alfvén waves in a rotating fluid. Find the relationships between the velocity and magnetic fluctuations for such waves. Sketch them graphically.
(2) Rotating magnetic fluids become unstable when Ω becomes a function of r, the radial distance from the origin. Study earlier work in this topic.
(3) Consider Eq. (20.25) that describes the eigenvalues for rotating magneto fluids. Expand λ for small ω_2 and describe the solution physically.

Chapter 21

Buoyant Flows Under Magnetic Field

In this chapter we describe the effects of magnetic field on Stably stratified flows and on RBC. We limit ourselves to the linear limit.

21.1 Stably stratified flows under magnetic field

We start with Eqs. (20.12, 20.13) and insert the buoyancy in Eq. (20.12). We add Eq. (2.39) for the density fluctuations ρ. In this and the next chapter we choose ρ instead of b (Eq. (2.37)) to avoid notational conflict with magnetic field fluctuation. Hence, the linearised equations for the buoyant magnetofluid are

$$\frac{d}{dt}\mathbf{u}(\mathbf{k}) = -ik\sigma(\mathbf{k}) + i[\mathbf{B}_0 \cdot \mathbf{k}]\mathbf{b}(\mathbf{k}) - N\rho(\mathbf{k})\hat{z}, \tag{21.1}$$

$$\frac{d}{dt}\mathbf{b}(\mathbf{k}) = i[\mathbf{B}_0 \cdot \mathbf{k}]\mathbf{u}(\mathbf{k}), \tag{21.2}$$

$$\frac{d}{dt}\rho(\mathbf{k}) = N\mathbf{u} \cdot \hat{z}, \tag{21.3}$$

$$\mathbf{k} \cdot \mathbf{u}(\mathbf{k}) = \mathbf{k} \cdot \mathbf{b}(\mathbf{k}) = 0. \tag{21.4}$$

In the above equations, $\mathbf{B}_0$, $\mathbf{b}$, and ρ are in velocity units.

It is easier to express the above equations and their solution in Craya-Herring basis:

$$\frac{d}{dt}
\begin{pmatrix}
u_1(\mathbf{k}) \\
u_2(\mathbf{k}) \\
b_1(\mathbf{k}) \\
b_2(\mathbf{k}) \\
\rho(\mathbf{k})
\end{pmatrix}
=
\begin{pmatrix}
0 & 0 & i\omega_1 & 0 & 0 \\
0 & 0 & 0 & i\omega_1 & \omega_2 \\
i\omega_1 & 0 & 0 & 0 & 0 \\
0 & i\omega_1 & 0 & 0 & 0 \\
0 & -\omega_2 & 0 & 0 & 0
\end{pmatrix}
\begin{pmatrix}
u_1(\mathbf{k}) \\
u_2(\mathbf{k}) \\
b_1(\mathbf{k}) \\
b_2(\mathbf{k}) \\
\rho(\mathbf{k})
\end{pmatrix}
\tag{21.5}$$

where $\omega_1 = \mathbf{B}_0 \cdot \mathbf{k}$ and $\omega_2 = N\sin\zeta$. For the above matrix, the eigenvalues are

$$\lambda = 0, i\omega_1, -i\omega_1, i\sqrt{\omega_1^2 + \omega_2^2}, -i\sqrt{\omega_1^2 + \omega_2^2}, \tag{21.6}$$

and the corresponding eigenvectors are

$$\begin{pmatrix} 0 \\ 0 \\ 0 \\ i\omega_2/\omega_1 \\ 1 \end{pmatrix} ;
\begin{pmatrix} 1 \\ 0 \\ 1 \\ 0 \\ 0 \end{pmatrix} ;
\begin{pmatrix} -1 \\ 0 \\ 1 \\ 0 \\ 0 \end{pmatrix} ;
\begin{pmatrix} 0 \\ -i\tilde{\omega}/\omega_2 \\ 0 \\ -i\omega_1/\omega_2 \\ 1 \end{pmatrix} ;
\begin{pmatrix} 0 \\ i\tilde{\omega}/\omega_2 \\ 0 \\ -i\omega_1/\omega_2 \\ 1 \end{pmatrix} ,
\tag{21.7}$$

where $\tilde{\omega} = \sqrt{\omega_1^2 + \omega_2^2}$.

Since the aforementioned eigenvalues are pure imaginary and zero, the system admits multiple wave solutions and a stationary solution. Among the wave solutions, $\lambda = \pm i\omega_1$ correspond to Alfvén waves with u_1 and b_1 fluctuations. These waves do not couple with gravity since u_1 and b_1 are perpendicular to gravity ($\hat{z}$). The other two wave solutions corresponding to eigenvalues $\lambda = \pm i\sqrt{\omega_1^2 + \omega_2^2}$ are combination of Alfvén and gravity waves. The stationary solution (the first eigenvector of Eq. (21.7)) correspond to force balance between the Lorentz force and buoyancy:

$$i[\mathbf{B}_0 \cdot \mathbf{k}]\mathbf{b}(\mathbf{k}) = N\rho(\mathbf{k}) \tag{21.8}$$

Substitution of the above in Eq. (21.2) yields $\lambda = 0$ that corresponds to the stationary solution. In the next section, we will discuss magnetoconvection.

21.2   Magnetoconvection

Often, a magnetofluid is subjected to a temperature gradient that sets up thermal convection. This process is called *magnetoconvection*. Such systems are encountered in laboratory and engineering flows, as well as in planetary and stellar dynamos. For simplicity, we assume that $\mathbf{B}_0$ is aligned along $\hat{z}$.

To derive the equations for magnetoconvection, we start with RBC equations [Sec. 2.7] and add Lorentz force to it. We nondimensionalize the equations using d as the length scale, κ/d as the velocity scale, and B_0 as the scale for the magnetic field. Hence,

$$\mathbf{B} = B_0[\hat{z} + \mathbf{b}], \tag{21.9}$$

where $B_0\mathbf{b}$ is the magnetic field fluctuation. In this system of units, the equations for magnetoconvection are

$$\frac{\partial \mathbf{u}}{\partial t} + \mathbf{u} \cdot \nabla \mathbf{u} = -\nabla \sigma + Q\frac{\mathrm{Pr}^2}{\mathrm{Pm}}[\hat{z} \cdot \nabla \mathbf{b} + \mathbf{b} \cdot \nabla \mathbf{b}] + \mathrm{RaPr}\theta\hat{z} + \mathrm{Pr}\nabla^2\mathbf{u}, \tag{21.10}$$

$$\frac{\partial \mathbf{b}}{\partial t} + \mathbf{u} \cdot \nabla \mathbf{b} = B_0[\hat{z} + \mathbf{b}] \cdot \nabla \mathbf{u} + \frac{\mathrm{Pr}}{\mathrm{Pm}}\nabla^2\mathbf{b}, \tag{21.11}$$

$$\nabla \cdot \mathbf{u} = \nabla \cdot \mathbf{B} = 0, \tag{21.12}$$

where

$$Q = \frac{B_{0,\mathrm{CGS}}^2 d^2}{4\pi\rho\nu\eta} \tag{21.13}$$

is *Chandrasekhar number* [Chandrasekhar (2013)][1], and

$$\mathrm{Pm} = \frac{\nu}{\eta} \tag{21.16}$$

---

[1]Chandrasekhar number is ratio of the Lorenz force and the viscous force, i.e.

$$Q = \frac{B_0\nabla b}{\nu\nabla^2 u}. \tag{21.14}$$

For large η, $\mathbf{B}_0 \cdot \nabla\mathbf{u} + \eta\nabla^2\mathbf{b} = 0$, substitution of which in the above yields Q of the form of Eq. (21.13). In SI system,

$$Q = \frac{B_{0,\mathrm{CGS}}^2 d^2}{\mu_0\rho\nu\eta} \tag{21.15}$$

is the ratio of kinematic viscosity and magnetic diffusivity, and it is called *magnetic Prandtl number*.

Under the limit $u \ll B_0$ and $b \ll B_0$, the nonlinear terms of magnetoconvection can be dropped. In Fourier space, the linearised equations are

$$\frac{d}{dt}\mathbf{u}(\mathbf{k}) = -ik\sigma(\mathbf{k}) + iQk_z\frac{\mathrm{Pr}^2}{\mathrm{Pm}}\mathbf{b}(\mathbf{k}) + \mathrm{RaPr}\theta(\mathbf{k})\hat{z} - \mathrm{Pr}k^2\mathbf{u}(\mathbf{k}), \quad (21.17)$$

$$\frac{d}{dt}\mathbf{b}(\mathbf{k}) = ik_z\mathbf{u}(\mathbf{k}) - k^2(\mathrm{Pr}/\mathrm{Pm})\mathbf{b}(\mathbf{k}), \quad (21.18)$$

$$\frac{d}{dt}\theta(\mathbf{k}) = \mathbf{u}\cdot\hat{z} - k^2\theta(\mathbf{k}), \quad (21.19)$$

$$\mathbf{k}\cdot\mathbf{u}(\mathbf{k}) = \mathbf{k}\cdot\mathbf{b}(\mathbf{k}) = 0, \quad (21.20)$$

where $k_z = n\pi$ is the wavenumber along $\hat{z}$. The dynamical evolution for magnetoconvection are similar to Eq. (21.5), with matrix having more entries. It is simpler to decouple the equations for (u_1, b_1) and (u_2, b_2, θ) that yields

$$\frac{d}{dt}\begin{pmatrix} u_1(\mathbf{k}) \\ b_1(\mathbf{k}) \end{pmatrix} = \begin{pmatrix} -k^2\mathrm{Pr} & iQk_z\frac{\mathrm{Pr}^2}{\mathrm{Pm}} \\ ik_z & -k^2\mathrm{Pr}/\mathrm{Pm} \end{pmatrix}\begin{pmatrix} u_1(\mathbf{k}) \\ b_1(\mathbf{k}) \end{pmatrix} \quad (21.21)$$

and

$$\frac{d}{dt}\begin{pmatrix} u_2(\mathbf{k}) \\ b_2(\mathbf{k}) \\ \theta(\mathbf{k}) \end{pmatrix} = \begin{pmatrix} -k^2\mathrm{Pr} & iQk_z\frac{\mathrm{Pr}^2}{\mathrm{Pm}} & -\mathrm{RaPr}\sin\zeta \\ ik_z & -k^2\mathrm{Pr}/\mathrm{Pm} & 0 \\ -\sin\zeta & 0 & -k^2 \end{pmatrix}\begin{pmatrix} u_2(\mathbf{k}) \\ b_2(\mathbf{k}) \\ \theta(\mathbf{k}) \end{pmatrix} \quad (21.22)$$

where ζ is the angle between $\mathbf{k}$ and $\hat{z}$, and $k^2 = n^2\pi^2 + k_\perp^2$. Equation (21.21) yields damped Alfvén waves that do not couple to gravity since u_1 and b_1 are perpendicular to gravity. Gravity however couples to u_2 and b_2 that leads to thermal convection. Eigenanalysis of Eq. (21.22) however is quite complex, hence we only discuss the neutral mode and the critical Rayleigh number.

The critical Rayleigh number is obtained by setting the determinant of the matrix of Eq. (21.22) to zero, which yields

$$\frac{\mathrm{Pr}^2}{\mathrm{Pm}}\mathrm{Ra}_ck^2\sin^2\zeta = \frac{\mathrm{Pr}^2}{\mathrm{Pm}}k^6 + Q\frac{\mathrm{Pr}^2}{\mathrm{Pm}}k^2k_z^2 \quad (21.23)$$

or

$$\mathrm{Ra}_c = \frac{k^2}{k_\perp^2}\left[k^4 + Q(n\pi)^2\right]. \quad (21.24)$$

Note that $\sin\zeta = k_\perp/k$ (see Fig. 3.6). For a given Q, we can determine $\mathrm{Ra}_{c,\min}$ by minimising Ra_c with respect to $k_\perp$, as performed in Sec. 6.1.

The neutral convective mode is the eigenvector corresponding to $\lambda = 0$, which is

$$\begin{pmatrix} -k^3/k_\perp \\ -i(kn\pi/k_\perp)(\mathrm{Pm}/\mathrm{Pr}) \\ 1 \end{pmatrix}. \quad (21.25)$$

Following the same procedure as in Chapter 19, we deduce the neutral mode in real space as

$$\mathbf{u}(\mathbf{r}) = \begin{pmatrix} u_2 \frac{n\pi k_x}{k k_\perp} \cos(\mathbf{k}_\perp \cdot \mathbf{r}_\perp + \phi_{k2}) \cos(n\pi z) \\ u_2 \frac{n\pi k_y}{k k_\perp} \cos(\mathbf{k}_\perp \cdot \mathbf{r}_\perp + \phi_{k2}) \cos(n\pi z) \\ u_2 (k_\perp/k) \sin(\mathbf{k}_\perp \cdot \mathbf{r}_\perp + \phi_{k2}) \sin(n\pi z) \end{pmatrix}, \tag{21.26}$$

$$\mathbf{b}(\mathbf{r}) = \begin{pmatrix} b_2' \frac{n\pi k_x}{k k_\perp} \cos(\mathbf{k}_\perp \cdot \mathbf{r}_\perp + \phi_{k2}) \sin(n\pi z) \\ b_2' \frac{n\pi k_y}{k k_\perp} \cos(\mathbf{k}_\perp \cdot \mathbf{r}_\perp + \phi_{k2}) \sin(n\pi z) \\ -b_2' (k_\perp/k) \sin(\mathbf{k}_\perp \cdot \mathbf{r}_\perp + \phi_{k2}) \cos(n\pi z) \end{pmatrix}, \tag{21.27}$$

$$\theta(\mathbf{r}) = \sin(\mathbf{k}_\perp \cdot \mathbf{r}_\perp + \phi_{k2}) \sin(n\pi z), \tag{21.28}$$

where $u_2 = -k^3/k_\perp$ and $b_2' = (n\pi/k_\perp)(\mathrm{Pm/Pr})$ [from Eq. (21.25)]. The terms u_1 and b_1 have been dropped in the above expressions because they decay to zero asymptotically. Note that Eqs. (21.26, 21.27, 21.28) satisfy the linearised magneto-convection equations with $\partial_t = 0$, as well as $\nabla \cdot \mathbf{u} = \nabla \cdot \mathbf{b} = 0$.

Magnetoconvection is a vast area of research with major applications in geo- and solar dynamos, as well as in engineering applications involving liquid metals. Such systems exhibit very interesting behaviour, including patterns, chaos, and turbulence [Weiss and Proctor (2014)], but these topics are beyond the scope of this book.

Further reading

Refer to Chandrasekhar (2013) for introduction on linear stability analysis of mag- neto convection. Recent book by Weiss and Proctor (2014) cover magnetoconvection in great detail.

Exercises

(1) Consider Eq. (21.8) that corresponds to the zero eigenvalue of the stability ma- trix for Stably stratified magnetic flow. Graphically illustrate the force balance corresponding to this solution.
(2) Consider the waves in stratified magnetic flow. Sketch their amplitudes and wave propagation as in Fig. 20.1.
(3) Consider the stability matrix of magnetoconvection. Perform the following analysis.

    (a) What are the eigenvalues of the matrix? Study them near Ra = Ra$_c$.
    (b) Compute the neutral mode. Interpret it physically. Draw the flow profile corresponding to the neutral mode.
    (c) Study how $\lambda = 0$ becomes nonzero. Is the transition pitchfork of Hopf?
    (d) Study variation of Ra$_c$ as a function of Chandrasekhar and Prandtl numbers.

(4) For magnetoconvection, what are the vorticity and kinetic helicity of the flow near the onset of convection?
(5) Decompose the neutral mode in helical basis.
(6) Contrast RBC, magnetoconvection, and rotating convection.

Chapter 22

Buoyant and Rotating Flows Under Magnetic Field

In this chapter, we will perform instability analysis of magnetofluid which is under simultaneous influence of rotation and buoyancy.

22.1 Stably stratified and rotating flow under magnetic field

We start with Eqs. (21.1–21.4) and insert the Coriolis force in Eq. (21.1). Then, the resulting equations in linearised form are

$$\frac{d}{dt}\mathbf{u}(\mathbf{k}) = -ik\sigma(\mathbf{k}) + i[\mathbf{B}_0 \cdot \mathbf{k}]\mathbf{b}(\mathbf{k}) - N\rho(\mathbf{k})\hat{z} - 2\mathbf{\Omega} \times \mathbf{u}, \qquad (22.1)$$

$$\frac{d}{dt}\mathbf{b}(\mathbf{k}) = i[\mathbf{B}_0 \cdot \mathbf{k}]\mathbf{u}(\mathbf{k}), \qquad (22.2)$$

$$\frac{d}{dt}\rho(\mathbf{k}) = N\mathbf{u} \cdot \hat{z}, \qquad (22.3)$$

$$\mathbf{k} \cdot \mathbf{u}(\mathbf{k}) = \mathbf{k} \cdot \mathbf{b}(\mathbf{k}) = 0. \qquad (22.4)$$

In the above equations, for the mass density, we choose symbol ρ instead of b [Eq. (2.37)] to avoid notational conflict with magnetic field fluctuation. In Craya-Herring basis, the above equations become

$$\frac{d}{dt}\begin{pmatrix} u_1(\mathbf{k}) \\ u_2(\mathbf{k}) \\ b_1(\mathbf{k}) \\ b_2(\mathbf{k}) \\ \rho(\mathbf{k}) \end{pmatrix} = \begin{pmatrix} 0 & \omega_3 & i\omega_1 & 0 & 0 \\ -\omega_3 & 0 & 0 & i\omega_1 & \omega_2 \\ i\omega_1 & 0 & 0 & 0 & 0 \\ 0 & i\omega_1 & 0 & 0 & 0 \\ 0 & -\omega_2 & 0 & 0 & 0 \end{pmatrix}\begin{pmatrix} u_1(\mathbf{k}) \\ u_2(\mathbf{k}) \\ b_1(\mathbf{k}) \\ b_2(\mathbf{k}) \\ \rho(\mathbf{k}) \end{pmatrix} \qquad (22.5)$$

where $\omega_1 = \mathbf{B}_0 \cdot \mathbf{k}$, $\omega_2 = N\sin\zeta$, and $\omega_3 = 2\Omega\cos\zeta$ with ζ as the angle between $\mathbf{k}$ and $\hat{z}$. We assume all the external fields—$\mathbf{B}_0$, $\mathbf{g}$, and $\mathbf{\Omega}$—to be aligned along $\hat{z}$. The eigenvalues of the matrix of the above equation are 0 and

$$\lambda = \pm\frac{i}{\sqrt{2}}\left\{2\omega_1^2 + \omega_2^2 + \omega_3^2 \pm \sqrt{\omega_2^4 + \omega_3^4 + 4\omega_1^2\omega_3^2 + 2\omega_2^2\omega_3^2}\right\}^{1/2} \qquad (22.6)$$

It is easy to verify that the eigenvalues λ are purely imaginary. Therefore, the system admits only wave solutions. The eigenvectors, except for $\lambda = 0$, are too long and complex to be written here. The eigenvector for $\lambda = 0$ is

$$
\begin{pmatrix} 0 \\ 0 \\ 0 \\ iN \sin \zeta / \omega_1 \\ 1 \end{pmatrix}.
\tag{22.7}
$$

This solution corresponds to the situation when the buoyancy cancels the Lorentz force in Eq. (22.1).

22.2 Magnetoconvection with rotation

In Fourier space, the linearised equations for rotating magnetoconvection are

$$
\frac{d}{dt} \mathbf{u}(\mathbf{k}) = -ik\sigma(\mathbf{k}) + iQk_z \frac{\mathrm{Pr}^2}{\mathrm{Pm}} \mathbf{b}(\mathbf{k}) + \mathrm{RaPr}\theta(\mathbf{k})\hat{z} - \mathrm{Pr}\sqrt{\mathrm{Ta}}\hat{z} \times \mathbf{u}(\mathbf{k}) - \mathrm{Pr}k^2,
\tag{22.8}
$$

$$
\frac{d}{dt} \mathbf{b}(\mathbf{k}) = ik_z \mathbf{u}(\mathbf{k}) - k^2 \frac{\mathrm{Pr}}{\mathrm{Pm}} \mathbf{b}(\mathbf{k}),
\tag{22.9}
$$

$$
\frac{d}{dt} \theta(\mathbf{k}) = \mathbf{u} \cdot \hat{z} - k^2 \theta(\mathbf{k}),
\tag{22.10}
$$

$$
\mathbf{k} \cdot \mathbf{u}(\mathbf{k}) = \mathbf{k} \cdot \mathbf{b}(\mathbf{k}) = 0.
\tag{22.11}
$$

where $k_z = n\pi$ is the wavenumber along $\hat{z}$, Ta is the Taylor number, Q is the Chandrasekhar number, and $\mathrm{Pm} = \nu/\eta$ is the magnetic Prandtl number. See Eq. (21.13) for definition. In Craya-Herring basis, the above equations can be rewritten as

$$
\frac{d}{dt} \begin{pmatrix} u_1(\mathbf{k}) \\ u_2(\mathbf{k}) \\ b_1(\mathbf{k}) \\ b_2(\mathbf{k}) \\ \theta(\mathbf{k}) \end{pmatrix} = A \begin{pmatrix} u_1(\mathbf{k}) \\ u_2(\mathbf{k}) \\ b_1(\mathbf{k}) \\ b_2(\mathbf{k}) \\ \theta(\mathbf{k}) \end{pmatrix}
\tag{22.12}
$$

where

$$
A = \begin{pmatrix}
-k^2\mathrm{Pr} & \mathrm{Pr}\sqrt{\mathrm{Ta}}\cos\zeta & iQk_z\frac{\mathrm{Pr}^2}{\mathrm{Pm}} & 0 & 0 \\
-\mathrm{Pr}\sqrt{\mathrm{Ta}}\cos\zeta & -k^2\mathrm{Pr} & 0 & iQk_z\frac{\mathrm{Pr}^2}{\mathrm{Pm}} & -\mathrm{RaPr}\sin\zeta \\
ik_z & 0 & -k^2\mathrm{Pr}/\mathrm{Pm} & 0 & 0 \\
0 & ik_z & 0 & -k^2\mathrm{Pr}/\mathrm{Pm} & 0 \\
0 & -\sin\zeta & 0 & 0 & -k^2
\end{pmatrix}.
\tag{22.13}
$$

For $\mathrm{Ra} < \mathrm{Ra}_c$, the fluctuations of the system decay to zero. However, the fluctuations grow with time for $\mathrm{Ra} > \mathrm{Ra}_c$. The crossover occurs when the eigenvalue of the matrix A becomes zero, which yields the following equation for Ra_c:

$$
k_\perp^2 \mathrm{Ra}_c \left[k^2 + Q\frac{(n\pi)^2}{k^2} \right] = k^8 + \mathrm{Ta}k^2(n\pi)^2 + 2Qk^4(n\pi)^2 + Q^2(n\pi)^4.
\tag{22.14}
$$

The eigenvectors are too complex to be presented here.

All the planets and stars are rotating about their axis, and many of them are magnetic. Note that the magnetic field in such systems are generated by dynamo mechanism [Weiss and Proctor (2014); Choudhuri (1998)], whose equations are same as above. The physics of such flows are quite complex, and they are beyond the scope of this book. The reader can refer to Weiss and Proctor (2014) and Choudhuri (1998) for further reading.

With this discussion, we conclude our instability analysis for buoyant flows under rotation and magnetic field.

Further reading

Refer to Chandrasekhar (2013) for introduction on linear stability analysis of rotating and buoyant MHD flows. Refer to Weiss and Proctor (2014) and Choudhuri (1998) for further reading.

Exercises

(1) Consider Eq. (22.7) that corresponds to the zero eigenvalue of the stability matrix of stratified and rotating magnetic flow. Graphically illustrate the force balance corresponding to this solution.
(2) Describe the nature of the waves in a stably-stratified and rotating magnetic flow. Sketch amplitudes and propagation of these waves.
(3) Consider the stability matrix of magnetoconvection with rotation. Perform the following analysis.

    (a) What are the eigenvalues of the matrix? Study them near Ra = Ra_c. Choose appropriate Pr and Pm for the analysis.
    (b) Compute the neutral mode. Interpret it physically. Draw the flow profile corresponding to the neutral mode.
    (c) Study how $\lambda = 0$ becomes nonzero. Is the transition pitchfork of Hopf?
    (d) Study variation of Ra_c as a function of system parameters.

(4) Near the convective transition in magnetoconvection with rotation, compute the vorticity and kinetic helicity of the flow?
(5) Contrast RBC, magnetoconvection, rotating convection, and magnetoconvection with rotation.

Chapter 23

Double Diffusive Convection

In nature, we often encounter advection of two scalar fields. For example, in ocean, the two scalar fields are the temperature and salt concentration. In this chapter, we will briefly discuss linear instability of such systems.

23.1 Linear instability analysis of double diffusive convection

In this section we consider two scalar fields, temperature field θ and concentration field Ξ. We assume that the fluid with the above scalars is confined between two layers that are separated by a vertical distance of d. The average linear stratification of these scalars are

$$\frac{d\bar{T}(z)}{dz} = \frac{T_t - T_b}{d}, \tag{23.1}$$

$$\frac{d\bar{\Xi}(z)}{dz} = \frac{\Xi_t - \Xi_b}{d} \tag{23.2}$$

where T_t and T_b are the temperatures at the top and bottom layers, and Ξ_t and Ξ_b are the values of concentrations at the top and bottom layers. We assume these quantities, T and Ξ, to be constants at the layers. We also define the fluctuations of T and Ξ over the background stratification as

$$\theta(x,y,z) = T(x,y,z) - \bar{T}(z) = T(x,y,z) - [T_b + \frac{d\bar{T}}{dz}z], \tag{23.3}$$

$$\xi(x,y,z) = \Xi(x,y,z) - \bar{\Xi}(z) = \Xi(x,y,z) - [\Xi_b + \frac{d\bar{\Xi}}{dz}z]. \tag{23.4}$$

The dynamical equations for the fluctuations, θ, ξ, and the velocity field, $\mathbf{u}$, under Boussinesq approximation are

$$\frac{\partial}{\partial t}\mathbf{u} + \mathbf{u} \cdot \nabla \mathbf{u} = -\nabla\sigma + \hat{z}g\left[\alpha\theta - \beta\xi\right] + \nu\nabla^2\mathbf{u}(\mathbf{k}), \tag{23.5}$$

$$\frac{\partial}{\partial t}\theta + \mathbf{u} \cdot \nabla\theta = -\frac{d\bar{T}}{dz}u_z + \kappa\nabla^2\theta, \tag{23.6}$$

$$\frac{\partial}{\partial t}\xi + \mathbf{u} \cdot \nabla\xi = -\frac{d\bar{\Xi}}{dz}u_z + \kappa_\xi\nabla^2\xi. \tag{23.7}$$

We can nondimensional the above equations in a similar manner as done for RBC. If we choose d/κ as the velocity scale, d as the length scale, T_b-T_t and $\Xi_b-\Xi_t$ as the temperature and concentration scales. Under these nondimensionalization, the aforementioned equations get transformed to

$$\frac{\partial}{\partial t}\mathbf{u}+\mathbf{u}\cdot\nabla\mathbf{u}=-\nabla\sigma+\hat{z}\left[\mathrm{PrRa}\theta(\mathbf{k})-\mathrm{ScRa}_\xi\xi(\mathbf{k})\right]+\mathrm{Pr}\nabla^2\mathbf{u}(\mathbf{k}),\qquad(23.8)$$

$$\frac{\partial}{\partial t}\theta+\mathbf{u}\cdot\nabla\theta=-S_\theta u_z+\nabla^2\theta,\qquad(23.9)$$

$$\frac{\partial}{\partial t}\xi+\mathbf{u}\cdot\nabla\xi=-S_\xi u_z+\frac{\mathrm{Pr}}{\mathrm{Sc}}\nabla^2\xi,\qquad(23.10)$$

where

$$\mathrm{Ra}=\frac{\beta|d\bar{T}/dz|d^4}{\nu\kappa},\qquad(23.11)$$

$$\mathrm{Ra}_\xi=\frac{\beta|d\bar{\Xi}/dz|d^4}{\nu\kappa_\zeta},\qquad(23.12)$$

$$\mathrm{Sc}=\frac{\nu}{\kappa_\xi},\qquad(23.13)$$

$$S_\theta=\frac{d\bar{T}}{dz}\bigg/\frac{(T_b-T_t)}{d},\qquad(23.14)$$

$$S_\xi=\frac{d\bar{\Xi}}{dz}\bigg/\frac{(\Xi_b-\Xi_t)}{d}.\qquad(23.15)$$

In Fourier space, the linearised version of the above equations are

$$\frac{d}{dt}\mathbf{u}(\mathbf{k})=-i\mathbf{k}\sigma(\mathbf{k})+\hat{z}\left[\mathrm{PrRa}\theta(\mathbf{k})-\mathrm{ScRa}_\xi\xi(\mathbf{k})\right]-\mathrm{Pr}k^2\mathbf{u}(\mathbf{k}),\qquad(23.16)$$

$$\frac{d}{dt}\theta(\mathbf{k})=-S_\theta u_z(\mathbf{k})-k^2\theta(\mathbf{k}),\qquad(23.17)$$

$$\frac{d}{dt}\xi(\mathbf{k})=-S_\xi u_z(\mathbf{k})-\frac{\mathrm{Pr}}{\mathrm{Sc}}k^2\xi(\mathbf{k}).\qquad(23.18)$$

Following similar procedure as in Sec. 6.5, we employ Craya-Herring basis to decompose $\mathbf{u}(\mathbf{k})$. As shown in Sec. 6.5, $u_1(\mathbf{k})\to0$, hence we do not include $u_1(\mathbf{k})$ component of Craya-Herring basis in the instability analysis. The Craya-Herring component, $u_2(\mathbf{k})$, however is driven by buoyancy. Hence, the dynamical equations in a matrix form are

$$\frac{d}{dt}\begin{pmatrix}u_2(\mathbf{k})\\\theta(\mathbf{k})\\\xi(\mathbf{k})\end{pmatrix}=A\begin{pmatrix}u_2(\mathbf{k})\\\theta(\mathbf{k})\\\xi(\mathbf{k})\end{pmatrix}\qquad(23.19)$$

where

$$A=\begin{pmatrix}-\mathrm{Pr}k^2&-\mathrm{RaPr}\sin\zeta&\mathrm{Ra}_\xi\mathrm{Sc}\sin\zeta\\S_\theta\sin\zeta&-k^2&0\\S_\xi\sin\zeta&0&-(\mathrm{Pr}/\mathrm{Sc})k^2\end{pmatrix},\qquad(23.20)$$

where ζ is the angle between $\mathbf{k}$ and $\hat{z}$, and $\sin \zeta = k_\perp / k$. We can work out the dynamics of the above equation in detail. But, in this book we only describe the neutral instability condition, which is obtained by setting the $\det(A) = 0$ that yields

$$\left[\left(\frac{\mathrm{Sc}}{\mathrm{Pr}} \right)^2 S_\xi \mathrm{Ra}_{c,\xi} - S_\theta \mathrm{Ra}_c \right] = \frac{k^6}{k_\perp^2}. \tag{23.21}$$

Using the above condition we can deduce the following results:

(1) When both θ and ξ are stable, or $S_\theta = 1$ and $S_\xi = -1$: For this case, the condition for neutral stability is not satisfied because temperature increases with z, and concentration decreases with z. This system is stable, and it supports waves.
(2) When both θ and ξ are unstable, or $S_\theta = -1$ and $S_\xi = 1$: Here, the temperature decreases with z, and concentration increase with z. Hence, both the terms in the left-hand side of Eq. (23.21) are positive, and neutral condition yields

$$\left[\mathrm{Ra}_c + \left(\frac{\mathrm{Sc}}{\mathrm{Pr}} \right)^2 \mathrm{Ra}_{c,\xi} \right] = \frac{k^6}{k_\perp^2}. \tag{23.22}$$

The flow becomes unstable when the above condition is satisfied.
(3) When θ is unstable but ξ are stable, or $S_\theta = -1$ and $S_\xi = -1$: For this case, the condition for neutral stability is

$$\left[\mathrm{Ra}_c - \left(\frac{\mathrm{Sc}}{\mathrm{Pr}} \right)^2 \mathrm{Ra}_{c,\xi} \right] = \frac{k^6}{k_\perp^2}. \tag{23.23}$$

(4) When θ is stable but ξ are unstable, or $S_\theta = 1$ and $S_\xi = 1$: Here, the condition for neutral stability is

$$\left[-\mathrm{Ra}_c + \left(\frac{\mathrm{Sc}}{\mathrm{Pr}} \right)^2 \mathrm{Ra}_{c,\xi} \right] = \frac{k^6}{k_\perp^2}. \tag{23.24}$$

Thus, we can determine the stability of a double diffusive convection given the temperature and concentration gradients. Before closing the discussion on double-diffusive systems, we remark that the finger width of double-diffusive systems can be deduced using the arguments of Sec. 17.10 [Turner (2009); Sadhukhan *et al.* (2017)]:

$$\delta \sim \left[\frac{g(\alpha d\bar{T}/dz - \beta d\bar{\Xi}/dz)}{\nu \kappa} \right]^{-1/4}. \tag{23.25}$$

In the above expression, more buoyant term among $g\alpha d\bar{T}/dz$ and $\beta d\bar{\Xi}/dz$) determines the finger width. There is no fingering when both the terms are stabilising. See Turner (2009), Sreenivas *et al.* (2009), Sadhukhan *et al.* (2017) for further discussion.

We can observe host of phenomena such as pattern, chaos, and turbulence in double diffusive systems. However, these topics are beyond the scope of this book. With this we conclude our discussion on double diffusive convection.

Further reading

Refer to Kundu *et al.* (2015) and Tritton (1988) for an introduction to double diffusive convection, and Turner (2009) for more a more detailed discussion.

Exercise

(1) Apply the instability analysis discussed in this chapter to the oceanic flows.

Chapter 24

Conclusions

In this book we covered various aspects of buoyancy-driven flows—linear, weakly nonlinear, and turbulent. Some of these topics are well understood, while others are either unsolved or partially understood. Here we present a summary of the present status of the field.

The linearised stably stratified flows exhibits waves, while linearised unstably stratified flows show instability. Weakly nonlinear buoyancy-driven systems yield patterns and chaos, but strongly nonlinear ones are turbulent. As described in the book, the linearised and weakly-nonlinear buoyancy-driven flows are reasonably well understood. Note however that that the low-dimensional models based on Galerkin truncation of RBC equations exhibit a zoo of secondary bifurcations, and the exact characterisation of routes to chaos in these systems is still not very clear.

Recent theoretical models of Reynolds and Nusselt numbers successfully describe the numerical and experimental results up to Rayleigh numbers of 10^{12} for moderate and large Prandtl numbers. These models show that in RBC, the walls matter in subtle ways. For example, the viscous dissipation rate $\epsilon_u \approx (U^3/d)\mathrm{Ra}^{-0.20}$, with the term $\mathrm{Ra}^{-0.20}$ arising purely due to walls. Also, walls play a major role in the Nusselt number scaling, $\mathrm{Nu} \sim \mathrm{Ra}^{0.30}$, as well as for $\theta(0,0,2n) \approx -1/(2n\pi)$ that has major implications on the entropy spectrum and flux. In this book we show how the temperature fluctuation $T(\mathbf{r}) - T_m(z) = \theta_{\mathrm{res}}(\mathbf{r})$ plays a critical role in many aspects of RBC, e.g., scaling of Pe and Nu scaling. These scaling however would need revision at large Ra, say Ra beyond 10^{15}.

In this book we present a model for the energy spectrum and flux of turbulence convection. The model shows that the behaviour of turbulent thermal convection is very similar to 3D hydrodynamic turbulence. In particular, according to the model, the kinetic energy spectrum of thermal convection is Kolmogorov-like, and the kinetic energy flux is constant, analogous to 3D hydrodynamic turbulence. The entropy flux too is constant, but the entropy exhibits bi-spectrum whose origin is connected to the boundary layers and confining walls. Surprisingly, the turbulent thermal convection is nearly isotropic, though thermal plumes give an appearance of large anisotropy.

The physics of boundary layers in turbulent convection remains unsolved. At present, the questions of major interests are: (a) the nature of the boundary layers—whether Prandtl-Blasius theory can describe the boundary layer or not? (b) the spectra and fluxes of the velocity and temperature fields in the boundary layer; (c) Relative contributions of boundary layers to the viscous and thermal (entropy) dissipation rates; (d) Is there a transition from viscous boundary layer to turbulent layer? (e) Relation of the boundary layer to the ultimate regime. Further experiments and numerical simulations hopefully would resolve these issues

There is a reasonable convergence on the dynamics of flow reversals in RBC. But we do not have predictive models that could describe the flow reversals in full parameter regime of Rayleigh and Prandtl numbers, and aspect ratio. In particular, we do not have very good low-dimensional models that capture all the properties of flow reversals. Though there is a qualitative understanding of the probability distribution function of the interval distribution between two consecutive reversals, a quantitative understanding of the phenomena is still missing. It would be interesting to relate the flow reversals to magnetic field reversals in dynamo.

The Bolgiano-Obukhov spectral theory describes stably stratified turbulence under moderate density stratification (Richardson number Ri $\approx$ 1). The flow is nearly isotropic in this regime. For strongly buoyancy, or for Ri $\gg$ 1, the flow becomes quasi two-dimensional with $E_\perp(k_\perp) \sim k_\perp^{-5/3}$, $E_\perp(k_\parallel) \sim k_\parallel^{-3}$, and constant kinetic energy flux. Though several numerical simulations appear to validate the aforementioned models, further investigation are required in this regime.

This is a broad summary of the results presented in the book. However, due to lack of space, the following important topics have been either skipped or very sketchily presented in the book:

(1) Amplitude equations and pattern formation
(2) Weak turbulence theories of stably stratified flows
(3) Rotating convection
(4) Magnetocovection, and rotating magnetoconvection
(5) Double diffusive systems
(6) Horizontal convection
(7) Natural convection, e.g. convection induced by a heated cylinder embedded in the atmosphere
(8) Analysis of the probability distribution functions of the fluctuations

In this book we covered linear theories of items 3–5. The nonlinear theory, specially the turbulence regime, of such systems is much less understood than RBC.

With this we close our discussion on buoyancy-driven flows with a hope that future works would resolve the burning unsolved problems of this field. Needless to say that these solutions would help us better model the atmosphere, the interior flows of planets and stars, as well as the engineering flows.

Appendix A

Thermal Parameters of Common Fluids

Table A.1 Thermal properties of common fluids. Note that the parameters for the Sun's convection zone are valid in the upper part of the atmosphere.

Quantity	Density	Thermal Conductivity	Specific heat	Dynamic Viscosity	Kinematic Viscosity	Thermal diffusivity	Prandtl number
	kg/m^3	W/m K	J/kg K	N/m^2 s	m^2/s	m^2/s	
Air	1	0.028	1006	18×10^{-6}	18×10^{-6}	28×10^{-6}	0.7
Water	1000	0.61	4180	0.9×10^{-3}	0.9×10^{-6}	0.14×10^{-6}	7
Ethyl alcohol	785	0.16	2440	1.1×10^{-3}	1.4×10^{-6}	0.084×10^{-6}	17
Liquid sodium	927	84	1250	0.6×10^{-3}	0.7×10^{-6}	80×10^{-6}	0.008
Mercury	1360	8.3	139	1.5×10^{-3}	1.1×10^{-6}	44×10^{-6}	0.025
Glycerin	1260	0.29	2620	1	790×10^{-6}	0.088×10^{-6}	9000
Castor oil	956	0.18	1970	0.650	670×10^{-6}	0.096×10^{-6}	7000
Earth's mantle	4400	3	1250	$10^{19} - 10^{22}$	$10^{16} - 10^{19}$	2.5×10^{-6}	$10^{22} - 10^{25}$
Earth's outer core	9900	50	850	0.01	10^{-6}	5×10^{-7}	0.2
Sun's conv. zone	4400				10^{-4}	10^{6}	10^{-10}

Table A.2   Miscellaneous parameters for Earth's mantle, outer core, and Sun's convection zone.
Note that the parameter values for the solar convection zone are very uncertain.

Quantity	Earth's mantle Earth's mantle	Earth's outer core	Sun's convection zone (top)	Sun's convection zone (bottom)
Velocity scale	2 cm/year	10^{-3} m/s	10^2 m/s	10^2 m/s
Length scale	3×10^6 m	3×10^6 m	10^6 m	10^6 m
Rayleigh number	5×10^7	10^{27}	10^{26}	
Reynolds number	10^{-20}	10^8	10^{13}	10^{13}
Magnetic diffusivity	-	2 m^2/s		
Magnetic Prandtl number	-	10^{-6}	10^{-7}	10^{-1}
Magnetic Reynolds number	-	10^2	10^3	10^{12}
Ekman number	-	10^{-15}	10^{-17}	10^{-15}

Appendix B

Phase and Group Velocities of a Wave Packet

Consider a wave packet at an initial time ($t = 0$):

$$f(\mathbf{r}, t = 0) = \sum_{\mathbf{k}} f(\mathbf{k}, t = 0) \exp\{i\mathbf{k} \cdot \mathbf{r}\}, \qquad (B.1)$$

where $f(\mathbf{k}, t = 0)$ is the Fourier transform of $f(\mathbf{r}, t = 0)$. Let us assume that $f(\mathbf{k}, t = 0)$ peaks sharply at $\mathbf{k} = \mathbf{k}_0$. Now, let us compute the profile of the wave packet as it moves due to dynamical effects.

In linear systems, the time evolution of the above wave packet when it is moving to the right is described by

$$f(\mathbf{k}, t) = f(\mathbf{k}, 0) \exp(-i\omega t), \qquad (B.2)$$

hence

$$f(\mathbf{r}, t) = \sum_{\mathbf{k}} f(\mathbf{k}, 0) \exp\{i(\mathbf{k} \cdot \mathbf{r} - \omega t)\}. \qquad (B.3)$$

Let us assume that the dispersion relation is

$$\omega = \omega(k). \qquad (B.4)$$

Since the wave packet peaks at $\mathbf{k} = \mathbf{k}_0$, we expand $\omega(k)$ near the carrier wave $\mathbf{k} = \mathbf{k}_0$ as

$$
\begin{aligned}
\omega(k) &= \omega(\mathbf{k}_0) + (\nabla_{\mathbf{k}}\omega|_{\mathbf{k}=\mathbf{k}_0}) \cdot \mathbf{k}' + H.O.T. \\
&= \omega(\mathbf{k}_0) + \mathbf{c}_g \cdot \mathbf{k}' + H.O.T.,
\end{aligned}
\qquad (B.5)
$$

where $\mathbf{k}' = \mathbf{k} - \mathbf{k}_0$, $H.O.T.$ stands for the higher order terms, and

$$\mathbf{c}_g = \nabla_{\mathbf{k}}\omega|_{\mathbf{k}=\mathbf{k}_0} \qquad (B.6)$$

is the group velocity of the wave packet. Now we substitute the expansion of Eq. (B.5) without H.O.T. in Eq. (B.3) and deduce that

$$
\begin{aligned}
f(\mathbf{r}, t) &= \exp\{i(\mathbf{k}_0 \cdot \mathbf{r} - \omega(\mathbf{k}_0)t)\} \sum_{\mathbf{k}'} f(\mathbf{k}', 0) \exp\{i\mathbf{k}' \cdot (\mathbf{r} - \mathbf{c}_g t)\}, \\
&= \exp\{i(\mathbf{k}_0 \cdot \mathbf{r} - \omega(\mathbf{k}_0)t)\} f(\mathbf{r} - \mathbf{c}_g t, 0).
\end{aligned}
\qquad (B.7)
$$

Thus, the evolution of the wave packet has two components:

305

(1) The carrier wave propagates with a phase velocity

$$\mathbf{c} = \frac{\omega(\mathbf{k_0})}{k_0}\hat{k}_0.$$
(B.8)

Note that the wave propagates along $\mathbf{k_0}$.

(2) The wave packet moves with the group velocity $\mathbf{c}_g$. The energy of the wave too moves with a velocity of $\mathbf{c}_g$.

It is easy to verify that if the wave propagation is along $-\mathbf{k}$, i.e.,

$$f(\mathbf{r}, t) = \sum_{\mathbf{k}} f(\mathbf{k}, t) \exp\{i(\mathbf{k} \cdot \mathbf{r} + \omega t)\},$$
(B.9)

then the phase and group velocities are

$$\mathbf{c} = -\frac{\omega(\mathbf{k_0})}{k_0}\hat{k}.$$
(B.10)

$$\mathbf{c}_g = -\nabla_{\mathbf{k}}\omega(\mathbf{k})|_{\mathbf{k}=\mathbf{k_0}}$$
(B.11)

Appendix C

Proper Orthogonal Decomposition of RBC

Proper Orthogonal Decomposition (POD) is a popular tool used for extracting the large-scale features of a system, such as a set of images, flow profiles, sound streams, etc. In this appendix, we sketch the basic idea of POD, and compare it with Fourier transform. For details refer to Kosambi (1943) and Holmes *et al.* (2012).

C.1 Proper Orthogonal Decomposition (POD)

In this appendix, we focus on the snapshot method, first proposed by Sirovich (1987) and Sirovich (1989). In this method, we consider M snapshots of a system. These snapshots for example could be pictures of an individual's face from different angles. In POD, these snapshots are akin to components of an "abstract" vector that represents the three-dimensional face of the individual. The objective of POD is to capture the abstract vector optimally. For the same, for each snapshot, we construct a column vector whose entries could be the field variables at the pixel of the picture. We label them by $|f_i\rangle$ with $i = 1 : M$. These column vectors, which could be interpreted as the components of the required abstract vector, are in general nonorthogonal.

Now we compute a correlation matrix whose components are

$$R_{ij} = \langle f_i | f_j \rangle. \tag{C.1}$$

Clearly, the tensor is a $M \times M$ symmetric matrix. Now we can compute the eigenvalues and eigenvectors of the matrix R. The eigenvectors corresponding to the large eigenvalues represent the dominant structures of the abstract vector. Note however that some of the eigenvalues of R could be quite small, hence a more suitable method, *singular value decomposition*, is employed to compute the eigenvalues and eigenvectors of the singular or near-singular matrix R.

C.2 POD vs. Fourier modes

In many parts of the book, we employ Fourier modes to extract large-scale structures of buoyancy-driven flows, in particular RBC. In Sec. 3.4 we described the free-slip

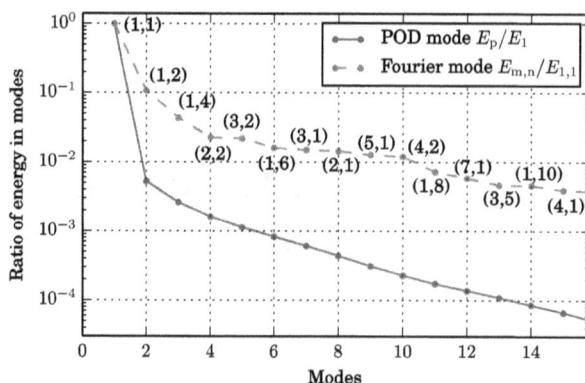

Fig. C.1   For RBC, plots of E_p/E_1, the ratio of the energies of the p-th POD and the first POD mode; and $E_{m,n}/E_{1,1}$, the ratio of the energies of Fourier mode (m, n) and the $(1,1)$ mode. The modes are ranked according to their energy contents. From Paul and Verma (2015). Reprinted with permission from World Scientific.

basis functions as Eqs. (3.47–3.49). These basis functions do not satisfy the no-slip boundary condition, yet they capture the large-scale structures of RBC quite well. See Chapters 8, 17, Chandra and Verma (2011), and Chandra and Verma (2013) for details. This is because the flow structures in the boundary layer are small in size, and they contribute to the large-wavenumber Fourier modes. Thus, the free-slip basis functions are quite useful in describing structures in thermal convection, and they have been employed on many occasions.

In the following discussion, we will perform POD and Fourier analysis on a set of flow pictures of 2D RBC in the turbulent regime away from reversal. We contrast the results obtained by these two methods.

Chandra and Verma (2011) and Chandra and Verma (2013) performed numerical simulation of 2D RBC for $Pr = 1$ and $Ra = 2 \times 10^7$ and observed flow reversals. See Sec. 17.2 for details. Paul and Verma (2015) took 1000 snapshots between two consecutive reversals and performed POD analysis on them. They constructed the column vector whose elements are $[u_x(x, y), u_y(x, y)]$ computed at the grid (x, y). The data analysis shows that ten most-energetic POD modes contain 97.5% of the total energy with the first POD mode sharing 88% of the total energy. In comparison, ten most-energetic Fourier modes contains 93% of the total energy. See Fig. C.1 for an illustration. Thus, both POD and Fourier analysis are able to capture the large-scale flow structures efficiently. POD however is more efficient than the Fourier transform because the POD modes contain larger fraction of energy than the corresponding Fourier modes.

In Fig. C.2 we illustrate the flow profiles of top three POD and Fourier modes [Paul and Verma (2015)]. The flow structures of these POD and Fourier modes appear quite similar. Based on these results we claim that the Fourier analysis provides an alternative to POD analysis. It is important to note that the

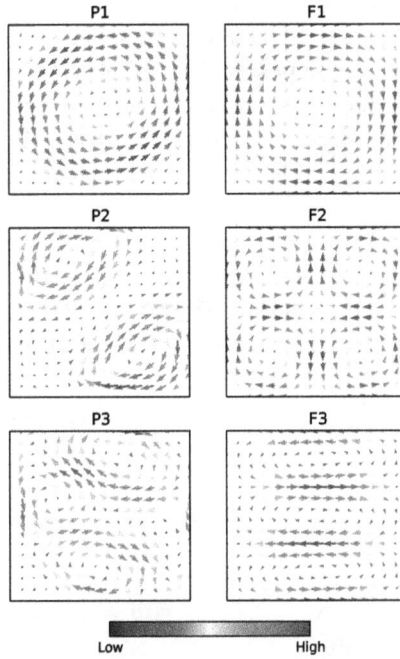

Fig. C.2 For 2D RBC, plots of the three most energetic POD modes (left column), and top three Fourier modes (right column). The Fourier modes are F1 = (1,1), F2 = (2,2), F3 = (1,3). From Paul and Verma (2015). Reprinted with permission from World Scientific.

Fourier analysis is performed for each snapshot independently, while POD analysis requires many snapshots. The POD analysis provides optimal basis functions that captures the large-scale structures from a set of snapshots.

Thus, the two methods, Fourier decomposition and POD, capture the large-scale flow structures quite well. However, these methods have their advantages and disadvantages that are listed below:

(1) Fourier modes can be computed for a single frame, but the POD computation requires a large number of frames.
(2) The Fourier modes have simpler visual interpretations than the POD modes.
(3) Low-dimensional models are constructed more easily using Fourier modes than using POD modes. Lorenz equation, discussed in Chapter 7 is an example of a low-dimensional model constructed using Fourier modes. The construction of low-dimensional models using POD modes requires simulation data [Podvin and Sergent (2015)], but those with Fourier modes can be constructed analytically, that is, without any numerical data.
(4) Fourier analysis is computationally less demanding than the POD computations. For extraction of the large-scale flow structures from three-dimensional datasets, specially in situ measurements, Fourier analysis may be more efficient than POD.

Thus, the two methods, Fourier decomposition and POD, capture the large-scale flow structures quite well. POD is quite popular among the fluid dynamists. We hope that the Fourier decomposition would also be used by researchers. We also remark that the other methods—principal component analysis (PCA) [Jackson (1991)] and dynamic mode decomposition (DMD) are similar in philosophy, but somewhat different in practice. With this, we end our discussion on POD.

Further reading

Refer to Holmes *et al.* (2012) for an introduction to POD. Paul and Verma (2015) compared the results of POD and Fourier analysis of 2D convective flows.

Bibliography

Ahlers, G., Grossmann, S., and Lohse, D. (2009). Heat transfer and large scale dynamics in turbulent Rayleigh-Bénard convection, *Rev. Mod. Phys.* **81**, 2, pp. 503–537.

Akula, B. and Ranjan, D. (2016). Dynamics of buoyancy-driven flows at moderately high Atwood numbers, *J. Fluid Mech.* **795**, pp. 313–355.

Arakeri, J. H., Avila, F. E., Dada, J. M., and Tovar, R. O. (2000). Convection in a long vertical tube due to unstable stratification-A new type of turbulent flow? *Curr. Sci.* **79**, pp. 859–866.

Araujo, F. F. and Grossmann, S. (2005). Wind reversals in turbulent Rayleigh-Bénard convection, *Phys. Rev. Lett.* **95**, p. 084502.

Ashkenazi, S. and Steinberg, V. (1999). Spectra and statistics of velocity and temperature fluctuations in turbulent convection, *Phys. Rev. Lett.* **83**, 23, pp. 4760–4763.

Bandyopadhyay, R. and Verma, M. K. (2017). Discrete symmetries in dynamo reversals, *Phys. Plasmas* **24**, 6, p. 062307.

Banerjee, A., Kraft, W. N., and Andrews, M. J. (2010). Detailed measurements of a statistically steady Rayleigh–Taylor mixing layer from small to high Atwood numbers, *J. Fluid Mech.* **659**, pp. 127–190.

Batchelor, G. K. (1959). Small-scale variation of convected quantities like temperature in turbulent fluid Part 1. General discussion and the case of small conductivity, *J. Fluid Mech.* **5**, pp. 113–133.

Behringer, R. P. (1985). Rayleigh-Bénard convection and turbulence in liquid helium, *Rev. Mod. Phys.* **57**, 3, pp. 657–687.

Benzi, R. (2005). Flow reversal in a simple dynamical model of turbulence, *Phys. Rev. Lett.* **95**, 2, p. 024502.

Berhanu, M., Monchaux, R., Fauve, S., Mordant, N., Pétrélis, F., Chiffaudel, A., Daviaud, F., Dubrulle, B., Marié, L., Ravelet, F., Bourgoin, M., Bourgoin, M., Odier, P., Pinton, J.-F., and Volk, R. (2007). Magnetic Field Reversals in an Experimental Turbulent Dynamo, *EPL* **77**, 5, p. 59001.

Bhattacharjee, J. K. (1987). *Convection and Chaos in Fluids* (World Scientific, Singapore).

Bhattacharjee, J. K. (2015). Kolmogorov argument for the scaling of the energy spectrum in a stratified fluid, *Phys. Lett. A* **379**, 7, pp. 696–699.

Bhattacharya, S., Pandey, A., Kumar, A., and Verma, M. K. (2018). Complexity of viscous dissipation in turbulent thermal convection, *Phys. Fluids*, 30, 031702.

Biferale, L. (2003). Shell models of energy cascade in turbulence, *Annu. Rev. Fluid Mech.* **35**, 1, pp. 441–468.

Bodenschatz, E., Pesch, W., and Ahlers, G. (2000). Recent developments in Rayleigh-Bénard convection, *Annu. Rev. Fluid Mech.* **32**, 1, pp. 709–778.

Boffetta, G., De Lillo, F., Mazzino, A., and Musacchio, S. (2011). Bolgiano scale in confined Rayleigh–Taylor turbulence, *J. Fluid Mech.* **690**, pp. 426–440.

Boffetta, G. and Mazzino, A. (2017). Incompressible Rayleigh–Taylor Turbulence, *Annu. Rev. Fluid Mech.* **49**, 1, pp. 119–143.

Bolgiano, R. (1959). Turbulent spectra in a stably stratified atmosphere, *J. Geophys. Res.* **64**, 12, pp. 2226–2229.

Borue, V. and Orszag, S. A. (1997). Turbulent convection driven by a constant temperature gradient, *J. Sci. Comput.* **12**, 3, pp. 305–351.

Boyd, J. P. (2003). *Chebyshev and Fourier Spectral Methods*, 2nd edn. (Dover Publications, New York).

Brandenburg, A. (1992). Energy-spectra in a model for convective turbulence, *Phys. Rev. Lett.* **69**, 4, pp. 605–608.

Breuer, M. and Hansen, U. (2009). Turbulent convection in the zero reynolds number limit, *EPL* **86**, p. 24004.

Brown, E. and Ahlers, G. (2006). Rotations and cessations of the large-scale circulation in turbulent Rayleigh–Bénard convection, *J. Fluid Mech.* **568**, pp. 351–386.

Bruneau, C.-H., Fischer, P., Kellay, H., and Xiong, Y.-L. (2017). Numerical simulations of thermal convection on a hemisphere, *Unpublished* , pp. 1–34.

Busse, F. H. and Whitehead, J. A. (1974). Oscillatory and collective instabilities in large Prandtl number convection, *J. Fluid Mech.* **66**, pp. 67–79.

Calzavarini, E., Lohse, D., and Toschi, F. (2005). Rayleigh and Prandtl number scaling in the bulk of Rayleigh–Bénard turbulence , *Phys. Fluids* **17**, p. 055107.

Calzavarini, E., Toschi, F., and Tripiccione, R. (2002). Evidences of Bolgiano-Obhukhov scaling in three-dimensional Rayleigh-Bénard convection, *Phys. Rev. E* **66**, p. 016304.

Canuto, C., Hussaini, M. Y., Quarteroni, A., and Zang, T. A. (1988). *Spectral Methods in Fluid Dynamics* (Springer-Verlag, Berlin Heidelberg).

Castaing, B. (1990). Scaling of turbulent spectra, *Phys. Rev. Lett.* **65**, 25, pp. 3209–3211.

Castaing, B., Gunaratne, G., Kadanoff, L. P., Libchaber, A., and Heslot, F. (1989). Scaling of hard thermal turbulence in Rayleigh-Bénard convection, *J. Fluid Mech.* **204**, pp. 1–30.

Chandra, M. and Verma, M. K. (2011). Dynamics and symmetries of flow reversals in turbulent convection, *Phys. Rev. E* **83**, p. 067303.

Chandra, M. and Verma, M. K. (2013). Flow Reversals in Turbulent Convection via Vortex Reconnections, *Phys. Rev. Lett.* **110**, 11, p. 114503.

Chandrasekhar, S. (2013). *Hydrodynamic and Hydromagnetic Stability* (Oxford University Press, Oxford).

Chassignet, E., Cenedese, C., and Verron, J. (eds.) (2012). *Buoyancy-Driven Flows* (Cambridge University Press, Cambridge).

Chatterjee, A. G., Verma, M. K., Kumar, A., Samtaney, R., Hadri, B., and Khurram, R. (2018). Scaling of a Fast Fourier Transform and a pseudo-spectral fluid solver up to 196608 cores, *J. Parallel Distrib. Comput.* **113**, pp. 77–91.

Chavanne, X., Chillà, F., Castaing, B., Hebral, B., Chabaud, B., and Chaussy, J. (1997). Observation of the ultimate regime in Rayleigh-Bénard convection, *Phys. Rev. Lett.* **79**, 19, pp. 3648–3651.

Chavanne, X., Chillà, F., Chabaud, B., Castaing, B., and Hebral, B. (2001). Turbulent Rayleigh-Bénard convection in gaseous and liquid He, *Phys. Fluids* **13**, 5, pp. 1300–1320.

Chertkov, M. (2003). Phenomenology of Rayleigh-Taylor turbulence, *Phys. Rev. Lett.* **91**, 11, p. 115001.

Chillà, F., Ciliberto, S., Innocenti, C., and Pampaloni, E. (1993). Boundary layer and scaling properties in turbulent thermal convection, *Nuovo Cimento D* **15**, 9, pp. 1229–1249.

Chillà, F. and Schumacher, J. (2012). New perspectives in turbulent Rayleigh-Bénard convection, *Eur. Phys. J. E* **35**, 7, p. 58.

Ching, E., Chui, K. W., Shang, X.-D., Qiu, X.-L., Tong, P., and Xia, K.-Q. (2004). Velocity and temperature cross-scaling in turbulent thermal convection, *J. Turbul.* **5**, pp. 1–12.

Ching, E. S. C. (2007). Scaling laws in the central region of confined turbulent thermal convection, *Phys. Rev. E* **75**, 5, p. 056302.

Ching, E. S. C. (2013). *Statistics and Scaling in Turbulent Rayleigh-Bénard Convection* (Springer, Berlin).

Ching, E. S. C. and Cheng, W. C. (2008). Anomalous scaling and refined similarity of an active scalar in a shell model of homogeneous turbulent convection, *Phys. Rev. E* **77**, 1, p. 015303.

Cholemari, M. R. and Arakeri, J. H. (2009). Axially homogeneous, zero mean flow buoyancy-driven turbulence in a vertical pipe, *J. Fluid Mech.* **621**, pp. 69–34.

Choudhuri, A. R. (1998). *The Physics of Fluids and Plasmas: An Introduction for Astrophysicists* (Cambridge University Press, Cambridge).

Cioni, S., Ciliberto, S., and Sommeria, J. (1995). Temperature structure functions in turbulent convection at low prandtl number, *EPL* **32**, 5, pp. 413–418.

Cioni, S., Ciliberto, S., and Sommeria, J. (1997). Strongly turbulent Rayleigh–Bénard convection in mercury: comparison with results at moderate Prandtl number, *J. Fluid Mech.* **335**, pp. 111–140.

Craya, A. (1958). *Contribution à l'analyse de la turbulence associée à des vitesses moyennes*, Ph.D. thesis, Université de Granoble.

Cross, M. and Greenside, H. (2009). *Pattern formation and dynamics in nonequilibrium systems* (Cambridge University Press, Cambridge).

Cross, M. C. and Hohenberg, P. C. (1993). Pattern formation outside of equilibrium, *Rev. Mod. Phys.* **65**, 3, pp. 851–1112.

Dar, G., Verma, M. K., and Eswaran, V. (2001). Energy transfer in two-dimensional magnetohydrodynamic turbulence: formalism and numerical results, *Physica D* **157**, 3, pp. 207–225.

Davidson, P. A. (2004). *Turbulence: An Introduction for Scientists and Engineers* (Oxford University Press, Oxford).

Davidson, P. A. (2013). *Turbulence in Rotating, Stratified and Electrically Conducting Fluids* (Cambridge University Press, Cambridge).

Davidson, P. A. (2014). The Dynamics and Scaling Laws of Planetary Dynamos Driven by Inertial Waves, *Geophys. J. Int.* **198**, 3, pp. 1832–1847.

Davidson, P. A. (2017). *An introduction to magnetohydrodynamics*, 2nd edn. (Cambridge University Press, Cambridge).

Ditlevsen, P. D. (2010). *Turbulence and Shell Models* (Cambridge University Press, Cambridge).

Doering, C. R., Otto, F., and Reznikoff, M. G. (2006). Bounds on vertical heat transport for infinite-Prandtl-number Rayleigh–Bénard convection, *J. Fluid Mech.* **560**, pp. 229–241.

Drazin, P. G. (2002). *Introduction to Hydrodynamic Stability* (Cambridge University Press, Cambridge).

Ecke, R. E. (1991). Quasiperiodicity, mode-locking, and universal scaling in Rayleigh-Bénard convection, in R. Artuso, P. Cvitanović, and G. Casati (eds.), *Chaos, Order, and Patterns* (Plenum Press, New York), pp. 77–108.

Ecke, R. E., Mainieri, R., and Sullivan, T. S. (1991). Universality in quasiperiodic Rayleigh-Bénard convection, *Phys. Rev. A* **44**, 12, pp. 8103–8118.

Emran, M. S. and Schumacher, J. (2008). Fine-scale statistics of temperature and its derivatives in convective turbulence, *J. Fluid Mech.* **611**, pp. 13–34.

Fauve, S. (1998). Pattern Forming Instabilities, in C. Godrèche and P. Manneville (eds.), *Hydrodynamics and nonlinear instabilities* (Cambridge University Press, Cambridge), pp. 387–492.

Fauve, S., Laroche, C., and Libchaber, A. (1981). Effect of a horizontal magnetic field on convective instabilities in mercury, *J. Physique Lett.* **42**, 21, p. L455.

Foroozani, N., Niemela, J. J., Armenio, V., and Sreenivasan, K. R. (2017). Reorientations of the large-scale flow in turbulent convection in a cube. *Phys. Rev. E* **95**, 3-1, p. 033107.

Frisch, U. (1995). *Turbulence: The Legacy of A. N. Kolmogorov* (Cambridge University Press, Cambridge).

Gage, K. S. and Nastrom, G. D. (1986). Theoretical interpretation of atmospheric wavenumber spectra of wind and temperature observed by commercial aircraft during GASP, *J. Atmos. Sci.* **43**, 7, pp. 729–740.

Gallet, B., Herault, J., Laroche, C., Pétrélis, F., and Fauve, S. (2012). Reversals of a large-scale field generated over a turbulent background, *Geophys. Astrophys. Fluid Dyn.* **106**, 4-5, pp. 468–492.

Gershuni, G. Z., Zhukhovitskii, E. M., and Nepomniashchii, A. A. (1989). *Stability of Convective Flows* (Izdatel'stvo Nauka, Moscow).

Gershuni, G. Z. and Zhukhovitskii, E. M. (1976). *Convective Stability of Incompressible Fluids* (Israel Program for Scientific Translations).

Getling, A. V. (1998). *Rayleigh-Bnard Convection: Structures and Dynamics* (World Scientific, Singapore).

Getling, A. V., Mazhorova, O. S., and Shcheritsa, O. V. (2013). Concerning the multiscale structure of solar convection, *Geomagn. Aeron.* **53**, 7, pp. 904–908.

Gledzer, E. B. (1973). System of hydrodynamic type allowing 2 quadratic integrals of motion , *Dokl Acad Nauk SSSR* **209**, pp. 1046–1048.

Godeferd, F. S. and Cambon, C. (1994). Detailed investigation of energy transfers in homogeneous stratified turbulence*, *Phys. Fluids* **6**, 6, pp. 2084–2100.

Gotoh, T., Watanabe, T., and Miura, H. (2014). Spectrum of Passive Scalar at Very High Schmidt Number in Turbulence, *Plasma and Fusion Research* **9**, 0, p. 3401019.

Gotoh, T. and Yeung, P. K. (2013). Passive scalar transport turbulence: a computational perspective , in P. A. Davidson, Y. Kaneda, and K. R. Sreenivasan (eds.), *Ten Chapters in Turbulence* (Cambridge University Press, Cambridge), pp. 87–131.

Greenspan, H. P. (1968). *The Theory of Rotating Fluids* (Cambridge University Press, Cambridge).

Grossmann, S. and Lohse, D. (1991). Fourier-Weierstrass mode analysis for thermally driven turbulence, *Phys. Rev. Lett.* **67**, 4, pp. 445–448.

Grossmann, S. and Lohse, D. (2000). Scaling in thermal convection: a unifying theory, *J. Fluid Mech.* **407**, pp. 27–56.

Grossmann, S. and Lohse, D. (2001). Thermal convection for large Prandtl numbers, *Phys. Rev. Lett.* **86**, 1, pp. 3316–3319.

Grossmann, S., Lohse, D., and Sun, C. (2016). High-Reynolds Number Taylor-Couette Turbulence, *Annu. Rev. Fluid Mech.* **48**, 1, pp. 53–80.

Guckenheimer, H. and Holmes, P. (2013). *Nonlinear Oscillations, Dynamical Systems, and Bifurcations of Vector Fields* (Springer, Berlin).

He, X., Funfschilling, D., Nobach, H., Bodenschatz, E., and Ahlers, G. (2012). Transition to the Ultimate State of Turbulent Rayleigh-Bénard Convection, *Phys. Rev. Lett.* **108**, 2, p. 024502.

Herring, J. R. (1974). Approach of axisymmetric turbulence to isotropy, *Phys. Fluids* **17**, 5, pp. 859–872.

Hilborn, R. C. (2001). *Chaos and Nonlinear Dynamics*, 2nd edn. (Oxford University Press, Oxford).

Holmes, P., Lumley, J. L., and Berkooz, G. (2012). *Turbulence, Coherent Structures, Dynamical Systems and Symmetry*, 2nd edn. (Cambridge University Press, Cambridge).

Hoyle, R. B. (2006). *Pattern formation* (Cambridge University Press, Cambridge).

Huisman, S. G., van Gils, D. P. M., Grossmann, S., Sun, C., and Lohse, D. (2012). Ultimate Turbulent Taylor-Couette Flow, *Phys. Rev. Lett.* **108**, 2, p. 024501.

Jackson, J. E. (1991). *A User's Guide to Principal Components* (Wiley, New Jersey).

King, E. M., Stellmach, S., and Aurnou, J. M. (2012). Heat transfer by rapidly rotating Rayleigh–Bénard convection, *J. Fluid Mech.* **691**, pp. 568–582.

King, E. M., Stellmach, S., Noir, J., and Hansen, U. (2008). Boundary layer control of rotating convection systems, *Nature* **457**, 7226, pp. 301–304.

Knobloch, E. (1992). Onset of zero prandtl number convection, *J. Phys. II France* **2**, 5, pp. 995–999.

Kolmogorov, A. N. (1941a). Dissipation of Energy in Locally Isotropic Turbulence, *Dokl Acad Nauk SSSR* **32**, 1, pp. 16–18.

Kolmogorov, A. N. (1941b). The local structure of turbulence in incompressible viscous fluid for very large Reynolds numbers, *Dokl Acad Nauk SSSR* **30**, 4, pp. 301–305.

Kosambi, D. D. (1943). Statistics in function space, *J. Ind. Math. Soc.* **7**, pp. 76–88.

Kraichnan, R. H. (1959). The structure of isotropic turbulence at very high Reynolds numbers, *J. Fluid Mech.* **5**, pp. 497–543.

Kraichnan, R. H. (1962). Turbulent thermal convection at arbitrary prandtl number, *Phys. Fluids* **5**, 11, pp. 1374–1389.

Kraichnan, R. H. (1967). Inertial ranges in two-dimensional turbulence, *Phys. Fluids* **10**, pp. 1417–1423.

Kraichnan, R. H. (1968). Small-scale structure of a scalar field convected by turbulence, *Phys. Fluids* **11**, pp. 945–953.

Kreyszig, E., Kreyszig, H., and Norminton, E. J. (2011). *Advanced Engineering Mathematics*, 10th edn. (Wiey, New York).

Krishnamurti, R. (1970a). On the transition to turbulent convection. Part 1. The transition from two- to three-dimensional flow, *J. Fluid Mech.* **42**, pp. 295–307.

Krishnamurti, R. (1970b). On the transition to turbulent convection. Part 2. The transition to time-dependent flow, *J. Fluid Mech.* **42**, pp. 309–320.

Kumar, A., Chatterjee, A. G., and Verma, M. K. (2014a). Energy spectrum of buoyancy-driven turbulence, *Phys. Rev. E* **90**, 2, p. 023016.

Kumar, A. and Verma, M. K. (2015). Shell model for buoyancy-driven turbulence. *Phys. Rev. E* **91**, 4, p. 043014.

Kumar, A. and Verma, M. K. (2017). Applicability of Taylor's hypothesis in thermally-driven turbulence, *arXiv* 1512.00959v2.

Kumar, A., Verma, M. K., and Sukhatme, J. (2017). Phenomenology of two-dimensional stably stratified turbulence under large-scale forcing, *J. Turbul.* **18**, 3, pp. 219–239.

Kumar, R., Verma, M. K., and Samtaney, R. (2014b). Energy transfers and magnetic energy growth in small-scale dynamo, *EPL* **104**, 5, p. 54001.

Kundu, P. K., Cohen, I. M., and Dowling, D. R. (2015). *Fluid Mechanics*, 6th edn. (Academic Press, San Diego).

Kunnen, R. P. J., Clercx, H. J. H., and Geurts, B. J. (2006). Heat flux intensification by vortical flow localization in rotating convection, *Phys. Rev. E* **74**, 5, p. 056306.

Kunnen, R. P. J., Clercx, H. J. H., and Geurts, B. J. (2010). Vortex statistics in turbulent rotating convection, *Phys. Rev. E* **82**, 3, p. 036306.

Kunnen, R. P. J., Clercx, H. J. H., Geurts, B. J., van Bokhoven, L. J. A., Akkermans, R. A. D., and Verzicco, R. (2008). Numerical and experimental investigation of structure-function scaling in turbulent Rayleigh-Bénard convection, *Phys. Rev. E* **77**, 1, p. 016302.

Lakkaraju, R., Stevens, R. J. A. M., Oresta, P., Verzicco, R., Lohse, D., and Prosperetti, A. (2013). Heat transport in bubbling turbulent convection. *PNAS* **110**, 23, pp. 9237–9242.

Landau, L. D. and Lifshitz, E. M. (1976). *Mechanics*, 3rd edn., Course of Theoretical Physics (Elsevier, Oxford).

Landau, L. D. and Lifshitz, E. M. (1987). *Fluid Mechanics*, 2nd edn., Course of Theoretical Physics (Elsevier, Oxford).

Lappa, M. (2010). *Thermal convection: Patterns, Evolution and Stability* (Wiley, Chichester).

Lappa, M. (2012). *Rotating Thermal Flows in Natural and Industrial Processes* (Wiley, Chichester).

Lesieur, M. (2008). *Turbulence in Fluids* (Springer-Verlag, Dordrecht).

Leslie, D. C. (1973). *Developments in the theory of turbulence* (Clarendon Press, Oxford).

Li, L., Shi, N., Du Puits, R., Resagk, C., Schumacher, J., and Thess, A. (2012). Boundary layer analysis in turbulent Rayleigh-Bénard convection in air: Experiment versus simulation, *Phys. Rev. E* **86**, 2, p. 026315.

Lindborg, E. (2006). The energy cascade in a strongly stratified fluid, *J. Fluid Mech.* **550**, pp. 207–242.

Lohse, D. and Toschi, F. (2003). Ultimate state of thermal convection, *Phys. Rev. Lett.* **90**, 3, p. 034502.

Lohse, D. and Xia, K.-Q. (2010). Small-scale properties of turbulent Rayleigh–Bénard convection, *Annu. Rev. Fluid Mech.* **42**, 1, pp. 335–364.

Lorenz, E. N. (1963). Deterministic nonperiodic flow, *J. Atmos. Sci.* **20**, pp. 130–141.

Lozhkin, S. A. and Frick, P. G. (1998). Inertial obukhov-bolgiano interval in shell models of convective turbulence, *Fluid Dyn.* **33**, 6, pp. 842–849.

L'vov, V. S. (1991). Spectra of velocity and temperature-fluctuations with constant entropy flux of fully-developed free-convective turbulence, *Phys. Rev. Lett.* **67**, 6, pp. 687–690.

L'vov, V. S. and Falkovich, G. (1992). Conservation laws and two-flux spectra of hydrodynamic convective turbulence, *Physica D* **57**, pp. 85–95.

L'vov, V. S., Podivilov, E., Pomyalov, A., Procaccia, I., and Vandembroucq, D. (1998). Improved shell model of turbulence, *Phys. Rev. E* **58**, 2, pp. 1811–1822.

Majumdar, S. and Sood, A. K. (2011). Universality and scaling behavior of injected power in elastic turbulence in wormlike micellar gel, *Phys. Rev. E* **84**, 1, p. 015302.

Malkus, W. V. R. (1954). The Heat Transport and Spectrum of Thermal Turbulence, *Proceedings of the Royal Society of London. Series A* **225**, 1, pp. 196–212.

Mannattil, M., Pandey, A., Verma, M. K., and Chakraborty, S. (2017). On the applicability of low-dimensional models for convective flow reversals at extreme Prandtl numbers, *Eur. Phys. J. B* **90**, 12, p. 259.

Manneville, P. (2004). *Instabilities, Chaos and Turbulence* (Imperial College Press, London).

Mashiko, T., Tsuji, Y., Mizuno, T., and Sano, M. (2004). Instantaneous measurement of velocity fields in developed thermal turbulence in mercury, *Phys. Rev. E* **69**, 3, p. 036306.

Mazzino, A. (2017). Two-dimensional turbulent convection, *Phys. Fluids* **29**, 11, p. 111102.

McComb, W. D. (1990). *The physics of fluid turbulence* (Clarendon Press, Oxford).

Meneveau, C. and Sreenivasan, K. R. (1987). Simple multifractal cascade model for fully developed turbulence, *Phys. Rev. Lett.* **59**, 13, pp. 1424–1427.

Messio, L., Morize, C., Rabaud, M., and Moisy, F. (2007). Experimental observation using particle image velocimetry of inertial waves in a rotating fluid, *Exp. Fluids* **44**, 4, pp. 519–528.

Mingshun, J. and Shida, L. (1997). Scaling behavior of velocity and temperature in a shell model for thermal convective turbulence, *Phys. Rev. E* **56**, 1, pp. 441–446.

Mishra, P. K., De, A. K., Verma, M. K., and Eswaran, V. (2011). Dynamics of reorientations and reversals of large-scale flow in Rayleigh–Bénard convection, *J. Fluid Mech.* **668**, pp. 480–499.

Mishra, P. K., Hérault, J., Fauve, S., and Verma, M. K. (2015). Dynamics of reversals and condensates in two-dimensional Kolmogorov flows. *Phys. Rev. E* **91**, 5, p. 053005.

Mishra, P. K. and Verma, M. K. (2010). Energy spectra and fluxes for Rayleigh-Bénard convection, *Phys. Rev. E* **81**, 5, p. 056316.

Mishra, P. K., Wahi, P., and Verma, M. K. (2010). Patterns and bifurcations in low–Prandtl-number Rayleigh–Bénard convection, *EPL* **89**, 4, p. 44003.

Nandukumar, Y. and Pal, P. (2016). Instabilities and chaos in low-Prandtl number rayleigh-Bénard convection, *Computers & Fluids* **138**, C, pp. 61–66.

Nath, D., Pandey, A., Kumar, A., and Verma, M. K. (2016). Near isotropic behavior of turbulent thermal convection, *Phys. Rev. Fluids* **1**, p. 064302.

Nath, D. and Verma, M. K. (2014). Annals of Nuclear Energy, *Ann. Nucl. Energy* **63**, pp. 51–58.

Ni, R., Huang, S.-D., and Xia, K.-Q. (2011). Local Energy Dissipation Rate Balances Local Heat Flux in the Center of Turbulent Thermal Convection, *Phys. Rev. Lett.* **107**, 17, p. 174503.

Niemela, J. J., Skrbek, L., Sreenivasan, K. R., and Donnelly, R. J. (2000). Turbulent convection at very high Rayleigh numbers, *Nature* **404**, pp. 837–840.

Niemela, J. J. and Sreenivasan, K. R. (2003). Confined turbulent convection, *J. Fluid Mech.* **481**, pp. 355–384.

Obukhov, A. M. (1959). On influence of buoyancy forces on the structure of temperature field in a turbulent flow, *Dokl Acad Nauk SSSR* **125**, p. 1246.

Pal, P., Wahi, P., Paul, S., Verma, M. K., Kumar, K., and Mishra, P. K. (2009). Bifurcation and chaos in zero-Prandtl-number convection, *EPL* **87**, 5, p. 54003.

Pandey, A., Kumar, A., Chatterjee, A. G., and Verma, M. K. (2016a). Dynamics of large-scale quantities in Rayleigh-Bénard convection, *Phys. Rev. E* **94**, 5, p. 053106.

Pandey, A. and Verma, M. K. (2016). Scaling of large-scale quantities in Rayleigh-Bénard convection, *Phys. Fluids* **28**, 9, p. 095105.

Pandey, A., Verma, M. K., Chatterjee, A. G., and Dutta, B. (2016b). Similarities between 2D and 3D convection for large Prandtl number, *Pramana-J. Phys.* **87**, 1, p. 13.

Pandey, A., Verma, M. K., and Mishra, P. K. (2014). Scaling of heat flux and energy spectrum for very large Prandtl number convection, *Phys. Rev. E* **89**, p. 023006.

Pao, Y.-H. (1965). Structure of Turbulent Velocity and Scalar Fields at Large Wavenumbers, *Phys. Fluids* **8**, 6, pp. 1063–1075.

Paul, S., Kumar, K., Verma, M. K., Carati, D., De, A. K., and Eswaran, V. (2010). Chaotic travelling rolls in Rayleigh–Bénard convection, *Pramana-J. Phys.* **74**, 1, pp. 75–82.

Paul, S. and Verma, M. K. (2015). Proper Orthogonal Decomposition vs. Fourier Analysis for Extraction of Large-Scale Structures of Thermal Convection, in T. K. Sengupta, S. K. Lele, K. R. Sreenivasan, and P. A. Davidson (eds.), *Proceedings of the Advances in Computation, Modeling and Control of Transitional and Turbulent Flows*, pp. 433–441.

Paul, S., Wahi, P., and Verma, M. K. (2011). Bifurcations and chaos in large-prandtl number Rayleigh–Bénard convection, *Int. J. Non Linear Mech.* **46**, pp. 772–781.

Pawar, S. S. and Arakeri, J. H. (2016). Kinetic energy and scalar spectra in high Rayleigh number axially homogeneous buoyancy driven turbulence, *Phys. Fluids* **28**, 6, p. 065103.

Pharasi, H. K., Kannan, R., Kumar, K., and Bhattacharjee, J. K. (2011). Turbulence in rotating Rayleigh-Bénard convection in low-Prandtl-number fluids, *Phys. Rev. E* **84**, 4, p. 047301.

Podvin, B. and Sergent, A. (2015). A large-scale investigation of wind reversal in a square Rayleigh–Bénard cell, *J. Fluid Mech.* **766**, pp. 172–201.

Pope, S. B. (2000). *Turbulent Flows* (Cambridge University Press, Cambridge).

Prakash, V. N., Martínez Mercado, J., van Wijngaarden, L., Mancilla, E., Tagawa, Y., Lohse, D., and Sun, C. (2016). Energy spectra in turbulent bubbly flows, *J. Fluid Mech.* **791**, pp. 174–190.

Procaccia, I. and Zeitak, R. (1989). Scaling exponents in nonisotropic convective turbulence, *Phys. Rev. Lett.* **62**, 18, pp. 2128–2131.

Ranjan, A. (2017). Segregation of helicity in inertial wave packets, *Phys. Rev. Fluids* **2**, 3, p. 033801.

Reddy, K. S., Kumar, R., and Verma, M. K. (2014). Anisotropic energy transfers in quasi-static magnetohydrodynamic turbulence, *Phys. Plasmas* **21**, 10, p. 102310.

Reddy, K. S. and Verma, M. K. (2014). Strong anisotropy in quasi-static magnetohydrodynamic turbulence for high interaction parameters, *Phys. Fluids* **26**, p. 025109.

Reid, W. H. and Harris, D. L. (1958). Some Further Results on the Bénard Problem, *Phys. Fluids* **1**, 2, pp. 102–110.

Remmel, M., Sukhatme, J., and Smith, L. M. (2013). Nonlinear gravity-wave interactions in stratified turbulence, *Theor. Comput. Fluid Dyn.* **28**, 2, pp. 131–145.

Rensen, J., Luther, S., and Lohse, D. (2005). The effect of bubbles on developed turbulence, *J. Fluid Mech.* **538**, -1, pp. 153–35.

Roberts, P. H. and Glatzmaier, G. A. (2000). Geodynamo theory and simulations, *Rev. Mod. Phys.* **72**, pp. 1081–1123.

Rubinstein, R. (1994). Renormalization group theory of Bolgiano scaling in Boussinesq turbulence, Tech. Rep. ICOM-94-8; CMOTT-94-2.

Sadhukhan, S., Gupta, H., and Chakraborty, S. (2017). On the helium fingers in the intracluster medium, *Mon. Not. R. Astron. Soc.* **469**, 3, pp. 2595–2601.

Sagaut, P. and Cambon, C. (2008). *Homogeneous turbulence dynamics* (Cambridge University Press, Cambridge).

Scheel, J. D., Emran, M. S., and Schumacher, J. (2013). Resolving the fine-scale structure in turbulent Rayleigh–Bénard convection, *New J. Phys.* **15**, 11, p. 113063.

Scheel, J. D., Kim, E., and White, K. R. (2012). Thermal and viscous boundary layers in turbulent Rayleigh–Bénard convection, *J. Fluid Mech.* **711**, pp. 281–305.

Scheel, J. D. and Schumacher, J. (2014). Local boundary layer scales in turbulent Rayleigh–Bénard convection, *J. Fluid Mech.* **758**, pp. 344–373.

Schumacher, J., Bandaru, V., Pandey, A., and Scheel, J. D. (2016). Transitional boundary layers in low-Prandtl-number convection, *Phys. Rev. Fluids* **1**, 8, p. 084402.

Seychelles, F., Amarouchene, Y., Bessafi, M., and Kellay, H. (2008). Thermal Convection and Emergence of Isolated Vortices in Soap Bubbles, *Phys. Rev. Lett.* **100**, 1, p. 144501.

Shang, X.-D. and Xia, K.-Q. (2001). Scaling of the velocity power spectra in turbulent thermal convection, *Phys. Rev. E* **64**, 6, p. 65301.

She, Z.-S. and Leveque, E. (1994). Universal scaling laws in fully developed turbulence, *Phys. Rev. Lett.* **72**, 3, pp. 336–339.

Sherman, F. S. and Imberger, J. (1978). Turbulence and mixing in stably stratified waters, *Annu. Rev. Fluid Mech.* **10**, pp. 267–288.

Shi, N., Emran, M. S., and Schumacher, J. (2012). Boundary layer structure in turbulent Rayleigh–Bénard convection, *J. Fluid Mech.* **706**, pp. 5–33.

Shraiman, B. I. and Siggia, E. D. (1990). Heat transport in high-Rayleigh-number convection, *Phys. Rev. A* **42**, 6, pp. 3650–3653.

Siggia, E. D. (1994). High Rayleigh number convection, *Annu. Rev. Fluid Mech.* **26**, 1, pp. 137–168.

Silano, G., Sreenivasan, K. R., and Verzicco, R. (2010). Numerical simulations of Rayleigh–Bénard convection for Prandtl numbers between 10^{-1} and 10^4 and Rayleigh numbers between 10^5 and 10^9, *J. Fluid Mech.* **662**, pp. 409–446.

Sirovich, L. (1987). Turbulence and the dynamics of coherent structures. I. Coherent structures, *Quart. Appl. Math.* **45**, 3, pp. 561–571.

Sirovich, L. (1989). Chaotic dynamics of coherent Structures structures, *Physica D* **37**, 1-3, pp. 126–145.

Skandera, D., Busse, A., and Müller, W.-C. (2007). Scaling Properties of Convective Turbulence, in *High Performance Computing in Science and Engineering, Garching/Munich 2007, Springer Berlin Heidelberg* (Garching/Munich), pp. 387–396.

Spiegel, E. A. (1971). Convection in stars: I. Basic Boussinesq convection, *Annu. Rev. Astron. Astrophys.* **9**, pp. 323–352.

Spiegel, E. A. (ed.) (2010). *The Theory of Turbulence: Subrahmanyan Chandrasekhar's 1954 Lectures* (Springer, Berlin).

Sreenivas, K. R., Singh, O. P., and Srinivasan, J. (2009). On the relationship between finger width, velocity, and fluxes in thermohaline convection, *Phys. Fluids* **21**, 2, p. 026601.

Sreenivasan, K. R. (1991). Fractals and multifractals in fluid turbulence, *Annu. Rev. Fluid Mech.* **23**, 1, pp. 539–600.

Sreenivasan, K. R., Bershadskii, A., and Niemela, J. J. (2002). Mean wind and its reversal in thermal convection, *Phys. Rev. E* **65**, 5, p. 056306.

Stepanov, R. and Plunian, F. (2006). Fully developed turbulent dynamo at low magnetic Prandtl numbers, *J. Turbul.* **7**, 39, pp. 1–15.

Stevens, R. J. A. M., Lohse, D., and Verzicco, R. (2011). Prandtl and Rayleigh number dependence of heat transport in high Rayleigh number thermal convection, *J. Fluid Mech.* **688**, pp. 31–43.

Stevens, R. J. A. M., van der Poel, E. P., Grossmann, S., and Lohse, D. (2013). The unifying theory of scaling in thermal convection: the updated prefactors, *J. Fluid Mech.* **730**, pp. 295–308.

Stevens, R. J. A. M., Verzicco, R., and Lohse, D. (2010). Radial boundary layer structure and Nusselt number in Rayleigh–Bénard convection, *J. Fluid Mech.* **643**, pp. 495–507.

Strogatz, S. H. (2014). *Nonlinear Dynamics and Chaos: With Applications to Physics, Biology, Chemistry, and Engineering*, 2nd edn. (Perseus Books, Reading MA).

Sugiyama, K., Ni, R., Stevens, R. J. A. M., Chan, T. S., Zhou, S.-Q., Xi, H.-D., Sun, C., Grossmann, S., Xia, K.-Q., and Lohse, D. (2010). Flow reversals in thermally driven turbulence, *Phys. Rev. Lett.* **105**, 3, p. 034503.

Sun, C., Zhou, Q., and Xia, K.-Q. (2006). Cascades of velocity and temperature fluctuations in buoyancy-driven thermal turbulence, *Phys. Rev. Lett.* **97**, 1, p. 144504.

Teaca, B., Verma, M. K., Knaepen, B., and Carati, D. (2009). Energy transfer in anisotropic magnetohydrodynamic turbulence, *Phys. Rev. E* **79**, 4, p. 046312.

Thorpe, S. A. (1973). Turbulence in stably stratified fluids: A review of laboratory experiments, *Boundary-Layer Meteorology* **5**, 1, pp. 95–119.

Tritton, D. J. (1988). *Physical Fluid Dynamics* (Clarendon Press, Oxord).

turbulencehub.org (2017). turbulencehub.org: Videos, .

Turner, J. S. (1985). Multicomponent convection, *Annu. Rev. Fluid Mech.* **17**, 1, pp. 11–44.

Turner, J. S. (2009). *Buoyancy effects in fluids* (Cambridge University Press, Cambridge).

Urban, P., Musilová, V., and Skrbek, L. (2011). Efficiency of heat transfer in turbulent Rayleigh-Bénard convection, *Phys. Rev. Lett.* **107**, 1, p. 014302.

Vallgren, A., Deusebio, E., and Lindborg, E. (2011). Possible Explanation of the Atmospheric Kinetic and Potential Energy Spectra, *Phys. Rev. Lett.* **107**, 26, p. 268501.

Vallis, G. K. (2006). *Atmospheric and Oceanic Fluid Dynamics* (Cambridge University Press, Cambridge).

van der Poel, E. P., Ostilla-Mónico, R., Verzicco, R., Grossmann, S., and Lohse, D. (2015). Logarithmic mean temperature profiles and their connection to plume emissions in turbulent Rayleigh-Bénard convection, *Phys. Rev. Lett.* **115**, 15, p. 154501.

van Reeuwijk, M., Jonker, H. J. J., and Hanjalic, K. (2008). Wind and boundary layers in Rayleigh–Bénard convection. I. Analysis and modeling, *Phys. Rev. E* **77**, 3, p. 036311.

Vasil'ev, A. Y. and Frick, P. (2011). Reversals of large-scale circulation in turbulent convection in rectangular cavities, *JETP Lett.* **93**, 6, pp. 330–334.

Vasiliev, A., Sukhanovskii, A., Frick, P., Budnikov, A., Fomichev, V., Bolshukhin, M., and Romanov, R. (2016). High Rayleigh number convection in a cubic cell with adiabatic sidewalls, *Int. J. Heat Mass Transfer* **102**, pp. 201–212.

Verma, M. K. (2001). Field theoretic calculation of scalar turbulence, *Int. J. Mod. Phys. B* **15**, 26, pp. 3419–3428.

Verma, M. K. (2004). Statistical theory of magnetohydrodynamic turbulence: recent results, *Phys. Rep.* **401**, 5, pp. 229–380.

Verma, M. K. (2017). Anisotropy in Quasi-Static Magnetohydrodynamic Turbulence, *Rep. Prog. Phys.* **80**, 8, p. 087001.

Verma, M. K., Ambhire, S. C., and Pandey, A. (2015). Flow reversals in turbulent convection with free-slip walls, *Phys. Fluids* **27**, 4, p. 047102.

Verma, M. K., Chatterjee, A. G., Yadav, R. K., Paul, S., Chandra, M., and Samtaney, R. (2013). Benchmarking and scaling studies of pseudospectral code Tarang for turbulence simulations, *Pramana-J. Phys.* **81**, 4, pp. 617–629.

Verma, M. K. and Kumar, A. (2014). Sweeping effect and Taylor's hypothesis via correlation function, *arXiv* 1411.2693v3.

Verma, M. K., Kumar, A., Kumar, P., Barman, S., Chatterjee, A. G., and Samtaney, R. (2017a). Energy fluxes and spectra for turbulent and laminar flows, *arXiv* 1705.04917.

Verma, M. K., Kumar, A., and Pandey, A. (2017b). Phenomenology of buoyancy-driven turbulence: recent results, *New J. Phys.* **19**, p. 025012.

Verma, M. K., Kumar, K., and Kamble, B. (2006a). Mode-to-mode energy transfers in convective patterns, *Pramana-J. Phys.* **67**, 6, pp. 1129–1140.

Verma, M. K., Manna, S., Banerjee, J., and Ghosh, S. (2006b). Universal scaling laws for large events in driven nonequilibrium systems, *EPL* **76**, 6, pp. 1050–1056.

Verma, M. K., Mishra, P. K., Pandey, A., and Paul, S. (2012). Scalings of field correlations and heat transport in turbulent convection, *Phys. Rev. E* **85**, 1, p. 016310.

Verma, M. K., Pandey, A., Mishra, P. K., and Chandra, M. (2014). Role of bulk flow in turbulent convection, in *AIP Conference Proceedings Series*, Proceedings of theInternational Conference on Complex Processes in Plasmas and Nonlinear Dynamical Systemsheld at Gandhinagar, (AIP Publishing LLC), pp. 224–238.

Weiss, N. O. and Proctor, M. R. E. (2014). *Magnetoconvection* (Cambridge University Press, Cambridge).

Whitehead, J. and Doering, C. R. (2011). Ultimate state of two-dimensional Rayleigh-Bénard convection between free-slip fixed-temperature boundaries, *Phys. Rev. Lett.* **106**, 24, p. 244501.

Wu, X.-Z., Kadanoff, L. P., Libchaber, A., and Sano, M. (1990). Frequency power spectrum of temperature fluctuations in free convection, *Phys. Rev. Lett.* **64**, 18, pp. 2140–2143.

Xi, H.-D., Zhang, Y.-B., Hao, J.-T., and Xia, K.-Q. (2016). Higher-order flow modes in turbulent Rayleigh–Bénard convection, *J. Fluid Mech.* **805**, pp. 31–51.

Xi, H.-D., Zhou, Q., and Xia, K.-Q. (2006). Azimuthal motion of the mean wind in turbulent thermal convection, *Phys. Rev. E* **73**, 5, p. 056312.

Xia, K.-Q., Lam, S., and Zhou, S.-Q. (2002). Heat-flux measurement in high-Prandtl-number turbulent Rayleigh–Bénard convection, *Phys. Rev. Lett.* **88**, 6, p. 064501.

Xin, Y.-B. and Xia, K.-Q. (1997). Boundary layer length scales in convective turbulence, *Phys. Rev. E* **56**, 3, pp. 3010–3015.

Yakhot, V. and Orszag, S. A. (1986). Renormalization group analysis of turbulence. I. Basic theory, *J. Sci. Comput.* **1**, 1, pp. 3–51.

Yeung, P. K., Donzis, D. A., and Sreenivasan, K. R. (2004). Simulations of Three-Dimensional Turbulent Mixing for Schmidt Numbers of the Order 1000, *Flow Turbul. Combust.* **72**, pp. 333–347.

Yoshida, J. and Nagashima, H. (2003). Numerical experiments on salt-finger convection, *Prog. Oceanogr.* **56**, 3-4, pp. 435–459.

Young, Y.-N., Tufo, H., Dubey, A., and Rosner, R. (2001). On the miscible Rayleigh–Taylor instability: two and three dimensions, *J. Fluid Mech.* **447**, pp. 377–408.

Zangwill, A. (2013). *Modern Electrodynamics* (Cambridge University Press, Cambridge).

Zhang, J., Wu, X. L., and Xia, K.-Q. (2005). Density Fluctuations in Strongly Stratified Two-Dimensional Turbulence, *Phys. Rev. Lett.* **94**, 17, p. 174503.

Zhou, Q., Liu, B.-F., Li, C.-M., and Zhong, B.-C. (2012). Aspect ratio dependence of heat transport by turbulent Rayleigh–Bénard convection in rectangular cells, *J. Fluid Mech.* **710**, pp. 260–276.

Zhou, S.-Q. and Xia, K.-Q. (2001). Scaling properties of the temperature field in convective turbulence, *Phys. Rev. Lett.* **87**, 6, p. 064501.

Zimin, V. and Frick, P. (1989). Turbulent Convection (Izdatel'stvo Nauka, Moscow).

Index

www.ingramcontent.com/pod-product-compliance
Lightning Source LLC
Chambersburg PA
CBHW081509190326
41458CB00015B/5323